Kazumasa Mizumura

Kanazawa Institute of Technology, Japan

Applied Mathematics in

HYDRAULIC ENGINEERING

An Introduction to Nonlinear Differential Equations

World Scientific

NEW JERSEY · LONDON · SINGAPORE · BEIJING · SHANGHAI · HONG KONG · TAIPEI · CHENNAI

Published by

World Scientific Publishing Co. Pte. Ltd.

5 Toh Tuck Link, Singapore 596224

USA office: 27 Warren Street, Suite 401-402, Hackensack, NJ 07601

UK office: 57 Shelton Street, Covent Garden, London WC2H 9HE

British Library Cataloguing-in-Publication Data
A catalogue record for this book is available from the British Library.

Cover image: Peter Moore

ISBN-13 978-981-4299-55-8
ISBN-10 981-4299-55-3

Typeset by Stallion Press
Email: enquiries@stallionpress.com

Preface

There are few books today on applied or engineering mathematics for civil or environmental engineers. In our daily lives, we live in or near bodies of fluid, such as rivers, seas, and the atmosphere. Even within the human body, blood and lymph flow. Because of this, there exist research fields to treat fluid motion under the gravitational action in hydraulic engineering, coastal engineering, hydrology, geophysics, and environmental engineering. Such fluid motions include turbulent motions, diffusion, density currents, rainfall-runoff analysis, sediment transport, water waves, littoral drift of sand, coastal changes, groundwater flow, etc.

When we plan or design structures, design conditions of the structure must first be predetermined for environmental assessment. For example, to analyze water pollution problems in the field, we have to consider the fluid motion beforehand. In fluid mechanics, fluid motions are originally and theoretically studied, thus studies of these motions in rivers, seas, groundwater, or atmosphere, are practically solved in civil engineering, and their theoretical bases are dependent on the Navier–Stokes equations in fluid mechanics.

In fluid mechanics there are many types of mathematics, such as nonlinear partial differential equations. Applied mathematics are classified into four categories. They are: analysis of linear lumped systems, analysis of linear distributed systems, analysis of nonlinear lumped and distributed systems, and problems involving statistical methods and the probability theory. Because of space limitations in this text, statistical methods and the probability theory are not included.

To understand experimental results (or the results of field surveys) in hydraulic engineering, coastal engineering, and hydrology, we are required to use or deal with mathematics. Most of the governing equations in fluid mechanics are nonlinear. As a result, many researchers have been transforming nonlinear partial differential equations to linear ones in order to obtain their solutions. Thus, many linear partial differential equations can be analytically solved by mathematics today. Although good graduate students or engineers are often able to derive nonlinear partial differential equations from surrounding phenomena based on experiments or field observations, they are still unable to obtain analytical solutions. Thus, it is important that we must first learn linearization methods for these nonlinear partial differential equations. I would like to show some of them in this book. Knowledge is power and much knowledge produces originality.

I thank Drs. Michael Strintzis, Mikio Hino, James A. Liggett, and the American Society of Civil Engineers (ASCE) Journal reviewers — Dr. Strintzis gave me a chance to study nonlinear systems analysis in electrical engineering, Dr. Hino led me to become interested in nonlinear differential equations of hydraulic, hydrologic, and coastal engineering, Dr. Liggett gave me warm advice and helpful suggestions when I submitted papers to the ASCE Journal, and the anonmyous ASCE Journal reviewers who criticized and commented on my papers led me to improve and polish them. Finally, I would like to thank Mr. Gregory Lee from World Scientific Publishing, for all his tremendous efforts in making the publication of this book possible.

Kazumasa Mizumura

Contents

Applied Mathematics in

HYDRAULIC
ENGINEERING

An Introduction to Nonlinear Differential Equations

Chapter 1

Introduction

We use various mathematics in engineering. Differential equations are mostly used in engineering to analyze various phenomena. To understand them, we are required to study calculus first. To master calculus, we are required to study other mathematics such as quadratic curves (conic curves), trigonometric functions, and infinite series. In this chapter, we study them. Readers who have sufficient knowledge about the above are encouraged to go straight to Chap. 2.

1.1. Quadratic Curves

The theory of quadratic curves was developed by the observation of stars in the sky. Planets or comets such as the Earth, the Mars, etc., form an orbit around the sun. The orbits are often an ellipse. The equation of the ellipse is in the Cartesian coordinate system given by

$$\frac{x^2}{a^2} + \frac{y^2}{b^2} = 1, \qquad a > b > 0 \tag{1.1}$$

in which $2a$ and $2b$ are called the long and short axes of the ellipse. The ellipse has two focuses. The positions of the focuses, $\pm\sqrt{b^2 - a^2}$ are coordinates on the x axis. The sun is located at one of the focuses of the orbits of the planets. When the sum of the two distances between a point on the ellipse and the focuses is constant, the orbit of the point forms the ellipse. The area of the ellipse is πab. Optically, the tangent line on a point P on the ellipse intersects the two lines between the point P and the focuses. A ray of light from a focus reflects at a point on the ellipse and reaches another focus. The orbit of some comets are expressed by a parabola or hyperbola.

1

The equation of the parabola is in the Cartesian coordinate system given by

$$y^2 = 4px, \qquad p > 0. \tag{1.2}$$

The parabola has a focus and its coordinate is p on the x axis. When the distance between a point P on the parabola and the focus is equal to the distance between the point P and a parallel line to the y axis that does not pass the focus, the locus of the point P forms the parabola. Optically, a ray of light parallel to the x axis from $x = +\infty$ reflects inside the parabola and passes through the focus.

A hyperbola is expressed by the Cartesian coordinate system as follows:

$$\frac{x^2}{a^2} - \frac{y^2}{b^2} = 1, \qquad a > 0, \quad b > 0. \tag{1.3}$$

Its focuses are $\pm\sqrt{a^2 + b^2}$ on the x axis. When the difference of the distances between a point P on the hyperbola and two focuses is constant, the locus forms the hyperbola. When $a = b$, it is called an orthogonal hyperbola. Optically, the tangent line at the point P equally divides the angle between the point P and the focuses. The asymptotic lines of Eq. (1.3) are obtained by assuming that the right side of Eq. (1.3) is zero.

$$\frac{x^2}{a^2} = \frac{y^2}{b^2} \quad \text{namely} \quad y = \frac{b}{a}x \quad \text{and} \quad y = -\frac{b}{a}x. \tag{1.4}$$

The equation

$$\frac{x^2}{a^2} - \frac{y^2}{b^2} = -1 \tag{1.5}$$

is called a conjugate hyperbola of Eq. (1.3). This is orthogonal to the original hyperbola (Eq. (1.3)). The quadratic curves are generally expressed by

$$f(x, y) = ax^2 + 2hxy + by^2 + 2gx + 2fy + c = 0. \tag{1.6}$$

Example 1.1. Show the optical property of the parabola.

Solution. Consider the parabola for $y > 0$. The equation of the parabola is given by

$$y = 2\sqrt{px}. \tag{1}$$

The derivative (Eq. (2.1)) at a point P at $x = x_0$ is

$$\frac{dy}{dx} = \sqrt{p/x_0}.\tag{2}$$

When an angle of the derivative is α, we have

$$\tan \alpha = \sqrt{p/x_0}.\tag{3}$$

The slope S of the line between the point P and the focus is

$$S = \frac{2\sqrt{px_0}}{x_0 - p}.\tag{4}$$

When an angle of the line is θ, we get

$$\tan \theta = \frac{2\sqrt{px_0}}{x_0 - p}.\tag{5}$$

In order that all light beam goes to the focus, the following equation is formed:

$$\tan (\theta - \alpha) = \tan \alpha.\tag{6}$$

This indicates $\theta = 2\alpha$. Thus, from Eq. (1.20) we get

$$\tan (2\alpha) = \frac{2 \tan \alpha}{1 - \tan^2 \alpha} = \frac{2\sqrt{p/x_0}}{x_0 - p}.\tag{7}$$

Equation (7) is the same as Eq. (4). This shows $\theta = 2\alpha$.

Problem 1.1. Show the optical property of the hyperbola.

1.2. Trigonometric Functions

Many formulas in trigonometric functions are very useful in engineering and mathematics. These are used for the derivation of governing equations from geometrical considerations and coordinate transformations. Readers must be familiar with the computations of these functions. The addition and subtraction of phases are transformed as

$$\sin(\alpha \pm \beta) = \sin \alpha \cos \beta \pm \cos \alpha \sin \beta, \tag{1.7}$$

$$\cos(\alpha \pm \beta) = \cos \alpha \cos \beta \mp \sin \alpha \sin \beta, \tag{1.8}$$

$$\tan(\alpha \pm \beta) = \frac{\tan \alpha \pm \tan \beta}{1 \mp \tan \alpha \tan \beta}. \tag{1.9}$$

The products of sin and cos functions are given as

$$\sin \alpha \cos \beta = \frac{1}{2}[\sin(\alpha + \beta) + \sin(\alpha - \beta)], \tag{1.10}$$

$$\cos \alpha \sin \beta = \frac{1}{2}[\sin(\alpha + \beta) - \sin(\alpha - \beta)], \tag{1.11}$$

$$\cos \alpha \cos \beta = \frac{1}{2}[\cos(\alpha + \beta) + \cos(\alpha - \beta)], \tag{1.12}$$

$$\sin \alpha \sin \beta = \frac{1}{2}[- \cos(\alpha + \beta) + \cos(\alpha - \beta)]. \tag{1.13}$$

Transforming Eqs. (1.10), (1.11), (1.12), and (1.13), we obtain

$$\sin \alpha + \sin \beta = 2 \sin \frac{\alpha + \beta}{2} \cos \frac{\alpha - \beta}{2}, \tag{1.14}$$

$$\sin \alpha - \sin \beta = 2 \cos \frac{\alpha + \beta}{2} \sin \frac{\alpha - \beta}{2}, \tag{1.15}$$

$$\cos \alpha + \cos \beta = 2 \cos \frac{\alpha + \beta}{2} \cos \frac{\alpha - \beta}{2}, \tag{1.16}$$

$$\cos \alpha - \cos \beta = -2 \sin \frac{\alpha + \beta}{2} \sin \frac{\alpha - \beta}{2}. \tag{1.17}$$

Substituting $\alpha = \beta$ in Eqs. (1.7), (1.8), and (1.9), and considering $+$ sign, we get

$$\sin 2\alpha = 2 \sin \alpha \cos \alpha, \tag{1.18}$$

$$\cos 2\alpha = \cos^2 \alpha - \sin^2 \alpha = 2 \cos^2 \alpha - 1 = 1 - 2 \sin^2 \alpha, \tag{1.19}$$

$$\tan 2\alpha = \frac{2 \tan \alpha}{1 - \tan^2 \alpha}. \tag{1.20}$$

The relationship $\sin^2 \alpha + \cos^2 \alpha = 1$ is used in Eq. (1.19). Using $3\alpha = 2\alpha + \alpha$ and Eqs. (1.7), (1.8), and (1.9), we have

$$\sin 3\alpha = 3 \sin \alpha - 4 \sin^3 \alpha, \tag{1.21}$$

$$\cos 3\alpha = 4 \cos^3 \alpha - 3 \cos \alpha, \tag{1.22}$$

$$\tan 3\alpha = \frac{3 \tan \alpha - \tan^3 \alpha}{1 - 3 \tan^2 \alpha}. \tag{1.23}$$

Using $4\alpha = 3\alpha + \alpha$, Eqs. (1.21), (1.22), and (1.23), we can derive these equations for 4α. Transforming Eqs. (1.8), (1.9), (1.20), (1.21), (1.22), and (1.23), we obtain

$$\sin^2 \frac{\alpha}{2} = \frac{1 - \cos \alpha}{2}, \tag{1.24}$$

$$\cos^2 \frac{\alpha}{2} = \frac{1 + \cos \alpha}{2}, \tag{1.25}$$

$$\sin^2 \alpha = \frac{1 - \cos 2\alpha}{2}, \tag{1.26}$$

$$\cos^2 \alpha = \frac{1 + \cos 2\alpha}{2}, \tag{1.27}$$

$$\sin^3 \alpha = \frac{3 \sin \alpha - \sin 3\alpha}{4}, \tag{1.28}$$

$$\cos^3 \alpha = \frac{3 \cos \alpha + \cos 3\alpha}{4}. \tag{1.29}$$

When $\tan \frac{x}{2} = t$, we have

$$\sin x = \frac{2t}{1 + t^2}, \tag{1.30}$$

$$\cos x = \frac{1 - t^2}{1 + t^2}, \tag{1.31}$$

$$\tan x = \frac{2t}{1 - t^2}. \tag{1.32}$$

Equations (1.30), (1.31), and (1.32) are often used in substitution integrals in Chap. 2.

Example 1.2. An Euler equation in Chap. 2 shows

$$e^{i\alpha} = \cos\alpha + i\sin\alpha, \tag{1}$$

$$e^{i\beta} = \cos\beta + i\sin\beta. \tag{2}$$

Using Eqs. (1) and (2), show

$$\sin(\alpha+\beta) = \sin\alpha\cos\beta + \cos\alpha\sin\beta. \tag{3}$$

Solution. Multiplying Eqs. (1) and (2), we obtain

$$e^{i(\alpha+\beta)} = \cos\alpha\cos\beta - \sin\alpha\sin\beta + i(\sin\alpha\cos\beta + \cos\alpha\sin\beta). \tag{4}$$

The imaginary part of Eq. (4) is

$$\sin(\alpha+\beta) = \sin\alpha\cos\beta + \cos\alpha\sin\beta. \tag{5}$$

Problem 1.2. When $\tan x/2 = t$, show Eq. (1.30).

1.3. Infinite Series

A sequence, a succession of terms, is formed according to some fixed rule or law. For example,

$$1, 4, 9, 16, 25, 36, \ldots \tag{1.33}$$

$$1, \frac{x}{1!}, \frac{x^2}{2!}, \frac{x^3}{3!}, \frac{x^4}{4!}, \ldots \tag{1.34}$$

are sequences. A series is the indicated sum of the terms of a sequence. That is, from the foregoing sequences we obtain the series

$$1 + 4 + 9 + 16 + 25 + 36 + \cdots \tag{1.35}$$

and

$$1 + \frac{x}{1!} + \frac{x^2}{2!} + \frac{x^3}{3!} + \frac{x^4}{4!} + \cdots. \tag{1.36}$$

If the number of terms is limited, the sequence or series is said to be finite. If the number of terms is unlimited, the sequence or series is said to be an infinite sequence or series. The general terms for the proceeding illustrations are

$$n^2 \quad \text{and} \quad \frac{x^n}{n!} \tag{1.37}$$

in which $n!$ is the fractional number given by

$$n! = 1 \times 2 \times 3 \times \cdots \times (n-1) \times n. \tag{1.38}$$

Consider the following series

$$S_n = a + ar + ar^2 + \cdots + ar^{n-1}. \tag{1.39}$$

This series is called a geometric series. Multiplying Eq. (1.39) by r, we get

$$rS_n = ar + ar^2 + ar^3 + \cdots + ar^n. \tag{1.40}$$

Let us now subtract Eq. (1.39) from Eq. (1.40). This gives

$$rS_n - S_n = ar^n - a, \tag{1.41}$$

hence

$$S_n = \frac{a(1-r^n)}{1-r}. \tag{1.42}$$

Now if $|r| < 1$, then r^n decreases in absolute value as n increases so that we have

$$\lim_{n \to} r^n = 0. \tag{1.43}$$

From Eq. (1.42) we then have

$$\lim_{n \to \infty} S_n = \frac{a}{1-r} = S. \tag{1.44}$$

Here, if $|r| < 1$, the sum S_n of the geometric series approaches a limit as the number of terms is increased indefinitely. In this case, the infinite series $a + ar + ar^2 + ar^3 + \cdots$ is said to be convergent. If $|r| > 1$, then r^n will become infinite as n increases indefinitely. Hence, from Eq. (1.42), $|S_n|$ will increase without a limit. In this case the series is said to be divergent. If $r = -1$, the geometric series becomes

$$a - a + a - a + \cdots \tag{1.45}$$

In this case, if n is even, S_n is zero. If n is odd, S_n is a. As n increases indefinitely, the absolute value of S_n does not increase indefinitely but still S_n does not approach a limit. A series of this sort is called an oscillating series. It is a divergent series. For example, let us consider the series

$$1 + 1 + \frac{1}{1 \times 2} + \frac{1}{1 \times 2 \times 3} + \cdots + \frac{1}{n!} + \cdots = \sum_{n=0}^{\infty} \frac{1}{n!}. \qquad (1.46)$$

If we temporarily neglect the first term, we may write

$$S_n = 1 + \frac{1}{1 \times 2} + \frac{1}{1 \times 2 \times 3} + \cdots + \frac{1}{n!}. \qquad (1.47)$$

Now let us consider the series defined by

$$U_n = 1 + \frac{1}{2} + \frac{1}{2 \times 2} + \cdots + \frac{1}{2^{n-1}}. \qquad (1.48)$$

Since the corresponding terms of the series S_n are less than the corresponding terms of the series U_n with the exception of the first two terms, it is obvious that

$$S_n < U_n. \qquad (1.49)$$

Now the series U_n is the geometric series with $a = 1$ and $r = 1/2$. Hence, $U_n < 2$ no matter how large n may be. The series S_n always increases as n increases but remains less than 2. Hence S_n approaches a limit as n becomes infinite, and this limit is less than 2. It is thus apparent that the series Eq. (1.46) is convergent and that its value is less than 3. It will be shown later that the sum of the infinite series Eq. (1.46) is less than 3. It will be shown later that the sum of the infinite series Eq. (1.46) is the constant $e = 2.71828\cdots$, the base of the natural-logarithm system. The following equation is very important in mathematics:

$$\lim_{n \to \infty} \left(1 + \frac{1}{n}\right)^n = e \quad \text{or} \quad \lim_{n \to \infty} \left(1 - \frac{1}{n}\right)^{-n} = e. \qquad (1.50)$$

By the use of the comparison principle, it is possible to test a series for divergence. Let

$$U = u_1 + u_2 + u_3 + \cdots \qquad (1.51)$$

be a series of positive terms to be tested which is never less than the corresponding terms of a series of positive terms

$$W = w_1 + w_2 + w_3 + \cdots \qquad (1.52)$$

which is known to be divergent. Then, Eq. (1.51) is a divergent series. By the use of this principle, we may prove that the harmonic series

$$U = 1 + \frac{1}{2} + \frac{1}{3} + \frac{1}{4} + \cdots + \frac{1}{n} + \cdots \qquad (1.53)$$

is divergent. This may be done by rewriting Eq. (1.53) as

$$U = \left(1 + \frac{1}{3} + \frac{1}{5} + \frac{1}{7} + \cdots\right) + \left(\frac{1}{2} + \frac{1}{4} + \frac{1}{6} + \frac{1}{8} + \cdots\right)$$
$$= \left(1 + \frac{1}{3} + \frac{1}{5} + \frac{1}{7} + \cdots\right) + \frac{U}{2}. \qquad (1.54)$$

Equation (1.54) shows

$$\frac{U}{2} = 1 + \frac{1}{3} + \frac{1}{5} + \frac{1}{7} + \cdots. \qquad (1.55)$$

If the harmonic series, Eq. (1.53) converges, Eqs. (1.54) and (1.55) indicate

$$1 + \frac{1}{3} + \frac{1}{5} + \frac{1}{7} + \cdots = \frac{1}{2} + \frac{1}{4} + \frac{1}{6} + \frac{1}{8} + \cdots. \qquad (1.55a)$$

Equation (1.55a) represents that the sum of reciprocals of odd number is the same as that of even number. This is a contradiction. Hence the series Eq. (1.54) is divergent. Euler in 1735 showed the following series converges:

$$1 + \frac{1}{2^2} + \frac{1}{3^2} + \frac{1}{4^2} + \frac{1}{5^2} + \cdots = \frac{\pi^2}{6}. \qquad (1.56)$$

A similar series is also convergent as

$$1 + \frac{1}{2^4} + \frac{1}{3^4} + \frac{1}{4^4} + \frac{1}{5^4} + \cdots = \frac{\pi^4}{90}. \qquad (1.57)$$

Thus, the Zeta function $\zeta(\cdot)$ is generally defined as

$$\zeta(2) = \sum_{k=1}^{\infty} \frac{1}{k^2} = \frac{\pi^2}{6}, \qquad (1.58)$$

$$\zeta(3) = \sum_{k=1}^{\infty} \frac{1}{k^3} = \text{not found}, \qquad (1.59)$$

$$\zeta(4) = \sum_{k=1}^{\infty} \frac{1}{k^4} = \frac{\pi^4}{90}, \tag{1.60}$$

$$\zeta(5) = \sum_{k=1}^{\infty} \frac{1}{k^5} = \text{not found}, \tag{1.61}$$

$$\zeta(6) = \sum_{k=1}^{\infty} \frac{1}{k^6} = \frac{\pi^6}{945}, \tag{1.62}$$

$$\vdots$$

$$\zeta(2n) = \frac{(2\pi)^{2n}}{2(2n)!} |B_{2n}|, \quad \text{for even number} \tag{1.63}$$

in which B_n = a Bernoulli number. The right side of Eq. (1.63) is called an Euler product. This indicates the relationship between the ζ function and prime numbers. For odd numbers $\zeta(s)$ is not found. This induced the theory of Riemann's Zeta function in 1859. It is defined by

$$\zeta(s) = \sum_{k=1}^{\infty} \frac{1}{k^s} = \prod_{\text{prime number}} \frac{1}{1 - \frac{1}{p^s}}. \tag{1.64}$$

In the infinite series,

$$u_1 + u_2 + u_3 + \cdots + u_n + u_{n+1} + \cdots \tag{1.65}$$

the ratio of the consecutive general terms u_n and u_{n+1} is called

$$\left| \frac{u_{n+1}}{u_n} \right| = \text{test ratio.} \tag{1.66}$$

Now find the limit of this ratio when n becomes infinite. Let this be

$$\rho = \lim_{n \to \infty} \left| \frac{u_{n+1}}{u_n} \right|, \tag{1.67}$$

provided that the limit exists. We then have

1. When $\rho < 1$, the series is convergent.
2. When $\rho > 1$, the series is divergent.
3. When $\rho = 1$, the test gives no information.

This is called the Cauchy ratio test. When the infinite series is convergent and the function of a variable, the derivative in Chap. 2 may be obtained by a term-by-term differentiation of the infinite series. When the infinite series is convergent, the integral may be found by integrating the infinite series term-by-term. The theory of the infinite series is applicable to the integrals of nonlinear differential equations and the perturbation theory. The following formulas for finite series are very familiar with readers:

$$\sum_{k=1}^{n} k = \frac{1}{2}n(n+1), \tag{1.68}$$

$$\sum_{k=1}^{n} k^2 = \frac{1}{6}n(n+1)(2n+1), \tag{1.69}$$

$$\sum_{k=1}^{n} k^3 = \left[\frac{1}{2}n(n+1)\right]^2. \tag{1.70}$$

These equations are used for the computation of integrals from the sum of infinite series.

Example 1.3. When

$$\sin x = x - \frac{x^3}{3!} + \frac{x^5}{5!} - \cdots + (-1)^{n-1}\frac{x^{2n-1}}{(2n-1)!} + \cdots, \tag{1}$$

show that

$$\frac{d}{dx}\sin x = \cos x. \tag{2}$$

Solution.

$$\frac{d}{dx}\sin x = 1 - \frac{x^2}{2!} + \frac{x^4}{4!} - \cdots + (-1)^{n-1}\frac{x^{2n-2}}{(2n-2)!} + \cdots = \cos x. \tag{3}$$

Problem 1.3. When

$$\sinh x = x + \frac{x^3}{3!} + \frac{x^5}{5!} + \cdots + \frac{x^{2n-1}}{(2n-1)!} + \cdots, \tag{4}$$

show that

$$\frac{d}{dx}\sinh x = \cosh x. \tag{5}$$

Chapter 2

Differentiations

We study differentiable functions and its applications. To read and understand books in fluid mechanics or hydraulic engineering, we must know differentials of functions, because the governing equations in fluid mechanics or hydraulic engineering contain differential operations.

2.1. Derivatives and Differentiations

When $f(x)$ is the function of x in a range and a real number a belongs to the range, if the limit

$$\lim_{\Delta x \to 0} \frac{f(a + \Delta x) - f(a)}{\Delta x} = \lim_{\Delta x \to 0} \frac{f(a) - f(a - \Delta x)}{\Delta x} = f'(a) \quad (2.1)$$

exists, $f'(a)$ is a differential coefficient of the function $f(x)$ at a. The sign "′" indicates d/dx. When a is a real variable, it is defined x. Then, the differential coefficient $f'(x)$ is called a derivative. Let us consider the derivative of $f(x) = \sin x$.

$$f'(x) = \lim_{\Delta x \to 0} \frac{\sin(x + \Delta x) - \sin x}{\Delta x}$$

$$= \lim_{\Delta x \to 0} \cos\left(x + \frac{\Delta x}{2}\right) \frac{\sin \frac{\Delta x}{2}}{\frac{\Delta x}{2}} = \cos x. \quad (2.2)$$

In the derivation of Eq. (2.2), Eq. (1.7) is used. The increment Δy of $y = f(x)$ is a complicated function of the increment of Δx. But when Δx is infinitesimally small, an approximation $\Delta y \cong f'(x)\Delta x$ is formed. When the differentiable function $y = f(x)$ and $z = g(y)$

exist, the derivative of the composite function $g(f(x))$ is given by

$$\frac{dz}{dx} = \frac{dz}{dy}\frac{dy}{dx} = g'(y)f'(x). \tag{2.3}$$

Equation (2.3) is often used in the analytical derivation of differential equations. As independent variables, real numbers are continuous.

Example 2.1. There is an open channel flow of which discharge is constant. When the discharge per unit width and the water depth are q and h, respectively, the specific energy is given by $E = q^2/(2gh^2) + h$, in which $g =$ the gravitational acceleration. If q is constant, obtain the critical (water) depth by differentiating the specific energy E by h. What do you get if the specific energy is differentiated by x?

Solution. When the specific energy is differentiated by the water depth h, we have

$$\frac{dE}{dh} = -\frac{q^2}{gh^3} + 1 = 0. \tag{1}$$

Then, from Eq. (1) we get

$$h = \left(\frac{q^2}{g}\right)^{1/3}. \tag{2}$$

This is the critical depth. The differentiation of the specific energy by x gives

$$\frac{dE}{dx} = \left(-\frac{q^2}{gh^3} + 1\right)\frac{dh}{dx} = 0. \tag{3}$$

The solution of the above equation indicates the critical depth if $dh/dx \neq 0$. The specific energy is minimum at the critical depth about the space or the water depth.

Problem 2.1. Differentiate x^x by x for $x > 0$.

2.2. Mean Value Theorem and Taylor Series

If the function $f(x)$ is continuous in a range $[a, b]$ and differentiable in (a, b), the mean value theorem is expressed as follows:

$$f(b) - f(a) = f'(\xi)(b - a), \quad \text{for } a < \xi < b, \tag{2.4}$$

in which ξ is a constant to satisfy the above equation in (a, b). When the function $f(x)$ is differentiable many times and expressed by a power series, we have

$$f(x) = a_0 + a_1 x + a_2 x^2 + a_3 x^3 + \cdots , \tag{2.5}$$

in which $a_0 = f(0), a_1 = f'(0)/1!, a_2 = f''(0)/2!, a_3 = f'''(0)/3!, \ldots$. Equation (2.5) is called a Maclaurin series. The application of the function $f(x + \Delta x)$ to Eq. (2.5) leads to

$$f(x + \Delta x) = a_0 + a_1(x + \Delta x) + a_2(x + \Delta x)^2 + a_3(x + \Delta x)^3 + \cdots . \tag{2.6}$$

Using Eq. (2.5) and expressing it in the power of Δx, we obtain

$$f(x + \Delta x) = f(x) + f'(x)\Delta x + \frac{f''(x)}{2!}(\Delta x)^2 + \frac{f'''(x)}{3!}(\Delta x)^3 + \cdots . \tag{2.7}$$

This is a Taylor series. Applying the Maclaurin series to $e^{i\theta}$, let us derive an Euler equation $e^{i\theta} = \cos\theta + i\sin\theta$.

$$\begin{aligned}
e^{i\theta} &= 1 + \frac{i\theta}{1!} + \frac{(i\theta)^2}{2!} + \frac{(i\theta)^3}{3!} + \frac{(i\theta)^4}{4!} + \frac{(i\theta)^5}{5!} + \cdots \\
&= 1 + \frac{i\theta}{1!} - \frac{\theta^2}{2!} - \frac{i\theta^3}{3!} + \frac{\theta^4}{4!} + \frac{i\theta^5}{5!} - \frac{\theta^6}{6!} - \cdots \\
&= \left(1 - \frac{\theta^2}{2!} + \frac{\theta^4}{4!} - \frac{\theta^6}{6!} + \cdots \right) + i\left(\frac{\theta}{1!} - \frac{\theta^3}{3!} + \frac{\theta^5}{5!} - \frac{\theta^7}{7!} + \cdots \right) \\
&= \cos\theta + i\sin\theta.
\end{aligned} \tag{2.8}$$

This indicates that

$$\cos\theta = 1 - \frac{\theta^2}{2!} + \frac{\theta^4}{4!} - \frac{\theta^6}{6!} + \cdots , \tag{2.9}$$

$$\sin\theta = \frac{\theta}{1!} - \frac{\theta^3}{3!} + \frac{\theta^5}{5!} - \frac{\theta^7}{7!} + \cdots . \tag{2.10}$$

Example 2.2. Derive one-dimensional continuity equation of flow.

Solution. As the input at x and output at $x + \Delta x$, discharge to the control volume are $Q(x)$ and $Q(x + \Delta x)$, respectively. The continuity relation shows that $Q(x) - Q(x + \Delta x) = 0$. The Taylor series approximately leads to

$$Q(x + \Delta x) \cong Q(x) + \Delta x \frac{dQ(x)}{dx}. \tag{1}$$

The substitution of the Taylor series into the continuity relation gives

$$\frac{dQ(x)}{dx} = 0. \tag{2}$$

Problem 2.2. Calculate the n-th derivative of $x/(x^2 - 1)$, in which n is an integer.

2.3. Partial Derivatives and Applications

When the function $f(x, y)$ is defined in a domain D, $f(x, y)$ has a finite limit at a point (x, y) in D, the following equation holds:

$$\lim_{\Delta x \to 0} \frac{f(x + \Delta x, y) - f(x, y)}{\Delta x} = \lim_{\Delta x \to 0} \frac{f(x, y) - f(x - \Delta x, y)}{\Delta x}$$

$$= \frac{\partial f(x, y)}{\partial x} = f_x(x, y). \tag{2.11}$$

Equation (2.11) is called a partial derivative of the function $f(x, y)$ about x. In the same way, a partial derivative of $f(x, y)$ about y is given by $\frac{\partial f(x,y)}{\partial y}$ or $f_y(x, y)$. The partial derivative of $f(x, y)$ is the derivative about x when y is assumed to be constant. When the increments of independent variables x and y are dx and dy, respectively, the total derivative of the function $f(x, y)$ is defined by

$$df(x, y) = f_x(x, y)dx + f_y(x, y)dy. \tag{2.12}$$

In general, there exists a function $f(x, y)$. It is differentiable and when the total derivative is given by $p(x, y)dx + q(x, y)dy$, the following relation holds:

$$df(x, y) = p(x, y)dx + q(x, y)dy, \tag{2.13}$$

in which $p(x, y)$ and $q(x, y)$ = differentiable functions of x and y. This is called a complete differentiation. The Taylor series for two independent variables is written as:

$$f(x + \Delta x, y + \Delta y) = f(x, y) + \frac{1}{1!}\left(\Delta x \frac{\partial}{\partial x} + \Delta y \frac{\partial}{\partial y}\right) f(x, y)$$

$$+ \frac{1}{2!}\left(\Delta x \frac{\partial}{\partial x} + \Delta y \frac{\partial}{\partial y}\right)^2 f(x, y) + \cdots$$

$$+ \frac{1}{(n-1)!}\left(\Delta x \frac{\partial}{\partial x} + \Delta y \frac{\partial}{\partial y}\right)^{n-1} f(x, y) + \cdots.$$

$$(2.14)$$

Next, we consider a tangent plain and a normal. When the equation of a curved plain is $z = f(x, y)$, the equation of the plain that intersects three points $P(a, b, f(a, b))$, $Q(a + \Delta x, b, f(a + \Delta x, b))$, and $R(a, b + \Delta y, f(a, b + \Delta y))$ is given by

$$\begin{vmatrix} x & y & z & 1 \\ a & b & f(a, b) & 1 \\ a + \Delta x & b & f(a + \Delta x, b) & 1 \\ a & b + \Delta y & f(a, b + \Delta y) & 1 \end{vmatrix} = 0. \qquad (2.15)$$

This is also written as:

$$z - f(a, b) = \frac{f(a + \Delta x, b) - f(a, b)}{\Delta x}(x - a)$$

$$+ \frac{f(a, b + \Delta y) - f(a, b)}{\Delta y}(y - b). \qquad (2.16)$$

As $\Delta x \to 0$ and $\Delta y \to 0$, the above equation approaches

$$z - f(a, b) = f_x(a, b)(x - a) + f_y(a, b)(y - b). \qquad (2.17)$$

This denotes that two vectors $(x - a, y - b, z - f(a, b))^T$ and $(f_x(a, b), f_y(a, b), -1)^T$ are orthogonal. The superscript T defines a transposed vector. Thus, Eq. (2.17) is the tangent plain of the function $z = f(x, y)$ at a point P. The perpendicular line to the tangent plain at the point P is called a normal. When a point (x, y, z) is on the normal, the two vectors to express Eq. (2.17) is parallel. Thus, we obtain the following equation for the normal:

$$\frac{x-a}{f_x(a,b)} = \frac{y-b}{f_y(a,b)} = \frac{z-f(a,b)}{-1}. \tag{2.18}$$

This is the equation of the normal to the given plane.

Example 2.3. When the stream function ψ is used, the velocities in the x and y directions are $u = \psi_y$ and $v = -\psi_x$, respectively. Show that the stream function ψ satisfies the equation of the streamline $dx/u = dy/v$.

Solution. When $f(x,y) = \psi$, Eq. (2.12) leads to $d\psi = \psi_x dx + \psi_y dy$. On the streamline, $d\psi = 0$ is formed. This denotes that $-vdx + udy = 0$ or $dx/u = dy/v$.

Problem 2.3. Transform the two-dimensional Laplace equation

$$\frac{\partial^2 \phi}{\partial x^2} + \frac{\partial^2 \phi}{\partial y^2} = 0 \tag{1}$$

in the Cartesian coordinate system (x, y) to the polar coordinate system as

$$\frac{\partial^2 \phi}{\partial r^2} + \frac{1}{r}\frac{\partial \phi}{\partial r} + \frac{1}{r^2}\frac{\partial^2 \phi}{\partial \theta^2} = 0, \tag{2}$$

in which $x = r\cos\theta$ and $y = r\sin\theta$.

2.4. Finite Difference Method and Newton Method

We call the finite difference when the derivative du/dx is expressed by the finite increments Δx and Δu. The forward finite difference scheme is defined by

$$\frac{du}{dx} \cong \frac{u(x+\Delta x) - u(x)}{\Delta x} = \frac{\Delta u}{\Delta x}. \tag{2.19}$$

The function $u(x + \Delta x)$ is also written as

$$u(x + \Delta x) \cong u(x) + (\Delta x)u'(x) + \frac{1}{2}(\Delta x)^2 u''(x + \theta \Delta x), \qquad (2.20)$$

in which $0 < \theta < 1$ and "$'$" indicates d/dx. The error in Eq. (2.19) is given by

$$\frac{u(x + \Delta x) - u(x)}{\Delta x} - u'(x) = O(\Delta x), \qquad (2.21)$$

in which O denotes an order. This means that

$$\lim_{\Delta x \to 0} \frac{O(\Delta x)}{\Delta x} < \infty \quad \text{(finite)} . \qquad (2.22)$$

As the increment Δx decreases, the accuracy of the finite difference scheme generally increases by Eq. (2.21). But since a round error induces the instability of the computation, Δx is not selected to be infinitesimal. In Eq. (2.21), the accuracy in the usage of Eq. (2.21) is 1. When the error is $O((\Delta x)^p)$ in Eq. (2.21), the accuracy is p. As the value of p increases, the accuracy increases. The accuracy of a Runge–Kutta method is 4 when it is often used in the numerical computations of nonlinear ordinary differential equations. To increase the accuracy of the computation, the higher derivatives are required in Eq. (2.20). When $\theta = 0$, Eq. (2.20) is written as

$$u(x + \Delta x) = u(x) + (\Delta x)u'(x) + \frac{1}{2}(\Delta x)^2 u''(x)$$

$$+ \frac{1}{6}(\Delta x)^3 u'''(x) + \frac{1}{24}(\Delta x)^4 u''''(x) + \cdots . \qquad (2.23)$$

Let us consider the numerical solution of a nonlinear equation, $f(x) = 0$. Assume that the function is convex upward. The equation of the tangent of the function $f(x)$ at $x = x_0$ is written as

$$y - f(x_0) = f'(x_0)(x - x_0). \qquad (2.24)$$

An intersection point x_1 between the x axis and the tangent of $f(x)$ at $x = x_0$ is given by

$$x_1 = x_0 - \frac{f(x_0)}{f'(x_0)}, \quad f''(x_0) > 0. \qquad (2.25)$$

An intersection point x_2 between the x axis and the tangent of $f(x)$ at x_1 is given by

$$x_2 = x_1 - \frac{f(x_1)}{f'(x_1)}, \quad f''(x_1) > 0. \tag{2.26}$$

In this way, we get approximate solutions $x_1, x_2, \ldots$, successively. We call these successive processes $x_0, x_1, x_2, \ldots$ the Newton method (or Newton–Raphson method). The applications of the Newton method are also discussed in Chap. 15.

Example 2.4. Solve $x^2 - 2 = 0$ using the Newton method.

Solution. Since $f'(x) = 2x$, the basic equation to calculate x_1 is $x_1 = x_0 - (x_0^2 - 2)/(2x_0)$. Assuming that $x_0 = 10$, we obtain $x_1 = 5.1$. For $x_1 = 5.1$, we get $x_2 = 2.746$. From $x_2 = 2.746$, we have $x_3 = 1.737, x_4 = 1.444, x_5 = 1.4145, x_6 = 1.4142$. This converges $\sqrt{2}$.

Problem 2.4. Solve $x^3 - 2 = 0$ by the Newton method.

2.5. Various Slopes of Unconfined Groundwater Flow

In open channel flows there generally exist several slopes such as energy slope, friction slope, free water surface slope, and bottom slope. In the unconfined groundwater flow, hydraulic slope (gradient) and bottom slope control the groundwater flow. When the flow among soil particles is laminar, a Darcy law is formed. It is written as

$$v = -k\frac{dH}{dx}, \tag{2.27}$$

in which $v =$ mean cross-sectional velocity, $k =$ hydraulic conductivity, and $H =$ piezometric head. When the unconfined groundwater flows on the uniform slope θ, the piezomeric head H at a point P in Fig. 2.1 is defined by

$$H = z + h\cos\theta, \tag{2.28}$$

in which $z =$ vertical distance above the base line. The slope $-dH/dx$ is also called by the hydraulic gradient. Then, the Darcy law is

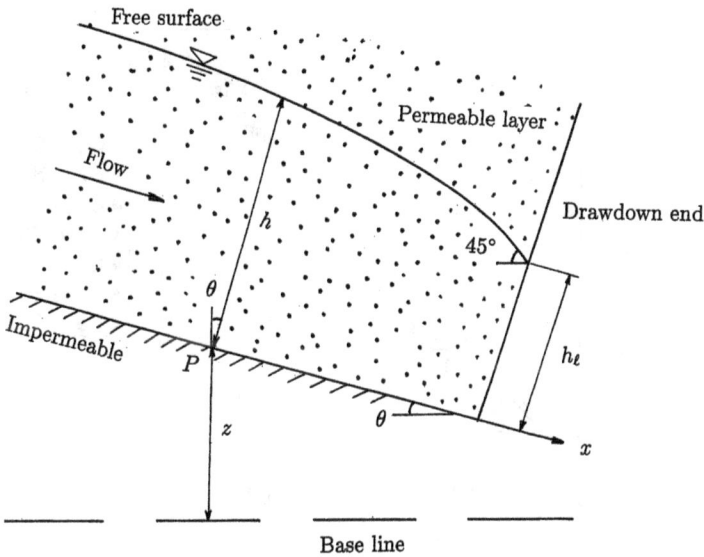

Figure 2.1. Definition of piezometric head and drawdown flow in the porous media.

written as

$$v = -k\left(\frac{dz}{dx} + \frac{dh}{dx}\cos\theta\right). \qquad (2.29)$$

The hydraulic gradient indicates that the piezometric head decreases along the flow direction x and its maximum hydraulic gradient is 1 from the definition of the hydraulic gradient (Mizumura, 2009b). As an example, the steepest hydraulic gradient is 1 at the vertical flow or the drawdown flow. As shown in Fig. 2.1, the hydraulic gradient is 1 when the slope of the seepage face is equal to or more than 90°. The seepage flow at the drawdown end in Fig. 2.1 is vertical. Therefore, when the unconfined groundwater flows from the vertical permeable wall, the hydraulic gradient is 1 or the water surface slope is 45° as shown in Fig. 2.1. This corresponds to the occurrence condition of piping or quick sand phenomena. Whenever the bottom slope of the unconfined groundwater flow is horizontal or not, the water surface slope is 45° or the hydraulic gradient is 1. Figure 2.2 represents the experimental relationship between fluid layer thickness and discharge

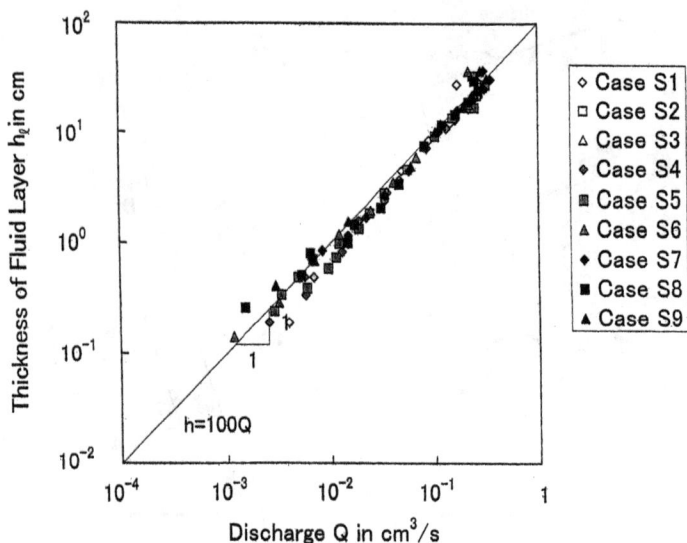

Figure 2.2.   Relationship between discharge and fluid layer thickness at draw-down end.

at the drawdown end (Mizumura, 2006a, 2009b). This relationship indicates

$$h_\ell = 100Q, \tag{2.30}$$

in which h_ℓ = the fluid layer thickness at the drawdown end and Q = discharge at the drawdown end. Using the Darcy law and $k = 1\,\mathrm{mm/s}$, Eq. (2.28) is rewritten as

$$\frac{dh_\ell}{dx} = -1. \tag{2.31}$$

This shows that the the fluid surface slope is theoretically $45°$ or the hydraulic gradient is 1 (Mizumura, 2006a, 2009b). This is the boundary condition at the drawdown end for the unconfined groundwater flow.

Example 2.5. Why is the hydraulic gradient 1 when the flow direction is vertical?

Solution. When the fluid vertically flows a distance, the piezometric head decreases in the same distance. Thus, the ratio between them is 1.

Problem 2.5. When the fluid layer thickness is $h_\ell/2$ at the downstream side of the drawdown end in Fig. 2.1, obtain the hydraulic gradient at the drawdown end.

Chapter 3

Integrations

Integration is an opposite operation of the differentiation. We can differentiate any differentiable functions. But we cannot integrate all functions. Since the governing equations in fluid mechanics contain derivatives (differential equations), they must be integrated to obtain the functional relationship. So we must be familiar with the operations of integrals.

3.1. Indefinite Integrals

When the indefinite integral of $f(x)$ is $F(x)$, the derivative of $F(x)$ is $f(x)$. That is

$$F(x) = \int f(x)dx. \tag{3.1}$$

If the indefinite integral of $f(x)$ is $F(x)$, all indefinite integrals are written by

$$\int f(x)dx = F(x) + C, \tag{3.2}$$

in which C is called an integral constant. When indefinite integrals of $f_1(x)$ and $f_2(x)$ are $F_1(x)$ and $F_2(x)$, respectively, an indefinite integral of $f_1(x)f_2(x)$ is given by

$$\int f_1(x)f_2(x)dx = F_1(x)f_2(x) - \int F_1(x)f_2'(x)dx + C. \tag{3.3}$$

This is called a partial integral. When we substitute $x = g(t)$ into Eq. (3.1), it is transformed as follows:

$$\int f(x)dx = \int f(g(t))g'(t)dt. \tag{3.4}$$

This is called a substitution integral. When the function $f(x)$ in Eq. (3.1) is a rational function, Eq. (3.4) is integrable by a fraction form. When the function in Eq. (3.1) is a surd function, Eq. (3.4) may be integrable by the substitution integral. The order of the function in a square root is more than 2, the function $f(x)$ is called an elliptic function. In this case Eq. (3.1) is not integrable. The elliptic functions are discussed in Chap. 9. For the trigonometric functions of x, the substitution integrals such as $\tan(x/2) = t$ or $\tan x = t$ lead to the integral of rational functions.

Example 3.1. Integrate $\int \frac{dx}{\sin x}$.

Solution. The substitution $\tan(x/2) = t$ into $\sin x$ gives

$$\sin x = 2 \sin \frac{x}{2} \cos \frac{x}{2} = 2 \tan \frac{x}{2} \cos^2 \frac{x}{2} = \frac{2t}{1+t^2}$$

and $\frac{dx}{2\cos^2(x/2)} = dt$ induces $dx = \frac{2}{1+t^2} dt$. Thus, we get

$$\int \frac{dx}{\sin x} = \int \frac{1+t^2}{2t} \cdot \frac{2}{1+t^2} dt = \int \frac{dt}{t} = \log \left| \tan \frac{x}{2} \right| + C.$$

Problem 3.1. Integrate $\int \frac{dx}{\sqrt{x^2-1}}$. (Hint: $x = \cosh y$).

3.2. Definite Integrals

A function $f(x)$ is assumed to be finite in an interval $[a, b]$. Then, the interval $[a, b]$ is subdivided into n subintervals and the coordinates are given as $a = x_0, x_1, \ldots, x_{k-1}, x_k, \ldots, x_n = b$. The length of the subinterval $[x_{k-1}, x_k]$ becomes Δx_k. When the arbitrary coordinate in the subinterval $[x_{k-1}, x_k]$ is ξ_k, the following summation

$$f(\xi_1)\Delta x_1 + f(\xi_2)\Delta x_2 + \cdots + f(\xi_n)\Delta x_n = \sum_{k=1}^{n} f(\xi_k)\Delta x_k \qquad (3.5)$$

is defined. As n becomes infinity, if Eq. (3.5) converges a finite value S, the function $f(x)$ is integrable in the interval $[a, b]$ and written by

$$\int_a^b f(x)dx = S. \qquad (3.6)$$

This is the definition of integral due to Riemann. When there are finite numbers of discontinuous points in $[a, b]$, the integral is defined. The function is not integrable if it is not continuous at irrational numbers in $[a, b]$. Then, the integral must be done by a Lebegue integral. The definite integral of $f(x)$ is obtained by using the indefinite integral $F(x)$ as follows:

$$\int_a^b f(x)dx = F(b) - F(a). \qquad (3.7)$$

In the definite integral of $f(x)$, when the function $f(x)$ is not finite at a point c in $[a, b]$, the following integral of $f(x)$ is called an improper integral as

$$\int_a^b f(x)dx = \lim_{\delta_1 \to 0} \int_a^{c-\delta_1} f(x)dx + \lim_{\delta_2 \to 0} \int_{c+\delta_2}^b f(x)dx. \qquad (3.8)$$

When Eq. (3.8) exists, the function $f(x)$ is integrable. Next, consider the length of a curve in a two-dimensional plane. The coordinate along the curve is s and the infinitesimally small linear element becomes $ds = \sqrt{(dx)^2 + (dy)^2}$. Thus, we get

$$s = \int_0^s \sqrt{(dx)^2 + (dy)^2} = \int_0^x \sqrt{1 + \left(\frac{dy}{dx}\right)^2} \, dx. \qquad (3.9)$$

In Eq. (3.9), $y = f(x)$ and $s = \int_0^x \sqrt{1 + [f'(x)]^2} dx$ is derived.

Example 3.2. There is water in a water tank and the specific weight of water is γ. When a vertical side wall of the water tank is rectangular, calculate the total pressure exerted on the side wall of the water tank. The width and the water depth of the vertical side wall are b and h, respectively. The pressure at the water depth y is given γy.

Solution. The total pressure P is given by the integral of pressure over the depth. It is obtained by

$$P = b \int_0^h \gamma y dy = \frac{b}{2}\gamma h^2. \qquad (1)$$

Problem 3.2. The velocity distribution in the vertical is given by $u = ay^2 + by$, when the water depth is h and $y = 0$ indicates the flow bottom. When a and b are constant, obtain the average velocity.

3.3. Double and Triple Integrals

There are continuous functions $\phi_1(x)$ and $\phi_2(x)$ and the following relation $\phi_1(x) < \phi_2(x)$ is formed. When the integral in a domain A which is bounded by $a \leq x \leq b$ and $\phi_1(x) \leq y \leq \phi_2(x,y)$, a double integral of $f(x,y)$ is defined as

$$\iint_A f(x,y)dxdy = \int_a^b \left[\int_{\phi_1(x)}^{\phi_2(x)} f(x,y)dy \right] dx, \qquad (3.10)$$

in which the function $f(x,y)$ has a finite number of discontinuous points in domain A. In the same way, a triple integral of $f(x,y,z)$ in a domain V in $a \leq x \leq b, \phi_1(x) \leq y \leq \phi_2(x)$, and $\psi_1(x,y) \leq z \leq \psi_2(x,y)$ is defined as

$$\iiint_V f(x,y,z)dxdydz = \int_a^b dx \int_{\phi_1(x)}^{\phi_2(x)} dy \int_{\psi_1(x,y)}^{\psi_2(x,y)} f(x,y,z)dz,$$

$$(3.11)$$

in which the function $f(x,y,z)$ is a finite number of discontinuous points in the domain V. When functions $x = x(u,v)$ and $y = y(u,v)$ which map one-to-one from a closed domain E on the uv-plane to a closed domain D on the xy-plane exist (Riemann theorem), the following relationship is derived:

$$\iint_D f(x,y)dxdy = \iint_E f[x(u,v),y(u,v)]\frac{\partial(x,y)}{\partial(u,v)}dudv, \qquad (3.12)$$

in which $\frac{\partial(x,y)}{\partial(u,v)}$ is the Jacobian. With Eq. (3.12), a volume V between the function $f(x,y)$ and the xy-plane on the domain D is given by

$$V = \iint_D f(x,y)dxdy. \qquad (3.13)$$

By using the triple integral, the mass M is defined by

$$M = \iint_D \left[\int_0^{f(x,y)} \rho(x,y,z)dz \right] dxdy = \iiint_W \rho(x,y,z)dxdydz,$$

$$(3.14)$$

in which $\rho(x, y, z)$ is density and W indicates the three-dimensional shape of the integral. Inertial moments about the $x, y,$ and z axes are, respectively

$$I_x = \iiint_W \rho(x, y, z)(y^2 + z^2)dxdydz, \qquad (3.15)$$

$$I_y = \iiint_W \rho(x, y, z)(z^2 + x^2)dxdydz, \qquad (3.16)$$

$$I_z = \iiint_W \rho(x, y, z)(x^2 + y^2)dxdydz. \qquad (3.17)$$

Since Eq. (2.18) represents the normal to $z = f(x, y)$, the angle θ is between the normal and the vertically upward to the xy plane. Then, the direction cosine is given by

$$\cos\theta \cong \left[1 + f_x^2(x, y) + f_y^2(x, y)\right]^{-1/2}. \qquad (3.18)$$

The real area of $z = f(x, y)$ on a horizontal area $dx \times dy$ is

$$\frac{dxdy}{\cos\theta} = \sqrt{1 + f_x^2(x, y) + f_y^2(x, y)}dxdy. \qquad (3.19)$$

The surface area S of $z = f(x, y)$ is calculated by

$$S = \iint_D \sqrt{1 + f_x^2(x, y) + f_y^2(x, y)}dxdy, \qquad (3.20)$$

in which D is the domain of the integral.

Example 3.3-1. Compute $\int_0^\infty \exp(-x^2)dx$.

Solution. Define $I = \int_0^\infty \exp(-x^2)dx$.
Thus, we have

$$I^2 = \int_0^\infty \exp(-x^2)dx \int_0^\infty \exp(-y^2)dy$$

$$= \int_0^\infty \int_0^\infty \exp(-x^2 - y^2)dxdy. \qquad (1)$$

The coordinate transformations of $x = r\cos\theta$ and $y = r\sin\theta$ show

$$I^2 = \int_0^{\pi/2} d\theta \int_0^\infty \exp(-r^2)rdr = \frac{\pi}{2}\left[-\frac{-\exp(-r^2)}{2}\right]_0^\infty = \frac{\pi}{4}, \qquad (2)$$

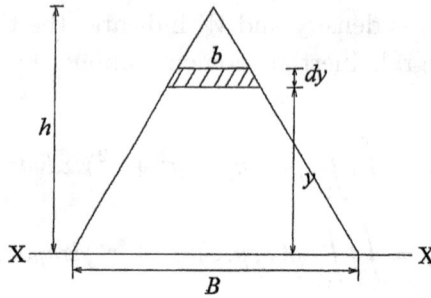

Figure 3.1.   Triangle with height h and base length B.

in which the Jacobian $\frac{\partial(r,\theta)}{\partial(x,y)} = rdrd\theta$ is used. Therefore, we get $I = \sqrt{\pi}/2$.

Example 3.3-2. Compute the second-order moment of a triangle of which height h and base length B about the base line (Fig. 3.1).

Solution. Define the y axis from the base line of the triangular normal to the base line. Since the width b of the triangular at y from the base line is given by $b = B - By/h$, the second-order cross-sectional moment is given by

$$I = \int_0^h by^2 dy = B \int_0^h \left(y^2 - \frac{y^3}{h} \right) dy = \frac{Bh^3}{12}. \tag{3}$$

Problem 3.3. Compute the volume of a sphere of which radius is R. Hint: Integrate a small cylindrical element of which radius is $r \sin \theta$ and the cross-sectional area is $rd\theta\, dr$ in the range $0 \leq \theta \leq \pi$ and $0 \leq r \leq R$.

3.4. Taylor's Diffusion Theory

G. I. Taylor derived the famous diffusion theory using calculus. Calculus includes differentiations and integrations. Assume that there is a uniform flow in x direction and its velocity is U, as shown in Fig. 3.2. Let us consider the turbulent diffusion of fluid particles from a source in the uniform flow. The turbulence in the uniform flow

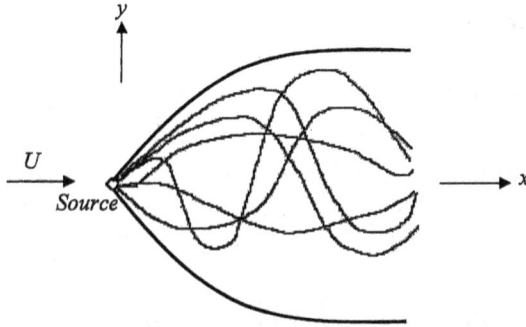

Figure 3.2.   Explanation of turbulent diffusion.

is assumed to be probabilistically stationary. This is the stationary stochastic process in the wide sense. This denotes that the correlation function or coefficient are dependent only on the time difference. The turbulent diffusion is theoretically defined by the ensemble average of the resultant motion of many fluid particles. The fluid particles or tracers with the disturbed velocity v_ℓ in the y direction at $t = 0$ mix with surrounding fluid particles and lose the properties of the initial turbulence. The turbulence component v_ℓ is the Lagrangian velocity component of a fluid particle. The distance $Y(T)$ which this fluid particle moves from $t = 0$ to $t = T$ in the y direction is given by

$$Y(T) = \int_0^T v_\ell(t)dt. \tag{3.21}$$

The transformation of an integral variable t to ξ is as follows:

$$t = T - \xi. \tag{3.22}$$

Using the new integral variable ξ, we give the distance $Y(T)$ as

$$Y(T) = \int_0^T v_\ell(T - \xi)d\xi. \tag{3.23}$$

The derivative of Eq. (3.21) about T gives

$$\frac{dY(T)}{dT} = v_\ell(T). \tag{3.24}$$

With Eq. (3.23), the product of $v_\ell(T)$ and $Y(T)$ shows

$$v_\ell(T)Y(T) = v_\ell(T) \int_0^T v_\ell(T - \xi)d\xi. \tag{3.25}$$

Since the variable T is independent of ξ, Eq. (3.25) is written as

$$v_\ell(T)Y(T) = \int_0^T v_\ell(T)v_\ell(T - \xi)d\xi. \tag{3.26}$$

The product of Eq. (3.24) and $Y(T)$ indicates

$$v_\ell(T)Y(T) = \frac{dY(T)}{dT}Y(T) = \frac{1}{2}\frac{dY^2(T)}{dT}. \tag{3.27}$$

The ensemble average of Eqs. (3.26) and (3.27) are written as

$$\overline{v_\ell(T)Y(T)} = \frac{1}{2}\frac{\overline{dY^2(T)}}{dT} = \int_0^T \overline{v_\ell(T)v_\ell(T - \xi)}d\xi. \tag{3.28}$$

The Lagrangian correlation coefficient which is wide sense stationary is defined as

$$R_\ell(\xi) = \frac{\overline{v_\ell(T)v_\ell(T - \xi)}}{\overline{v_\ell^2}}, \tag{3.29}$$

in which $\overline{v_\ell^2}$ is statistically constant and $R_\ell(\xi)$ is independent of T, since the flow is uniform. Thus, Eq. (3.28) becomes

$$\overline{Y^2(T)} = 2\overline{v_\ell^2} \int_0^T \int_0^\eta R_\ell(\xi)d\xi d\eta. \tag{3.30}$$

This is Taylor's diffusion theory. The physical characteristics of the Taylor's diffusion theory is explained in the following example and problem.

Example 3.4. For the turbulent flow, the Lagrangian correlation coefficient, Eq. (3.29) is almost 1 near $\xi = 0$ and the integration of Eq. (3.29) for large ξ is approximately constant. That is

$$\int_0^\xi R_\ell(\xi)d\xi = T_*, \tag{1}$$

in which $T_* = $ const. Thus, show that for $T \cong 0$,

$$\sqrt{\overline{Y^2(T)}} \cong \sqrt{2\overline{v_\ell^2} \cdot T} \tag{2}$$

and for $T \to \infty$

$$\sqrt{\overline{Y^2(T)}} \cong \sqrt{2\overline{v_\ell^2} T_*} \sqrt{T}. \tag{3}$$

Solution. If $R_\ell(\xi) = 1$ in Eq. (3.30),

$$\overline{Y^2(T)} = 2\overline{v_\ell^2} \int_0^T \int_0^\eta R_\ell(\xi) d\xi d\eta \cong 2\overline{v_\ell^2} \int_0^T \eta d\eta = \overline{v_\ell^2} T^2. \tag{4}$$

The width of the turbulent diffusion for $T \cong 0$ is

$$\sqrt{\overline{Y^2(T)}} \cong \sqrt{\overline{v_\ell^2} \cdot T}. \tag{5}$$

This indicates that the width of the turbulent diffusion increases linearly near the source. For a large T in Eq. (3.30), we have

$$\int_0^\eta R_\ell(\xi) d\xi = T_*. \tag{6}$$

Thus,

$$\overline{Y^2(T)} = 2\overline{v_\ell^2} \int_0^T T_* d\eta = 2\overline{v_\ell^2} T_* T. \tag{7}$$

The width of the turbulent diffusion for $T \to \infty$ is given by

$$\sqrt{\overline{Y^2(T)}} = \sqrt{2\overline{v_\ell^2} T_*} \cdot \sqrt{T}. \tag{8}$$

This explains why the width of the turbulent diffusion parabolically increases for the large T. Therefore, the Taylor's diffusion theory indicates that the diffusion range linearly expands near the source and parabolically expands some distance from the source.

Problem 3.4. Using Fig. 3.3, show that Eq. (3.30) is expressed by

$$\overline{Y^2(T)} = 2\overline{v_\ell^2} \int_0^T (T - \xi) R_\ell(\xi) d\xi. \tag{9}$$

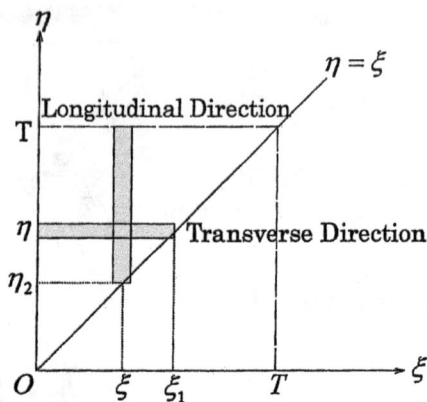

Figure 3.3.   Transform of integral range.

Hint: Eq. (3.30) becomes

$$\overline{Y^2(T)} = 2\overline{v_\ell^2} \int_0^T \int_0^\eta R_\ell(\xi)d\xi d\eta$$

$$= 2\overline{v_\ell^2} \int_0^T \left[\int_0^\eta R_\ell(\xi)d\xi \right] d\eta$$

$$= 2\overline{v_\ell^2} \int_\xi^T \left[\int_0^T R_\ell(\xi)d\xi \right] d\eta$$

$$= 2\overline{v_\ell^2} \int_0^T \left[\int_\xi^T R_\ell(\xi)d\eta \right] d\xi. \qquad (10)$$

The Lagrangian correlation $R_\ell(\xi)$ is not a function of η. The integral variable η is not a function of ξ. Thus, we obtain

$$\text{Eq. (3.30)} = 2\overline{v_\ell^2} \int_0^T (T - \xi) R_\ell(\xi)d\xi. \qquad (11)$$

Figure 3.3 indicates the geometrical mean for the transformation of the integral region.

Chapter 4

Vector and Tensor

A quantity which specifies magnitude like temperature, elevation, height, weight, etc., is called a scalar. On the other hand, a vector contains more than two components — force, velocity, etc. Force and velocity contain directions and magnitudes. The scalar and vector belong to tensors. The tensors of order zero and one are the scalar and vector, respectively. The differential equations in the higher dimensions, which are not visually recognizable, are easily expressed using vectors or tensors. The applications of many formulas in vector and tensor analysis make computations easy in fluid mechanics and hydraulic engineering.

4.1. Vector Algebra

When there are 10 persons in a group, the group has a total of 10 weights. The weight is a scalar variable but the 10 weights of the group form a vector of ranks 10. The vector is usually a column vector as expressed by

$$\vec{X} \qquad \mathbf{X} \qquad \underline{X}$$

The selection of the above definitions of the vector is dependent on the field of science to be treated. When the magnitude of the vector is 1, it is called a unit vector. An arbitrary vector $\vec{X}$ which has three components in the Cartesian space is written as $\vec{X} = x\vec{i} + y\vec{j} + z\vec{k}$. The vectors $\vec{i}$, $\vec{j}$, and $\vec{k}$ which are orthogonal to each other are unit vectors in the x, y, and z directions, respectively. When the magnitudes of the vectors $\vec{A}$ and $\vec{B}$ are a and b, respectively, and the angle between the vectors $\vec{A}$ and $\vec{B}$ is θ, the dot or scalar product of the vectors $\vec{A}$ and $\vec{B}$ is defined by

$$\vec{A} \cdot \vec{B} = ab\cos\theta. \tag{4.1}$$

The cross or vector product is also defined by

$$\vec{A} \times \vec{B} = ab \sin \theta' \, \vec{e}, \tag{4.2}$$

in which the vector $\vec{e}$ is a unit vector perpendicular to the plane of the two vectors in the sense of advance of a right-handed screw rotated from the first to second of the given vectors through the smaller angle θ' between their positive directions. Using the unit vectors $\vec{i}$, $\vec{j}$, and $\vec{k}$, we write the cross product of the vectors $\vec{A}$ and $\vec{B}$ by

$$\begin{aligned}
\vec{A} \times \vec{B} &= (A_x \vec{i} + A_y \vec{j} + A_z \vec{k}) \times (B_x \vec{i} + B_y \vec{j} + B_z \vec{k}) \\
&= (A_y B_z - A_z B_y)\vec{i} + (A_z B_x - A_x B_z)\vec{j} + (A_x B_y - A_y B_x)\vec{k} \\
&= \begin{vmatrix} \vec{i} & \vec{j} & \vec{k} \\ A_x & A_y & A_z \\ B_x & B_y & B_z \end{vmatrix},
\end{aligned} \tag{4.3}$$

in which A_x, A_y, and A_z = the x, y, and z components of the vector $\vec{A}$, respectively, and B_x, B_y, and B_z = the x, y, and z components of the vector $\vec{B}$, respectively.

The three vectors $\vec{A}$, $\vec{B}$, and $\vec{C}$ of whose origins are the same, form a parallelepiped whose edges are the vectors $\vec{A}$, $\vec{B}$, and $\vec{C}$. The volume of the parallelepiped is given by the absolute value of Eq. (4.4).

$$\vec{A} \cdot (\vec{B} \times \vec{C}) = \vec{B} \cdot (\vec{C} \times \vec{A}) = \vec{C} \cdot (\vec{A} \times \vec{B}) = [\vec{A}\vec{B}\vec{C}]. \tag{4.4}$$

Equation (4.4) is positive when the angle which $\vec{A}$ and $\vec{B} \times \vec{C}$ forms is less than $90°$. Otherwise, it is negative. When a vector $\vec{C} = C_x \vec{i} + C_y \vec{j} + C_z \vec{k}$ is defined, using two vectors $\vec{A}$ and $\vec{B}$, we have

$$[\vec{A}\vec{B}\vec{C}] = \vec{A} \cdot (\vec{B} \times \vec{C}) = \begin{vmatrix} A_x & A_y & A_z \\ B_x & B_y & B_z \\ C_x & C_y & C_z \end{vmatrix}. \tag{4.5}$$

Thus, we have

$$[\vec{A}\vec{B}\vec{C}]^2 = \begin{vmatrix} (\vec{A} \cdot \vec{A}) & (\vec{A} \cdot \vec{B}) & (\vec{A} \cdot \vec{C}) \\ (\vec{B} \cdot \vec{A}) & (\vec{B} \cdot \vec{B}) & (\vec{B} \cdot \vec{C}) \\ (\vec{C} \cdot \vec{A}) & (\vec{C} \cdot \vec{B}) & (\vec{C} \cdot \vec{C}) \end{vmatrix}. \tag{4.6}$$

When an arbitrary vector $\vec{v}$ is expressed by three vectors $\vec{e}_1$, $\vec{e}_2$, and $\vec{e}_3$ which are linearly independent and in the positive direction, the vector $\vec{v}$ is written as

$$\vec{v} = v^1\vec{e}_1 + v^2\vec{e}_2 + v^3\vec{e}_3. \tag{4.7}$$

The word "linearly independent" indicates that the three vectors are not on the same plane. The positive direction shows that the angle between $\vec{e}_1$ and $\vec{e}_2 \times \vec{e}_3$ is less than 90°. These three vectors $\vec{e}_1$, $\vec{e}_2$, and $\vec{e}_3$ form base vectors. Equation (4.7) is rewritten in a short form as

$$\vec{v} = \sum_{\kappa=1}^{3} v^\kappa \vec{e}_\kappa. \tag{4.8}$$

In Eq. (4.8), the index number κ appears more than twice and Eq. (4.8) shows the sum of changing quantity of the index number 1, 2, and 3. Then, Eq. (4.8) is written by a simplified form

$$\vec{v} = v^\kappa \vec{e}_\kappa. \tag{4.9}$$

This is called Einstein's rule. With Eq. (4.9), the square of the vector $\vec{v}$ is written as

$$(\vec{v})^2 = v^\lambda \vec{e}_\lambda \cdot v^\kappa \vec{e}_\kappa = (\vec{e}_\lambda \cdot \vec{e}_\kappa)v^\lambda v^\kappa. \tag{4.10}$$

If $\vec{e}_\lambda \cdot \vec{e}_\kappa = g_{\lambda\kappa} = g_{\kappa\lambda}$ is defined, Eq. (4.10) is written as

$$\vec{v}^2 = g_{\lambda\kappa}v^\lambda v^\kappa, \tag{4.11}$$

in which $g_{\lambda\kappa}$ is called a metric tensor (Sokolnikoff, 1951). We define

$$[\vec{e}_1\vec{e}_2\vec{e}_3]^2 = \begin{vmatrix} (\vec{e}_1 \cdot \vec{e}_1) & (\vec{e}_1 \cdot \vec{e}_2) & (\vec{e}_1 \cdot \vec{e}_3) \\ (\vec{e}_2 \cdot \vec{e}_1) & (\vec{e}_2 \cdot \vec{e}_2) & (\vec{e}_2 \cdot \vec{e}_3) \\ (\vec{e}_3 \cdot \vec{e}_1) & (\vec{e}_3 \cdot \vec{e}_2) & (\vec{e}_3 \cdot \vec{e}_3) \end{vmatrix} = g. \tag{4.12}$$

Since $\vec{e}_1$, $\vec{e}_2$, and $\vec{e}_3$ are in the positive direction and $[\vec{e}_1\vec{e}_2\vec{e}_3]$ is a scalar, $[\vec{e}_1\vec{e}_2\vec{e}_3] > 0$. Thus, we define

$$[\vec{e}_1\vec{e}_2\vec{e}_3] = \sqrt{g}. \tag{4.13}$$

The element of the matrix of which determinant g is defined by $g_{\lambda\kappa}$ and the matrix $g_{\lambda\kappa}$ is symmetric. The element of the inverse matrix of $g_{\lambda\kappa}$ is $g^{\lambda\kappa}$. They are defined as follows:

$$g^{\lambda\kappa} = \begin{pmatrix} g^{11} & g^{12} & g^{13} \\ g^{21} & g^{22} & g^{23} \\ g^{31} & g^{32} & g^{33} \end{pmatrix}. \tag{4.14}$$

From the definition of the matrix computation we have

$$g_{\mu\lambda}g^{\lambda\kappa} = \delta_\mu^\kappa, \tag{4.15}$$

in which δ_μ^κ is the Kronecker delta. When $\vec{e}^\kappa = g^{\kappa\lambda}\vec{e}_\lambda$ is defined, we have

$$\vec{e}_\lambda = g_{\lambda\kappa}\vec{e}^\kappa. \tag{4.16}$$

Therefore, we derive

$$\vec{e}^\kappa \cdot \vec{e}_\mu = (g^{\kappa\lambda}\vec{e}_\lambda)\vec{e}_\mu = g^{\kappa\lambda}(\vec{e}_\lambda \cdot \vec{e}_\mu) = g^{\kappa\lambda}g_{\lambda\mu} = \delta_\mu^\kappa. \tag{4.17}$$

The dot product of two vectors $\vec{e}^\lambda$ and $\vec{e}^\kappa$ is given by

$$\begin{aligned} \vec{e}^\lambda \cdot \vec{e}^\kappa &= (g^{\lambda\rho}\vec{e}_\rho)(g^{\kappa\sigma}\vec{e}_\sigma) = g^{\lambda\rho}g^{\kappa\sigma}(\vec{e}_\rho \cdot \vec{e}_\sigma) \\ &= g^{\lambda\rho}g^{\kappa\sigma}g_{\rho\sigma} = g^{\lambda\rho}\delta_\rho^\kappa = g^{\lambda\kappa}. \end{aligned} \tag{4.18}$$

The three vectors $\vec{e}_1$, $\vec{e}_2$, and $\vec{e}_3$ are called a set of triples and the vectors $\vec{e}^1$, $\vec{e}^2$, and $\vec{e}^3$ are also called an inverse set of triples. Since the vector $\vec{e}^1$ is orthogonal to the vectors $\vec{e}_2$ and $\vec{e}_3$ from Eq. (4.17), we have

$$\vec{e}^1 = \frac{1}{\sqrt{g}}(\vec{e}_2 \times \vec{e}_3), \quad \vec{e}^2 = \frac{1}{\sqrt{g}}(\vec{e}_3 \times \vec{e}_1), \quad \vec{e}^3 = \frac{1}{\sqrt{g}}(\vec{e}_1 \times \vec{e}_2). \tag{4.19}$$

Let us define

$$\vec{v} \cdot \vec{e}_\lambda = v_\lambda, \tag{4.20}$$

we have

$$v_\lambda = v^\kappa\vec{e}_\kappa \cdot \vec{e}_\lambda = v^\kappa g_{\kappa\lambda}. \tag{4.21}$$

Since $v_\lambda g^{\kappa\lambda} = v^\kappa g_{\kappa\lambda}g^{\kappa\lambda} = v^\kappa\delta_\kappa^\kappa = v^\kappa$, the following equation is derived:

$$v^\kappa = v_\lambda g^{\kappa\lambda}. \tag{4.22}$$

The vector components v^λ and v_λ are called a contravariant and covariant component, respectively.

Example 4.1. Compute the volume of the parallelepiped which the following three vectors form:

$$\vec{A} = \begin{pmatrix} 1 \\ 0 \\ 0 \end{pmatrix}, \quad \vec{B} = \begin{pmatrix} 0 \\ 1 \\ 0 \end{pmatrix}, \quad \vec{C} = \begin{pmatrix} 0 \\ 0 \\ 1 \end{pmatrix}. \tag{1}$$

Solution.

$$\vec{A} \cdot (\vec{B} \times \vec{C}) = \begin{pmatrix} 1 \\ 0 \\ 0 \end{pmatrix} \cdot \begin{vmatrix} \vec{i} & \vec{j} & \vec{k} \\ 0 & 1 & 0 \\ 0 & 0 & 1 \end{vmatrix} = 1. \tag{2}$$

Problem 4.1. Show that the cosine of the two vectors $\vec{u}$ and $\vec{v}$ is given by the following equation when $\vec{u} = u^\lambda \vec{e}_\lambda$ and $\vec{v} = v^\kappa \vec{e}_\kappa$.

$$\cos \theta = \frac{g_{\lambda\kappa} u^\lambda v^\kappa}{\sqrt{g_{\lambda\kappa} u^\lambda u^\kappa} \sqrt{g_{\lambda\kappa} v^\lambda v^\kappa}}. \tag{3}$$

4.2. Tensor

From a set of triples in the positive direction $\vec{e}_1$, $\vec{e}_2$, and $\vec{e}_3$, the following equation

$$\vec{e}_{\lambda'} = A^\lambda_{\lambda'} \vec{e}_\lambda \tag{4.23}$$

leads to a new set of triples $\vec{e}_{\lambda'}$. Then, the new set of $\vec{e}_{1'}$, $\vec{e}_{2'}$, and $\vec{e}_{3'}$ also triples. When the inverse matrix of $A^\lambda_{\lambda'}$ is $A^\kappa_{\kappa'}$,

$$A^{\kappa'}_\kappa A^\kappa_{\lambda'} = \delta^{\kappa'}_{\lambda'}, \quad \text{and} \quad A^{\lambda'}_\lambda A^\kappa_{\lambda'} = \delta^\kappa_\lambda \tag{4.24}$$

are derived. The vector $\vec{e}_\lambda$ in Eq. (4.23) is

$$\vec{e}_\lambda = A^{\lambda'}_\lambda \vec{e}_{\lambda'}. \tag{4.25}$$

If a vector $\vec{v}$ is expressed by the triples $\vec{e}_1$, $\vec{e}_2$, and $\vec{e}_3$, the following equation is obtained:

$$\vec{v} = v^\kappa \vec{e}_\kappa. \tag{4.26}$$

If we use different triples $\vec{e}_{1'}$, $\vec{e}_{2'}$, and $\vec{e}_{3'}$ the vector $\vec{v}$ is given by

$$\vec{v} = v^{\kappa'} \vec{e}_{\kappa'}. \tag{4.27}$$

Therefore,

$$v^{\kappa'}\vec{e}_{\kappa'} = v^{\kappa}\vec{e}_{\kappa}. \tag{4.28}$$

The substitution of $\vec{e}_{\kappa} = A_{\kappa}^{\kappa'}\vec{e}_{\kappa'}$ into Eq. (4.28) shows

$$v^{\kappa'}\vec{e}_{\kappa'} = v^{\kappa}A_{\kappa}^{\kappa'}\vec{e}_{\kappa'}. \tag{4.29}$$

Since $\vec{e}_{\kappa'}$ is linearly independent, we have

$$v^{\kappa'} = A_{\kappa}^{\kappa'}v^{\kappa}. \tag{4.30}$$

This indicates that v^{κ} component in the vector $\vec{v}$ is inversely transformed from Eq. (4.23). Thus, the component v^{κ} is called a contravariant component of $\vec{v}$ about the triples $\vec{e}_1$, $\vec{e}_2$, and $\vec{e}_3$. Since $\vec{v}\cdot\vec{e}_{\lambda} = v_{\lambda}$ and $\vec{v}\cdot\vec{e}_{\lambda'} = v_{\lambda'}$ are formed, the dot product of $\vec{e}_{\lambda'} = A_{\lambda'}^{\lambda}\vec{e}_{\lambda}$ and $\vec{v}$ leads to

$$v_{\lambda'} = A_{\lambda'}^{\lambda}v_{\lambda}. \tag{4.31}$$

This is the same transformation of Eq. (4.23). Thus, v_{λ} is the covariant component of $\vec{v}$. Going further, if we put

$$g_{\lambda\kappa} = \vec{e}_{\lambda}\cdot\vec{e}_{\kappa}, \qquad g_{\lambda'\kappa'} = \vec{e}_{\lambda'}\cdot\vec{e}_{\kappa'} \tag{4.32}$$

and substitute $\vec{e}_{\lambda'} = A_{\lambda'}^{\lambda}\vec{e}_{\lambda}$ and $\vec{e}_{\kappa'} = A_{\kappa'}^{\kappa}\vec{e}_{\kappa}$ into Eq. (4.32),

$$g_{\lambda'\kappa'} = \vec{e}_{\lambda'}\cdot\vec{e}_{\kappa'} = (A_{\lambda'}^{\lambda}\vec{e}_{\lambda})\cdot(A_{\kappa'}^{\kappa}\vec{e}_{\kappa}) = A_{\lambda'}^{\lambda}A_{\kappa'}^{\kappa}(\vec{e}_{\lambda}\cdot\vec{e}_{\kappa}) = A_{\lambda'}^{\lambda}A_{\kappa'}^{\kappa}g_{\lambda\kappa}. \tag{4.33}$$

There are the quantity $T_{\mu\lambda}^{\cdots\kappa}$ about the triples $\vec{e}_1$, $\vec{e}_2$, and $\vec{e}_3$ and the quantity $T_{\mu'\lambda'}^{\cdots\kappa'}$ about the triples $\vec{e}_{1'}$, $\vec{e}_{2'}$, and $\vec{e}_{3'}$. When the triples

$$\vec{e}_{\lambda'} = A_{\lambda'}^{\lambda}\vec{e}_{\lambda}, \quad \vec{e}_{\lambda} = A_{\lambda}^{\lambda'}\vec{e}_{\lambda'} \tag{4.34}$$

are given, $T_{\mu\lambda}^{\cdots\kappa}$ and $T_{\mu'\lambda'}^{\cdots\kappa'}$ are connected as follows:

$$T_{\mu'\lambda'}^{\cdots\kappa'} = A_{\mu'}^{\mu}A_{\lambda'}^{\lambda}A_{\kappa}^{\kappa'}T_{\mu\lambda}^{\cdots\kappa}. \tag{4.35}$$

The quantity which is expressed by $T_{\mu\lambda}^{\cdots\kappa}$ and $T_{\mu'\lambda'}^{\cdots\kappa'}$ is called a tensor of the first order contravariant and the second-order covariant. The quantities $T_{\mu\lambda}^{\cdots\kappa}$ and $T_{\mu'\lambda'}^{\cdots\kappa'}$ are tensor components of the triple system. A tensor of the p-th order contravariant and the zeroth-order covariant is called the contravariant tensor. A tensor of the zeroth-order contravariant and the zeroth-order covariant is called the covariant tensor. The vectors $\vec{v}^{\kappa}$, $\vec{v}_{\lambda}$, and the tensor $g_{\lambda\kappa}$ are the components of the contravariant vector, the covariant vector, and the covariant tensor, respectively.

Example 4.2. Show that $A^\lambda_\lambda = I$ is a unit matrix.

Solution. In Eq. (4.23), $\vec{e}_{\lambda'} = A^\lambda_{\lambda'}\vec{e}_\lambda$, when $\lambda = \lambda'$ is assumed, $\vec{e}_\lambda = A^\lambda_\lambda\vec{e}_\lambda$ is formed. Thus, $A^\lambda_\lambda = I$, in which I = the unit matrix.

Problem 4.2. Show that T^μ_μ is a scalar.
Hint: $T^{\mu'}_{\mu'} = A^\mu_{\mu'} A^{\mu'}_\mu T^\mu_\mu = T^\mu_\mu$.

4.3. Curvilinear Coordinate System

Three families of curved surfaces in the orthogonal coordinate system (Yano, 1957) are

$$F(x,y,z) = u^1, \quad G(x,y,z) = u^2, \quad H(x,y,z) = u^3. \qquad (4.36)$$

Let us assume that Eqs. (4.36) intersect at a point P in a domain D when the values of u^1, u^2, and u^3 are specified. The condition is that the Jacobian of Eqs. (4.36) is not zero as follows:

$$\frac{\partial(F,G,H)}{\partial(x,y,z)} = \begin{vmatrix} \dfrac{\partial F}{\partial x} & \dfrac{\partial G}{\partial x} & \dfrac{\partial H}{\partial x} \\[2mm] \dfrac{\partial F}{\partial y} & \dfrac{\partial G}{\partial y} & \dfrac{\partial H}{\partial y} \\[2mm] \dfrac{\partial F}{\partial z} & \dfrac{\partial G}{\partial z} & \dfrac{\partial H}{\partial z} \end{vmatrix} \neq 0. \qquad (4.37)$$

In this case, the specific values u^1, u^2, and u^3 determine three curved surfaces $F = u^1$, $G = u^2$, and $H = u^3$ and the intersection of three curved surfaces shows the point P. Inversely, if the point P is fixed, the coordinates (x,y,z) of the point P denote the values of u^1, u^2, and u^3 by using Eq. (4.36). That is, since the point P corresponds to a set of three numbers in the domain D, u^1, u^2, and u^3, the three numbers u^1, u^2, and u^3 are considered to be the coordinates of the point P. These three numerals u^1, u^2, and u^3 are called curvilinear coordinates. In this case, the numerical values of x, y, and z are computed as:

$$x = f(u^1, u^2, u^3), \quad y = g(u^1, u^2, u^3), \quad z = h(u^1, u^2, u^3). \qquad (4.38)$$

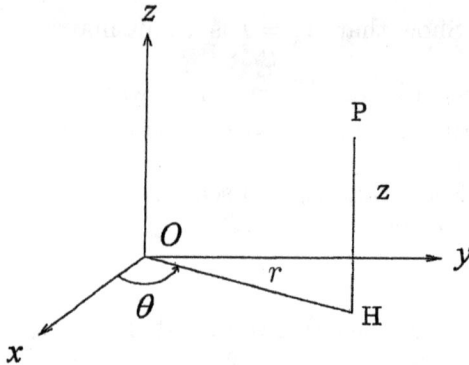

Figure 4.1.   Cylindrical coordinate system.

Equation (4.36) draws a curve for changing u^1 when u^2 and u^3 are fixed. This is called a u^1 curve. In the same way, a u^2 curve and u^3 curve are defined. These three are coordinate curves. When the coordinate system is linear (curved), the coordinate curves are linear (curved). The curved surfaces which are defined by $F(x, y, z) = u^1$, $G(x, y, z) = u^2$, and $H(x, y, z) = u^3$ are called a u^1 curved surface, a u^2 curved surface, and a u^3 curved surface, respectively. They are called the coordinate system of curved surfaces. As shown in Fig. 4.1, we define a perpendicular line which passes through the point P to the xy plane. The intersection is a point H on the xy plane. When we define $OH = r$, $\angle xOH = \theta$, and $HP = z$, the three variables r, θ, and z form the cylindrical coordinate system as $x = r\cos\theta$, $y = r\sin\theta$, and $z = z$, in which O indicates the origin of the coordinate system. The variables r, θ, and z are the cylindrical coordinate system which is curvilinear. When the point P is expressed by the curvilinear coordinates u^1, u^2, and u^3 and a vector from the origin O to the point P is $\vec{X}$, the vector $\vec{X}$ is the function of u^1, u^2, and u^3 as explained in Fig. 4.2. Then, we have

$$\vec{X}_1 = \frac{\partial \vec{X}}{\partial u^1}, \quad \vec{X}_2 = \frac{\partial \vec{X}}{\partial u^2}, \quad \vec{X}_3 = \frac{\partial \vec{X}}{\partial u^3}. \tag{4.39}$$

These vectors are tangent to the u^1 curve, the u^2 curve, and the u^3 curve and linearly independent of each other. These curves form

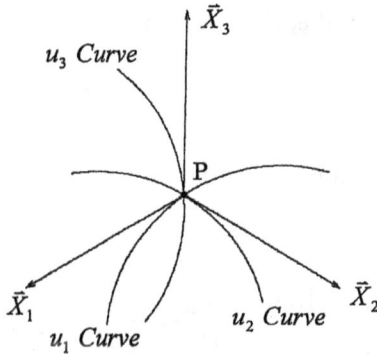

Figure 4.2.   Curvilinear coordinate system.

triples of vectors. The vectors $\vec{X}(u)$ and $\vec{X}(u) + d\vec{X}(u)$ which originate from the origin correspond to the curvilinear coordinates u^κ and $u^\kappa + du^\kappa$ for $(\kappa = 1, 2, 3)$. The curvilinear coordinates u^κ and $u^\kappa + du^\kappa$ also start from the origin O. Then, $d\vec{X}$ is given by

$$d\vec{X} = \vec{X}_1 du^1 + \vec{X}_2 du^2 + \vec{X}_3 du^3. \tag{4.40}$$

The squared distance $ds^2 = (d\vec{X})^2$ between u^κ and $u^\kappa + du^\kappa$ is shown by $ds^2 = d\vec{X} \cdot d\vec{X} = (\vec{X}_1 du^1 + \vec{X}_2 du^2 + \vec{X}_3 du^3) \cdot (\vec{X}_1 du^1 + \vec{X}_2 du^2 + \vec{X}_3 du^3)$. Thus, it is written as

$$g_{\mu\lambda} = \vec{X}'_\mu \cdot \vec{X}_\lambda. \tag{4.41}$$

Therefore, the following equation is derived:

$$ds^2 = g_{\mu\lambda}(u) du^\mu du^\lambda. \tag{4.42}$$

Equation (4.42) is called a fundamental metric form. The quantities g_{11}, g_{22}, and g_{33} are squares of the vectors $\vec{X}_1$, $\vec{X}_2$, and $\vec{X}_3$, respectively. When the angles between the vectors $\vec{X}_2$ and $\vec{X}_3$ or the coordinates u^2 and u^3 curves, the vectors $\vec{X}_3$ and $\vec{X}_1$ or the coordinates u^3 and u^1 curves, and the vectors $\vec{X}_1$ and $\vec{X}_2$ or the coordinates u^1 and u^2 curves are denoted by θ_{23}, θ_{31}, and θ_{12}, respectively. Then, their direction cosines are written as

$$\cos\theta_{23} = \frac{g_{23}}{\sqrt{g_{22}}\sqrt{g_{33}}}, \quad \cos\theta_{31} = \frac{g_{31}}{\sqrt{g_{33}}\sqrt{g_{11}}}, \quad \cos\theta_{12} = \frac{g_{12}}{\sqrt{g_{11}}\sqrt{g_{22}}}.$$

$$\tag{4.43}$$

Thus, the necessary and satisfactory conditions for the orthogonality of coordinate curves are

$$g_{23} = g_{31} = g_{12} = 0. \tag{4.44}$$

The coordinate system which satisfies Eq. (4.44) is called the orthogonal curvilinear coordinate system. For the volume element $dV = dxdydz$, we have

$$dV = dxdydz = \begin{vmatrix} \dfrac{\partial x}{\partial u^1} & \dfrac{\partial y}{\partial u^1} & \dfrac{\partial z}{\partial u^1} \\ \dfrac{\partial x}{\partial u^2} & \dfrac{\partial y}{\partial u^2} & \dfrac{\partial z}{\partial u^2} \\ \dfrac{\partial x}{\partial u^3} & \dfrac{\partial y}{\partial u^3} & \dfrac{\partial z}{\partial u^3} \end{vmatrix} du^1 du^2 du^3. \tag{4.45}$$

Using the Jacobian in Eq. (4.45), the following equation is obtained:

$$\begin{bmatrix} \dfrac{\partial x}{\partial u^1} & \dfrac{\partial y}{\partial u^1} & \dfrac{\partial z}{\partial u^1} \\ \dfrac{\partial x}{\partial u^2} & \dfrac{\partial y}{\partial u^2} & \dfrac{\partial z}{\partial u^2} \\ \dfrac{\partial x}{\partial u^3} & \dfrac{\partial y}{\partial u^3} & \dfrac{\partial z}{\partial u^3} \end{bmatrix} \cdot \begin{bmatrix} \dfrac{\partial x}{\partial u^1} & \dfrac{\partial x}{\partial u^2} & \dfrac{\partial x}{\partial u^3} \\ \dfrac{\partial y}{\partial u^1} & \dfrac{\partial y}{\partial u^2} & \dfrac{\partial y}{\partial u^3} \\ \dfrac{\partial z}{\partial u^1} & \dfrac{\partial z}{\partial u^2} & \dfrac{\partial z}{\partial u^3} \end{bmatrix} = \begin{vmatrix} g_{11} & g_{12} & g_{13} \\ g_{21} & g_{22} & g_{23} \\ g_{31} & g_{32} & g_{33} \end{vmatrix} = g. \tag{4.46}$$

Therefore, when the determinant of the matrix $g_{\mu\lambda}$ is g, the volume element $dV = \sqrt{g}\,du^1 du^2 du^3$. Since for the cylindrical coordinate system $x = r\cos\theta$, $y = r\sin\theta$, and $z = z$, the vectors $\vec{X}_1$, $\vec{X}_2$, and $\vec{X}_3$ for $r = u^1$, $\theta = u^2$, and $z = u^3$, respectively become

$$\begin{aligned} \vec{X}_1&: (\quad \cos\theta \qquad \sin\theta \quad\ 0\ )^T \\ \vec{X}_2&: (\ -r\sin\theta \ \ r\cos\theta \ \ 0\ )^T \\ \vec{X}_3&: (\quad 0 \qquad\quad 0 \qquad 1\ )^T \end{aligned}$$

in which the superscript T denotes the transpose. Thus, from Eq. (4.46) the elements $g_{\mu\lambda}$ are

$$g_{\mu\lambda} = \begin{pmatrix} 1 & 0 & 0 \\ 0 & r^2 & 0 \\ 0 & 0 & 1 \end{pmatrix}. \tag{4.47}$$

The squared distance ds^2 in space is

$$ds^2 = dr^2 + r^2 d\theta^2 + dz^2. \tag{4.48}$$

The volume element is also

$$dV = r dr d\theta dz. \tag{4.49}$$

From Eq. (4.47), the cylindrical coordinate system is the orthogonal curvilinear coordinate system.

Example 4.3. Obtain $g_{\mu\lambda}$ for the spherical coordinate system when $x = r \sin\theta \cos\phi$, $y = r \sin\theta \sin\phi$, and $z = r \cos\theta$.

Solution. If $r = u^1$, $\theta = u^2$, and $\phi = u^3$, we get

$$
\begin{array}{ll}
\vec{X}_1: & (\quad \sin\theta\cos\phi \qquad \sin\theta\sin\phi \qquad \cos\theta \quad)^T \\
\vec{X}_2: & (\ r\cos\theta\cos\phi \quad r\cos\theta\sin\phi \quad -r\sin\theta\ )^T. \\
\vec{X}_3: & (-r\sin\theta\sin\phi \quad r\sin\theta\cos\phi \qquad 0 \quad)^T
\end{array} \tag{1}
$$

Thus,

$$
g_{\mu\lambda} =
\begin{bmatrix}
\sin\theta\cos\phi & \sin\theta\sin\phi & \cos\theta \\
r\cos\theta\cos\phi & r\cos\theta\sin\phi & -r\sin\theta \\
-r\sin\theta\sin\phi & r\sin\theta\cos\phi & 0
\end{bmatrix}
$$

$$
\times
\begin{bmatrix}
\sin\theta\cos\phi & \sin\theta\sin\phi & \cos\theta \\
r\cos\theta\cos\phi & r\cos\theta\sin\phi & -r\sin\theta \\
-r\sin\theta\sin\phi & r\sin\theta\cos\phi & 0
\end{bmatrix}^T
$$

$$
=
\begin{bmatrix}
1 & 0 & 0 \\
0 & r^2 & 0 \\
0 & 0 & r^2\sin^2\theta
\end{bmatrix}, \tag{2}
$$

$$dV = \sqrt{g}\, du^1 du^2 du^3 = r^2 \sin\theta\ dr d\theta d\phi, \tag{3}$$

in which T in Eq. (2) indicates the transpose.

Problem 4.3. When the two-dimensional cylindrical coordinate system is defined by $x = r\cos\theta$ and $y = r\sin\theta$, calculate the metric tensor $g_{\mu\lambda}$.

4.4. Vector Analysis

When a vector $\vec{A}$ varies as a parameter τ changes, the derivative of $\vec{A}$ is defined as

$$\frac{d\vec{A}}{d\tau} = \lim_{\Delta\tau \to 0} \frac{\vec{A}(\tau + \Delta\tau) - \vec{A}(\tau)}{\Delta\tau}. \tag{4.50}$$

Practically, the derivative of $\vec{A}$ is obtained by differentiating each component of the vector $\vec{A}$. When there is a scalar field of the velocity potential ϕ, the derivative of ϕ about x, y, and z is called a gradient vector of ϕ and expressed by $grad \ \phi$ or $\nabla\phi$. Their components are $\partial\phi/\partial x$, $\partial\phi/\partial y$, and $\partial\phi/\partial z$, respectively. Using unit vectors $\vec{i}$, $\vec{j}$, and $\vec{k}$, we have

$$\nabla\phi = \frac{\partial\phi}{\partial x}\vec{i} + \frac{\partial\phi}{\partial y}\vec{j} + \frac{\partial\phi}{\partial z}\vec{k}. \tag{4.51}$$

Thus, the vector operator, a nabla ∇ is defined as

$$\nabla = \vec{i}\frac{\partial}{\partial x} + \vec{j}\frac{\partial}{\partial y} + \vec{k}\frac{\partial}{\partial z}. \tag{4.52}$$

The total derivative of ϕ as a scalar is

$$d\phi = \nabla\phi \cdot d\vec{r}, \tag{4.53}$$

in which $r = \sqrt{x^2 + y^2 + z^2}$. When a vector $\vec{A}$ is defined in space, the space is called a vector space. The x, y, and z components of the vector $\vec{A}$ are A_x, A_y, and A_z, respectively. Thus,

$$\nabla \cdot \vec{A} = \frac{\partial A_x}{\partial x} + \frac{\partial A_y}{\partial y} + \frac{\partial A_z}{\partial z} = div \ \vec{A}. \tag{4.54}$$

This is called the divergence of the vector $\vec{A}$. When the vector $\vec{A}$ shows velocity, the following equation is derived:

$$\nabla \cdot \vec{A} = 0. \tag{4.55}$$

This is the continuity equation in fluid mechanics. Since the nabla is the vector operator, Eq. (4.55) is a scalar. The gradient vector of the velocity potential ϕ is the velocity. The divergence of the gradient vector is

$$\nabla(\nabla\phi) = \nabla^2\phi = \frac{\partial^2\phi}{\partial x^2} + \frac{\partial^2\phi}{\partial y^2} + \frac{\partial^2\phi}{\partial z^2}, \tag{4.56}$$

in which ∇^2 is a Laplace operator and an equation

$$\nabla^2\phi = 0 \tag{4.57}$$

is called a Laplace equation and the velocity potential ϕ which satisfies Eq. (4.57) is the harmonic function. The rotation of the vector $\vec{A}$ is given by

$$rot\ \vec{A} = \begin{vmatrix} \vec{i} & \vec{j} & \vec{k} \\ \dfrac{\partial}{\partial x} & \dfrac{\partial}{\partial y} & \dfrac{\partial}{\partial z} \\ A_x & A_y & A_z \end{vmatrix} = \nabla \times \vec{A}. \tag{4.58}$$

This is called a rotation of the vector $\vec{A}$ and also written as *curl* $\vec{A}$. When the closed surface S forms the volume V, the following divergence theorem for the vector $\vec{A}$ is derived:

$$\iiint_V div\ \vec{A}\ dV = \iint_S \vec{A} \cdot \vec{n}\ dS, \tag{4.59}$$

in which $\vec{n}$ is a normal unit vector on the surface S which surrounds the volume V. The meaning of the closed surface is that the volume is completely surrounded by the surface. The right side of Eq. (4.59) denotes fluid discharge from the surface S in fluid mechanics. This is also called a Gauss theorem. Applying Eq. (4.59) in the two dimensions, the following equation is obtained:

$$\int_C (A_x dx + A_y dy) = \iint_S \left(\frac{\partial A_y}{\partial x} - \frac{\partial A_x}{\partial y} \right) dx dy, \tag{4.60}$$

in which C indicates the closed curve surrounding the surface S. This equation leads to the following Stokes theorem:

$$\int_C \vec{A} \cdot \vec{t}\ ds = \iint_S \vec{n} \cdot rot\ \vec{A}\ dS, \tag{4.61}$$

in which $\vec{t}$ is a tangent unit vector on the curve C and ds is an infinitesimal line element. In Eq. (4.59), when the scalar functions

f and g are assumed as

$$\vec{A} = f\nabla g, \quad \nabla g \cdot \vec{n} = \frac{\partial g}{\partial n}, \tag{4.62}$$

the following Green theorem is derived:

$$\iint_S f\frac{\partial g}{\partial n}\,dS = \iiint_V \left[f\nabla^2 g + (\nabla f)(\nabla g)\right]dV. \tag{4.63}$$

The computation of the vector analysis is necessary to understand fluid mechanics.

Example 4.4. Derive the following equation:

$$\iint_S \left(f\frac{\partial g}{\partial n} - g\frac{\partial f}{\partial n}\right)dS = \iiint_V (f\nabla^2 g - g\nabla^2 f)dV. \tag{1}$$

When the functions f and g are harmonic, show that

$$\iint_S \left(f\frac{\partial g}{\partial n} - g\frac{\partial f}{\partial n}\right)dS = 0. \tag{2}$$

Solution. Change the functions f and g in Eq. (4.63). Then, we get

$$\iint_S g\frac{\partial f}{\partial n}\,dS = \iiint_V \left[g\nabla^2 f + (\nabla g)(\nabla f)\right]dV. \tag{3}$$

Subtracting the above equation from Eq. (4.63), we obtain

$$\iint_S f\frac{\partial g}{\partial n}\,dS - \iint_S g\frac{\partial f}{\partial n}\,dS = \iiint_V (f\nabla^2 g - g\nabla^2 f)dV. \tag{4}$$

Since $\nabla^2 f = \nabla^2 g = 0$, we obtain

$$\iint_S \left(f\frac{\partial g}{\partial n} - g\frac{\partial f}{\partial n}\right)dS = 0. \tag{5}$$

Problem 4.4. For two vectors $\vec{A}$ and $\vec{B}$, show that

$$div(\vec{A} \times \vec{B}) = \vec{B}\,rot\,\vec{A} - \vec{A}\,rot\,\vec{B}. \tag{6}$$

4.5. Application to Navier–Stokes Equations

The equation to explain viscous fluid motion is the Navier–Stokes equations (Prager, 1961). Let us write these equations using a vector and tensor. According to Aris (1962), the equations of viscous fluid motion are given by

$$\rho \frac{dv_i}{dt} = \rho f_i + T_{ij,j},$$ (4.64)

in which ρ = fluid density, v_i = fluid velocity in $i-$ direction, t = time, f_i = body force in $i-$ direction, and $T_{ij,j}$ = partial derivative of stress tensor T_{ij} about x_j. The stress tensor T_{ij} of viscous fluid is expressed by

$$T_{ij} = (-p + \lambda\Theta)\delta^i_j + 2\mu e_{ij},$$ (4.65)

in which p = pressure, μ = dynamic viscosity, λ = a coefficient, Θ = divergence of strains, e_{ij} = deformation, and δ^i_j = Kronecker delta. The divergence Θ of strains is defined as follows:

$$\Theta = e_{11} + e_{22} + e_{33}.$$ (4.66)

The velocity gradient tensor can be written as the sum of symmetric and antisymmetric parts. It is

$$\frac{\partial v_i}{\partial x_j} = \frac{1}{2}\left(\frac{\partial v_i}{\partial x_j} + \frac{\partial v_j}{\partial x_i}\right) + \frac{1}{2}\left(\frac{\partial v_i}{\partial x_j} - \frac{\partial v_j}{\partial x_i}\right) = e_{ij} + \Omega_{ij},$$ (4.67)

in which $e_{ij} = \frac{1}{2}(\frac{\partial v_i}{\partial x_j} + \frac{\partial v_j}{\partial x_i})$ and $\Omega_{ij} = \frac{1}{2}(\frac{\partial v_i}{\partial x_j} - \frac{\partial v_j}{\partial x_i})$. The relationship between λ and μ is introduced by Stokes.

$$\lambda + \frac{2}{3}\mu = 0.$$ (4.68)

Equation (4.68) is formed for imcompressible fluid such as water and air. Thus, the effect of the coefficient λ is not found in the Navier–Stokes equations. The partial derivative of the symmetric strain e_{ij} is

$$e_{ij,j} = \frac{1}{2}\frac{\partial}{\partial x_j}\left(\frac{\partial v_i}{\partial x_j} + \frac{\partial v_j}{\partial x_i}\right) = \frac{1}{2}\nabla^2 v_i + \frac{1}{2}\frac{\partial}{\partial x_i}(\nabla \cdot \vec{v}).$$ (4.69)

The partial derivative of Eq. (4.65) about x_j shows

$$T_{ij,j} = -\frac{\partial p}{\partial x_i} + (\lambda + \mu)\frac{\partial}{\partial x_i}(\nabla \cdot \vec{v}) + \mu\nabla^2 v_i.$$ (4.70)

Thus, the Navier–Stokes equations are written in vector and tensor form as:

$$\rho\frac{dv_i}{dt} = \rho f_i - \frac{\partial p}{\partial x_i} + (\lambda + \mu)\frac{\partial}{\partial x_i}(\nabla \cdot \vec{v}) + \mu\nabla^2 v_i. \qquad (4.71)$$

Example 4.5. Differentiate e_{ij} about x_j.

Solution.

$$e_{ij,j} = \frac{1}{2}\frac{\partial}{\partial x_j}\left(\frac{\partial v_i}{\partial x_j} + \frac{\partial v_j}{\partial x_i}\right)$$

$$= \frac{1}{2}\frac{\partial^2 v_i}{\partial x_j \partial x_j} + \frac{1}{2}\frac{\partial}{\partial x_i}\frac{\partial v_j}{\partial x_j} = \frac{1}{2}\nabla^2 v_i + \frac{1}{2}\frac{\partial}{\partial x_i}(\nabla \cdot \vec{v}). \qquad (1)$$

Problem 4.5. Derive Eq. (4.70) from Eq. (4.65).

Chapter 5

Ordinary Differential Equations

Ordinary differential equations which are used in hydraulic, hydrologic, and coastal engineering are derived from the Euler equation of motion or the Navier–Stokes equations. Thus, the order of the ordinary differential equations is usually two. The continuity equation indicates the first- or the second-order. To investigate fluid motion, the ordinary differential equations can be solved analytically. If the ordinary differential equations are linear, the analytical solution can be obtained. This chapter gives skills to treat linear ordinary differential equations. The skills are very important to handle the following partial differential equations.

5.1. Method of Separation of Variables

When x and y are an independent and a dependent variable, respectively, the derivative is given by

$$\frac{dy}{dx} = f(x)g(y). \tag{5.1}$$

Then, Eq. (5.1) is called a type of the separation of variables. Functions $f(x)$ and $g(y)$ are given functions of x and y, respectively. The solution of Eq. (5.1) is determined by

$$\int \frac{dy}{g(y)} = \int f(x)dx. \tag{5.2}$$

The substitution of variables x and y reduces to the type of separation of variables as the homogeneous form

$$\frac{dy}{dx} = f\left(\frac{y}{x}\right). \tag{5.3}$$

The substitution of $y = xz$, shows that the dependent variable changes from y to z. Thus, Eq. (5.3) is written as

$$x\frac{dz}{dx} + z = f(z). \tag{5.4}$$

The solution of Eq. (5.4) is

$$\int \frac{dz}{f(z) - z} = \int \frac{dx}{x}. \tag{5.5}$$

These are elementary methods to solve ordinary differential equations of the first-order.

Example 5.1. There is a two-dimensional velocity field. When the velocity in the x-direction and y-direction are $u = \frac{1}{x}$ and $v = \frac{y}{x^2}$, respectively, obtain the equation of streamline.

Solution. The equation of streamline is

$$\frac{dx}{u} = \frac{dy}{v}, \tag{1}$$

in which $u = $ velocity in the x-direction and $v = $ velocity in the y-direction. Since $u = \frac{1}{x}$ and $v = \frac{y}{x^2}$,

$$\int \frac{dx}{x} = \int \frac{dy}{y} + C, \tag{2}$$

in which $C = $ an integral constant. The integral of Eq. (2) shows

$$\frac{x}{y} = e^C. \tag{3}$$

Problem 5.1. Transform the following equation to the form of separation of variables:

$$x\frac{dy}{dx} + y + \frac{y^2}{x} = 0. \tag{4}$$

5.2. Linear Differential Equations of the First-Order

Let us consider the following linear differential equations of the first-order:

$$\frac{dy}{dx} + f(x)y = g(x). \tag{5.6}$$

When $g(x) = 0$, Eq. (5.6) is written as

$$\frac{dy}{dx} + f(x)y = 0. \tag{5.7}$$

The solution of Eq. (5.7) is obtained by the separation of variables as

$$\int \frac{dy}{y} + \int f(x)dx = 0. \tag{5.8}$$

Thus, the solution of Eq. (5.7) is given by

$$y = \exp\left[-\int f(x)dx\right]. \tag{5.9}$$

When we define

$$F(x) = \exp\left[\int f(x)dx\right], \tag{5.10}$$

multiply $F(x)$ to Eq. (5.6). Then, we get

$$\exp\left[\int f(x)dx\right]\frac{dy}{dx} + \exp\left[\int f(x)dx\right]f(x)y$$

$$= \exp\left[\int f(x)dx\right]g(x). \tag{5.11}$$

Noting the following equation from Eq. (5.10),

$$F'(x) = \exp\left[\int f(x)dx\right]f(x), \tag{5.12}$$

we simplify Eq. (5.11) as

$$\frac{d}{dx}[F(x)y] = F(x)g(x). \tag{5.13}$$

Equation (5.13) becomes the fundamental form of the separation of variables. The solution is easily

$$y = \frac{1}{F(x)}\left[\int F(x)g(x)dx + C\right] = \exp\left[-\int f(x)dx\right]$$

$$\times \left\{\int \exp\left[\int f(x)dx\right]g(x)\right\}dx + C\exp\left[-\int f(x)dx\right], \tag{5.14}$$

in which C = an integral constant. It is very important that the exponent of y of the second term on the left side of Eq. (5.6) is 1. Otherwise, Eq. (5.6) becomes nonlinear and usually cannot be analytically solved. Let us consider the famous Bernoulli's differential equation which is transformed to the linear differential equation. It is

$$y' + P(x)y = Q(x)y^n, \quad n \neq 0, 1, \tag{5.15}$$

in which "$'$" $= d/dx$. Substituting $y^{1-n} = u$ into Eq. (5.15), we have

$$u' - (n-1)P(x)u = -(n-1)Q(x). \tag{5.16}$$

This is the same type as Eq. (5.6) and analytically solvable. Next is the following Clairaut's differential equation:

$$y = xy' + f(y'). \tag{5.17}$$

Differentiating Eq. (5.17) about x, we have

$$y' = y' + xy'' + f'(y')y''. \tag{5.18}$$

Equation (5.18) is written as

$$y''[x + f'(y')] = 0. \tag{5.19}$$

Equation (5.19) induces $y'' = 0$ or $x + f'(y') = 0$. The former derives $y = C_1 x + C_2$ ($y' = C_1$), in which C_1 and C_2 are integral constants. Substituting $y' = C_1$ into Eq. (5.17), we obtain

$$y = C_1 x + f(C_1). \tag{5.20}$$

Equation (5.20) is the general solution. For the latter the following two equations:

$$y = xy' + f(y') \quad \text{and} \quad x + f'(y') = 0 \tag{5.21}$$

lead to another solution. If y' is eliminated from two equations of Eq. (5.21), another solution is obtained. This solution is called a particular one which envelops Eq. (5.20).

Next, let us consider a problem in hydraulics. As shown in Fig. 5.1, water flow in a horizontal pipe AB of which cross-sectional area a is constant and length is ℓ is connected to an upstream water tank of which water level H is constant above pipe AB. There is a valve at the outlet of the point B and the valve is abruptly opened at the

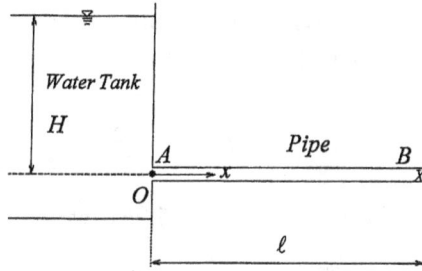

Figure 5.1.   Water flow from water flume.

time $t = 0$. The point A is defined as the origin of the x axis along the pipe AB. Then, the Euler equation of motion in the horizontal pipe AB is written as

$$\frac{\partial u}{\partial t} + u \frac{\partial u}{\partial x} = -\frac{1}{\rho}\frac{\partial p}{\partial x}, \tag{5.22}$$

in which $u = $ velocity, $\rho = $ water density, and $p = $ pressure. If ρ is constant and the pipe does not deform, the equation of continuity is given by

$$au = F(t), \tag{5.23}$$

in which $F(t)$ is an arbitrary function of time. Since u is the function of time t only, $\frac{\partial u}{\partial x} = 0$ is formed. Therefore, the Euler equation of motion Eq. (5.22) is simplified as

$$\frac{du}{dt} = -\frac{1}{\rho}\frac{\partial p}{\partial x}. \tag{5.24}$$

Since the pressure in the pipe is unknown, the pressure p is assumed to be linearly distributed in the pipe AB. When the pressure at the point A is $p_A(t)$ and $p_B = 0$ at the point B, the following equation for the pressure gradient is assumed:

$$\frac{dp}{dx} = -\frac{p_A(t)}{\ell}. \tag{5.25}$$

Applying the Bernoulli equation in the pipe between the point A and B, we have

$$H = \frac{u_A^2}{2g} + \frac{p_A}{\gamma}, \tag{5.26}$$

in which u_A = the velocity at the point A, g = the gravitational acceleration, and γ = the specific weight of water. The velocity is

$$u_A = \sqrt{2g\left(H - \frac{p_A}{\gamma}\right)}.$$ (5.27)

Substituting p_A in Eq. (5.26) to Eq. (5.25), we obtain

$$\frac{dp}{dx} = -\frac{\gamma}{\ell}\left(H - \frac{u^2}{2g}\right).$$ (5.28)

The velocity u_A at the point A is equal to the velocity in the pipe u. The subsitution of Eq. (5.28) into Eq. (5.24) gives the following nonlinear ordinary differential equation.

$$\frac{du}{dt} + \frac{u^2}{2\ell} - \frac{gH}{\ell} = 0.$$ (5.29)

Equation (5.29) is transformed to

$$\frac{du}{\frac{gH}{\ell} - \frac{u^2}{2\ell}} = dt.$$ (5.30)

The left side is transformed to the form of partial fraction. Thus, we get

$$\frac{1}{\sqrt{2gH}}\left(\frac{1}{\sqrt{2gH} - u} + \frac{1}{\sqrt{2gH} + u}\right)du = \frac{dt}{\ell}.$$ (5.31)

Integrating Eq. (5.31), we obtain

$$\frac{1}{\sqrt{2gH}}\log\frac{\sqrt{2gH} + u}{\sqrt{2gH} - u} = \frac{t}{\ell} + c_1.$$ (5.32)

When the valve is opened at $t = 0$, the velocity $u = 0$. Therefore, $c_1 = 0$. Thus, u in Eq. (5.32) becomes

$$u = \frac{e^{kt} - 1}{e^{kt} + 1}\sqrt{2gH},$$ (5.33)

in which $k = \frac{\sqrt{2gH}}{\ell}$. Equation (5.33) indicates that $u = 0$ at $t = 0$ and $u \to \sqrt{2gH}$ for $t \to \infty$. Mizumura (1995a) succeeded in the computation of rainfall and runoff process, using a simple water tank as Fig. 5.1. The coefficients for the water tank were determined from information of recession curves of the observed hydrograph. The resultant continuity equation is linear.

Example 5.2. When an inflow, outflow, and time are I, O, and t, respectively, the continuity equation is

$$I - O = \frac{dS}{dt}, \tag{1}$$

in which S is a storage function. When $S = KO$ is given and K is a constant in Eq. (1), solve Eq. (1).

Solution. Substituting $y = O$, $x = t$, $f(x) = 1/K$, and $g(x) = I/K$ into Eq. (5.14), we have

$$O = \exp(-t/K) \int \exp(t/K) \frac{I}{K} dt + C \exp(-t/K). \tag{2}$$

Problem 5.2. Solve $y = xy' + y'^2$.

5.3. Linear Differential Equations of the Second-Order

The following linear differential equation of the second-order (Ince, 1956):

$$\frac{d^2y}{dx^2} + f(x)\frac{dy}{dx} + g(x)y = h(x) \tag{5.34}$$

is not usually solved. The functions $f(x)$, $g(x)$, and $h(x)$ are differentiable. But the next types are solved.

(A) When $h(x) = 0$, Eq. (5.34) is homogeneous and a solution is known:

When a function $y = \phi(x)$ satisfies the homogeneous form of Eq. (5.34), define $y = \phi(x)z$ and substitute it into Eq. (5.34). Then, Eq. (5.34) is rearranged to

$$\frac{d^2}{dx^2}[\phi(x)z] + f(x)\frac{d}{dx}[\phi(x)z] + g(x)\phi(x)z = h(x). \tag{5.35}$$

Equation (5.35) is rewritten as

$$\frac{d^2z}{dx^2} + \left[2\frac{\phi'(x)}{\phi(x)} + f(x)\right]\frac{dz}{dx} = \frac{h(x)}{\phi(x)}, \tag{5.36}$$

in which "$'$" $= d/dx$.

Equation (5.36) is the linear ordinary differential equation of the first-order about dz/dx. This is the same as Eq. (5.6). To both sides of Eq. (5.36),

$$\phi^2(x)e^{\int f(x)dx} \qquad (5.37)$$

is multiplied. Then, Eq. (5.36) becomes

$$\frac{d}{dx}\left[\phi^2(x)e^{\int f(x)dx}\frac{dz}{dx}\right] = \phi(x)h(x)e^{\int f(x)dx}. \qquad (5.38)$$

Thus, Eq. (5.38) is

$$\frac{dz}{dx} = \frac{1}{\phi^2(x)}e^{-\int f(x)dx}\left[\int \phi(x)h(x)e^{\int f(x)dx}dx + C_1\right]. \qquad (5.39)$$

Integrating Eq. (5.39) again, we obtain

$$z = \int \frac{1}{\phi^2(x)}e^{-\int f(x)dx}\left[\int \phi(x)h(x)e^{\int f(x)dx}dx + C_1\right]dx + C_2, \qquad (5.40)$$

in which C_1 and C_2 are integration constants. The solution y is obtained by multiplying Eq. (5.40) and $\phi(x)$. For an example, a function $\phi(x) = x$ satisfies the homogeneous form of the following linear ordinary differential equation of the second-order:

$$\frac{d^2y}{dx^2} - x\frac{dy}{dx} + y = x^2. \qquad (5.41)$$

(B) When the left side of Eq. (5.34) is factorized.

The left side of Eq. (5.34) is factorized as follows:

$$\left[\frac{d}{dx} + p_1(x)\right]\left[\frac{dy}{dx} + p_2(x)y\right] = h(x). \qquad (5.42)$$

Equation (5.42) reduces to two linear ordinary differential equations of the first-order. They are

$$Y = \frac{dy}{dx} + p_2(x)y \qquad (5.43)$$

and

$$\frac{dY}{dx} + p_1(x)Y = h(x), \qquad (5.44)$$

in which $p_1(x) + p_2(x) = f(x)$ and $p_1(x)p_2(x) + \frac{dp_2(x)}{dx} = g(x)$ in Eq. (5.34). Since Eq. (5.44) is the linear ordinary differential equation of the first-order of Y, the solution is easily obtained as Eq. (5.6). When we substitute the solution Y into Eq. (5.43), the resultant equation becomes the linear ordinary differential equation of the first-order as Eq. (5.6). These solutions are easily obtained by Eq. (5.14).

Example 5.3. Solve the following linear ordinary differential equation of the second-order.

$$y'' + 5xy' + (6x^2 + 3)y = x, \tag{1}$$

in which $y(0) = 1$ and $y'(0) = 0$.

Solution. This corresponds to case (B). Using Eq. (5.42), Eq. (1) is factorized as:

$$\left(\frac{d}{dx} + 2x\right)(y' + 3xy) = x. \tag{2}$$

When we define

$$Y = y' + 3xy, \tag{3}$$

Eq. (2) becomes

$$\frac{dY}{dx} + 2xY = x. \tag{4}$$

Using Eq. (5.14), the solution of Eq. (4) is

$$Y = \frac{1}{2} + C_1 e^{-x^2}. \tag{5}$$

Thus, Eq. (3) becomes

$$\frac{dy}{dx} + 3xy = \frac{1}{2} + C_1 e^{-x^2}. \tag{6}$$

From Eq. (5.14) the solution of Eq. (6) is

$$y = e^{-3x^2/2} \int_0^x e^{3x^2/2}\left(\frac{1}{2} + C_1 e^{-x^2}\right)dx + C_2 e^{-3x^2/2}. \tag{7}$$

From $y(0) = 1$ we have $C_2 = 1$. Differentiating Eq. (7), we obtain

$$\frac{dy}{dx} = -3xe^{-3x^2/2} \int_0^x e^{3x^2/2}\left(\frac{1}{2} + C_1 e^{-x^2}\right) dx$$

$$+ \left(\frac{1}{2} + C_1 e^{-x^2}\right) - 3xe^{-3x^2/2}. \tag{8}$$

By the boundary condition $y'(0) = 0$ the substitution of $x = 0$ into Eq. (8) indicates $0 = 1/2 + C_1$. Thus, $C_1 = -1/2$ is obtained. Therefore, the solution is written as

$$y = e^{-3x^2/2} \int_0^x e^{3x^2/2}\left(\frac{1}{2} - \frac{1}{2}e^{-x^2}\right) dx + e^{-3x^2/2}. \tag{9}$$

Problem 5.3. Solve the following linear differential equation of the second-order:

$$y'' + 3xy' + (2x^2 + 1)y = \sin x, \quad y(0) = 1, \; y'(0) = 0. \tag{10}$$

5.4. Operational Method

The general linear ordinary differential equation with constant coefficients is written as

$$\frac{d^n y}{dx^n} + a_1 \frac{d^{n-1} y}{dx^{n-1}} + a_2 \frac{d^{n-2} y}{dx^{n-2}} + \cdots + a_n y = g(x), \tag{5.45}$$

in which $a_1, a_2, \ldots, a_n$ are constants. When the operation of differentiation about x is defined as D,

$$D = \frac{d}{dx}, \quad D^2 = \frac{d^2}{dx^2}, \ldots, \quad D^n = \frac{d^n}{dx^n}. \tag{5.46}$$

Thus, Eq. (5.45) is expressed by

$$(D^n + a_1 D^{n-1} + a_2 D^{n-2} + \cdots + a_n)y = g(x). \tag{5.47}$$

If we define

$$f(D) = D^n + a_1 D^{n-1} + a_2 D^{n-2} + \cdots + a_n, \tag{5.48}$$

Eq. (5.47) is written as

$$f(D)y = g(x). \tag{5.49}$$

When $f(D)$ is assumed to be factorized, the following equation is derived:

$$f(D) = (D - \alpha_1)(D - \alpha_2) \cdots (D - \alpha_{n-1})(D - \alpha_n). \qquad (5.50)$$

When D in Eq. (5.49) is assumed as a constant coefficient, Eq. (5.49) is formally solvable. The general solution of Eq. (5.49) is expressed by the summation of a particular integral y_p and a homogeneous solution y_h which satisfies $f(D)y_h = 0$. The particular integral satisfies Eq. (5.49). The particular integral is formally given by

$$y_p = \frac{1}{f(D)} g(x). \qquad (5.51)$$

The homogeneous solution is expressed as follows:

$$f(D)y_h = 0. \qquad (5.52)$$

When an algebraic equation of D has roots such as $\alpha_1, \alpha_2, \ldots, \alpha_m$ and their roots are assumed to be r_1 repeated, r_2 repeated, $\ldots$, r_m repeated, the homogeneous solution of Eq. (5.52) (Yoshida, 1950) is

$$y_h = (c_{11} + c_{12}x + \cdots + c_{1r_1}x^{r_1-1})e^{\alpha_1 x}$$
$$+ (c_{21} + c_{22}x + \cdots + c_{2r_2}x^{r_2-1})e^{\alpha_2 x} + \cdots$$
$$+ (c_{m1} + c_{m2}x + \cdots + c_{mr_m}x^{r_m-1})e^{\alpha_m x}. \qquad (5.53)$$

Next, consider the particular integral. When the meromorphic function of D, $1/f(D)$ is expanded in partial fractions as

$$\frac{1}{f(D)} = \frac{A_{11}}{D - \alpha_1} + \frac{A_{12}}{(D - \alpha_1)^2} + \cdots + \frac{A_{1r_1}}{(D - \alpha_1)^{r_1}}$$
$$+ \frac{A_{21}}{D - \alpha_2} + \frac{A_{22}}{(D - \alpha_2)^2} + \cdots + \frac{A_{2r_2}}{(D - \alpha_2)^{r_2}}$$
$$+ \cdots + \frac{A_{m1}}{D - \alpha_m} + \cdots + \frac{A_{mr_m}}{(D - \alpha_m)^{r_m}}, \qquad (5.54)$$

the particular integral of $f(D)y_p = g(x)$ is

$$y_p = \frac{1}{f(D)} g(x)$$
$$= \frac{A_{11}}{D - \alpha_1} g(x) + \frac{A_{12}}{(D - \alpha_1)^2} g(x) + \cdots + \frac{A_{1r_1}}{(D - \alpha_1)^{r_1}} g(x)$$

$$+ \frac{A_{21}}{D - \alpha_2}g(x) + \cdots + \frac{A_{2r_2}}{(D - \alpha_2)^{r_2}}g(x) + \cdots$$

$$+ \frac{A_{1r_m}}{D - \alpha_m}g(x) + \cdots + \frac{A_{mr_m}}{(D - \alpha_m)^{r_m}}g(x), \tag{5.55}$$

in which the particular integral of $(D - \alpha)y = g(x)$ is derived from Eq. (5.14) with $C = 0$ as

$$\frac{1}{D - \alpha}g(x) = e^{\alpha x} \int e^{-\alpha x} g(x) dx. \tag{5.56}$$

The particular integral of $(D - \alpha)^k y = g(x)$ is

$$\frac{1}{(D - \alpha)^k}g(x) = e^{\alpha x} \underbrace{\iint \cdots \int}_{k \text{ times}} e^{-\alpha x} g(x)\, dx \cdots dx. \tag{5.57}$$

Thus, the particular integral, Eq. (5.56) is

$$\frac{1}{f(D)}g(x) = A_{11}e^{\alpha_1 x} \int e^{-\alpha_1 x} g(x)dx + A_{12}e^{\alpha_1 x} \iint e^{-\alpha_1 x} g(x)\, dxdx$$

$$+ \cdots + A_{1r_1}e^{\alpha_1 x} \underbrace{\iint \cdots \int}_{r_1 \text{ times}} e^{-\alpha_1 x} g(x)dx \cdots dx$$

$$+ A_{21}e^{\alpha_2 x} \int e^{-\alpha_2 x} g(x)dx + A_{22}e^{\alpha_2 x} \iint e^{-\alpha_2 x}$$

$$\times g(x)\, dxdx + \cdots + A_{2r_2}e^{\alpha_2 x} \underbrace{\iint \cdots \int}_{r_2 \text{ times}} e^{-\alpha_2 x}$$

$$\times g(x)dx \cdots dx + \cdots + A_{1r_m}e^{\alpha_r m x} \int e^{-\alpha_m x} g(x)dx$$

$$+ \cdots + A_{mr_m}e^{\alpha_m x} \underbrace{\iint \cdots \int}_{r_m \text{ times}} e^{-\alpha_m x} g(x)dx \cdots dx. $$

$$\tag{5.58}$$

Therefore, in the case that $g(x)$ is an elementary function, the particular integral can be easily obtained as follows:

(i) Case of $g(x) = e^{\beta x}$ and $f(\beta) \neq 0$.

$$\frac{1}{f(D)} e^{\beta x} = \frac{1}{f(\beta)} e^{\beta x}. \qquad (5.59)$$

(ii) Case of $g(x) = e^{\beta x}$ and $f(\beta) = 0$.

When the roots of $f(D) = 0$ are $\alpha_1, \alpha_2, \ldots, \alpha_m$, assume that β is one of the roots. When

$$f(D) = (D - \beta)^r f_1(D), \qquad f_1(\beta) \neq 0, \qquad (5.60)$$

is given, the particular integral is

$$\frac{1}{f(D)} e^{\beta x} = \frac{1}{(D - \beta)^r} \frac{e^{\beta x}}{f_1(D)} = \frac{1}{f_1(\beta)} \frac{1}{(D - \beta)^r} e^{\beta x}$$

$$= \frac{1}{f_1(\beta)} \cdot \frac{1}{r!} x^r e^{\beta x}. \qquad (5.61)$$

(iii) Case of $g(x) = \cos bx$ or $\sin bx$.

After the computation of the particular integral of $g(x) = e^{ibx}$, obtain its real or imaginary part.

(iv) Case that $g(x)$ is a polynomial of x of order k.

If $f(D) = b_0 D^\ell (1 + b_1 D + b_2 D^2 + \cdots)$, the particular integral is

$$\frac{1}{f(D)} g(x) = \frac{1}{b_0 D^\ell} \frac{1}{1 + b_1 D + b_2 D^2 + \cdots} g(x)$$

$$= \frac{1}{b_0 D^\ell} [1 - (b_1 D + b_2 D^2 + \cdots)$$

$$+ (b_1 D + b_2 D^2 + \cdots)^2 - \cdots] g(x)$$

$$= \frac{1}{b_0 D^\ell} [g(x) - (b_1 D + b_2 D^2 + \cdots) g(x)$$

$$+ (b_1 D + b_2 D^2 + \cdots)^2 g(x) - \cdots]. \qquad (5.62)$$

Since the polynomial $g(x)$ is k-th order of x, the higher order terms than

$$(b_1 D + b_2 D^2 + \cdots)^{k+1} g(x) \qquad (5.63)$$

are 0. The remains in [ ] in Eq. (5.62) are

$$g(x) - (b_1 D + b_2 D^2 + \cdots) g(x) + (b_1 D + b_2 D^2 + \cdots)^2 g(x)$$

$$- \cdots + (b_1 D + b_2 D^2 + \cdots)^k g(x). \qquad (5.64)$$

This is $P_k(x)$ and the polynomial of x of k-th order. Thus, the particular integral is

$$\frac{1}{f(D)}g(x) = \frac{1}{b_0}\frac{1}{D^\ell}P_k(x) = \frac{1}{b_0}\underbrace{\int\!\!\int \cdots \int}_{\ell \text{ times}} P_k(x)\,dxdx\cdots dx. \quad (5.65)$$

The particular integral is expressed by $1/b_0$ times the ℓ times integration of $P_k(x)$ about x.

Example 5.4. Solve the following linear ordinary differential equation of the second-order:

$$(D^2 - 3D - 10)y = e^x \cos 3x. \quad (1)$$

Solution. The solution of the homogeneous ordinary differential equation

$$(D^2 - 3D - 10)y = 0 \quad (2)$$

is given by Eq. (5.58). The homogeneous solution y_h is obtained as:

$$y_h = C_1 e^{5x} + C_2 e^{-2x} \quad (3)$$

from

$$D^2 - 3D - 10 = (D-5)(D+2) = 0, \quad (4)$$

in which C_1 and C_2 are integral constants. The particular integral y_p is obtained from Eq. (5.59) and $e^x \cos 3x = \Re[e^{(1+3i)x}]$. Since $\beta = 1 + 3i$

$$y_p = \frac{1}{f(\beta)}e^{\beta x} = \Re\left[\frac{e^{(1+3i)x}}{(3i-4)(3i+3)}\right]$$

$$= -\frac{1}{150}e^x(7\cos 3x + \sin 3x), \quad (5)$$

in which $\Re$ indicates a real part of [ ] in Eq. (5). Thus, the general solution y is

$$y = y_h + y_p = C_1 e^{5x} + C_2 e^{-2x} - \frac{1}{150}e^x(7\cos 3x + \sin 3x). \quad (6)$$

Problem 5.4. Solve the following linear ordinary differential equation of the second-order.

$$(D^2 - 2D + 1)y = e^{-x}. \quad (7)$$

5.5. The Method of Undetermined Coefficients

Another method to get a particular integral is the method of undetermined coefficients. This may be illustrated by examples. Consider the linear ordinary differential equation

$$\frac{d^2y}{dx^2} + 3\frac{dy}{dx} + 2y = x^2 + x. \tag{5.66}$$

To obtain the particular integral, let us assume a general polynomial of the second degree of the form

$$y_p = ax^2 + bx + c. \tag{5.67}$$

Since the maximum degree of the polynomial on the right side is 2, the maximum degree of Eq. (5.67) is 2. Substituting this into Eq. (5.66), we obtain

$$2a + 3(2ax + b) + 2(ax^2 + bx + c) = x^2 + x. \tag{5.68}$$

Equating coefficients of like powers of x, we obtain

$$2a = 1, \quad 6a + 2b = 1, \quad 2a + 3b + 2c = 0. \tag{5.69}$$

Solving these equations, we obtain

$$a = 1/2, \quad b = -1, \quad c - 1. \tag{5.70}$$

Substituting these values into Eq. (5.67), we obtain the particular integral as

$$y_p = \frac{1}{2}x^2 - x + 1. \tag{5.71}$$

The method of undetermined coefficients for obtaining the particular integral is adopted when the function $g(x)$ is a sum of terms such as sines, cosines, exponentials, powers of x, and their products. As another example, consider the differential equation

$$\frac{d^2y}{dx^2} + y = e^{2x} \cos 3x. \tag{5.72}$$

In this case we assume

$$y_p = ae^{2x} \cos 3x + be^{2x} \sin 3x. \tag{5.73}$$

Substituting Eq. (5.73) into Eq. (5.72), we obtain

$$12b - 4a = 1, \quad 4b + 12a = 0. \tag{5.74}$$

Solving these equations, we obtain

$$a = -\frac{1}{40}, \quad b = \frac{3}{40}. \tag{5.75}$$

Substituting these values into Eq. (5.73), we obtain the particular integral as

$$y_p = -\frac{1}{40}e^{2x}\cos 3x + \frac{3}{40}e^{2x}\sin 3x. \tag{5.76}$$

When the driving function $g(x)$ contains terms of the homogeneous solution, the method must be altered in the following one. When the terms that are included in the assumed particular integral are contained in the homogeneous solution, the terms in the particular integral must be multiplied by x^m, where m is of a large enough order so that all of these particular terms are of the form xy_h. As an example, consider the following ordinary differential equation

$$\frac{d^2y}{dx^2} + y = \sin x. \tag{5.77}$$

According to the standard procedure, the particular integral is assumed to have the form

$$y_p = a\cos x + b\sin x. \tag{5.78}$$

However, the homogeneous solution is

$$y_h = c_1\cos x + c_2\sin x. \tag{5.79}$$

Thus, the assumed particular integral must be modified as

$$y_p = a\cos x + b\sin x + cx\cos x + dx\sin x. \tag{5.80}$$

Because the particular integral Eq. (5.78) is not independent of Eq. (5.79). Substituting Eq. (5.80) into Eq. (5.77), we obtain,

$$a = 0, \quad b = 0, \quad c = -\frac{1}{2}, \quad d = 0. \tag{5.81}$$

The particular integral is

$$y_p = -\frac{1}{2}x\cos x. \tag{5.82}$$

Example 5.5. Obtain the particular integral of following linear ordinary differential equation.

$$\frac{d^3y}{dx^3} + y = e^{2x}\cos 3x. \tag{1}$$

Solution. We assume

$$y_p = ae^{2x}\cos 3x + be^{2x}\sin 3x. \tag{2}$$

Substituting Eq. (2) into Eq. (1), we obtain

$$9b - 45a = 1, \quad 9a + 45b = 0. \tag{3}$$

Solving Eqs. (3) and substituting them into Eq. (2), we obtain

$$y_p = \frac{e^{2x}}{234}(\sin 3x - 5\cos 3x). \tag{4}$$

Problem 5.5. Obtain the particular integral of the following linear ordinary differential equation.

$$\frac{d^2y}{dx^2} + \frac{dy}{dx} = e^{-x}. \tag{5}$$

5.6. Linear Ordinary Differential Equations of the Higher Orders

A linear ordinary differential equation of the higher order with constant coefficients $a_1, a_2, \ldots, a_n$ is given by

$$\frac{d^ny}{dx^n} + a_1\frac{d^{n-1}y}{dx^{n-1}} + a_2\frac{d^{n-2}y}{dx^{n-2}} + \cdots + a_ny = g(x). \tag{5.83}$$

This can be solved by (B) in Sec. 5.3. When Eq. (5.83) is written by using the differential operator D, it is

$$D^n + a_1D^{n-1} + a_2D^{n-2} + \cdots + a_n = 0. \tag{5.84}$$

The roots of Eq. (5.84) are assumed to be $\alpha_1, \alpha_2, \ldots, \alpha_n$ Then, Eq. (5.83) is written as

$$\left(\frac{d}{dx} - \alpha_1\right)\underbrace{\left(\frac{d}{dx} - \alpha_2\right)\cdots\left(\frac{d}{dx} - \alpha_n\right)y}_{Y_1} = g(x). \tag{5.85}$$

If

$$\left(\frac{d}{dx} - \alpha_2\right)\left(\frac{d}{dx} - \alpha_3\right)\cdots\left(\frac{d}{dx} - \alpha_n\right) y = Y_1 \tag{5.86}$$

is defined,

$$\frac{dY_1}{dx} - \alpha_1 Y_1 = g(x) \tag{5.87}$$

is derived. The solution of Eq. (5.87) is given by

$$Y_1 = e^{\alpha_1 x}\left[\int e^{-\alpha_1 x} g(x)dx + C_1\right], \tag{5.88}$$

in which $C_1 =$ an integral constant. In the same way,

$$\left(\frac{d}{dx} - \alpha_3\right)\left(\frac{d}{dx} - \alpha_4\right)\cdots\left(\frac{d}{dx} - \alpha_n\right) y = Y_2 \tag{5.89}$$

is defined and Eq. (5.86) is

$$\frac{dY_2}{dx} - \alpha_2 Y_2 = Y_1. \tag{5.90}$$

Substituting Eq. (5.88) into Eq. (5.90), we obtain

$$Y_2 = e^{\alpha_2 x}\int e^{(\alpha_1 - \alpha_2)x}\int e^{-\alpha_1 x} g(x)\,dxdx$$

$$+ C_1 e^{\alpha_2 x}\int e^{(\alpha_1 - \alpha_2)x}dx + C_2 e^{\alpha_2 x}, \tag{5.91}$$

in which $C_2 =$ an integral constant. These processes of computations give the solution of Eq. (5.83).

Example 5.6. Solve the oscillation equation of water surface in a U-tube as shown in Fig. 5.2, when the cross-sectional area of the U-tube is constant. The symbols ℓ and ζ show the length of the U-tube and the water surface level, respectively.

Solution. When an s coordinate is defined along the U-tube, the unsteady Bernoulli equation along the s coordinate is given by

$$\frac{1}{g}\frac{\partial v}{\partial t} + \frac{\partial}{\partial s}\left(\frac{v^2}{2g} + z + \frac{p}{\gamma}\right) = 0, \tag{1}$$

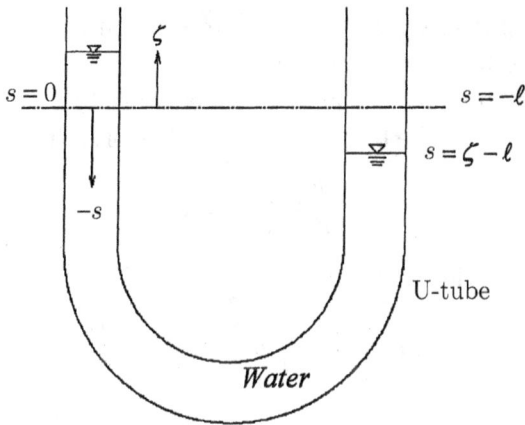

Figure 5.2. Example 5.5.

in which v = velocity, z = potential head, p = pressure, γ = specific weight of water, and s = the coordinate system along the U-tube as shown in Fig. 5.2. Since the cross-sectional area is constant along the s-coordinate, we obtain

$$\frac{\partial v}{\partial s} = 0. \tag{2}$$

Since the time variation of the water level ζ is equal to the velocity in the U-tube, we have

$$v = \frac{\partial \zeta}{\partial t}. \tag{3}$$

Substituting Eqs. (2) and (3) into Eq. (1), we obtain

$$\frac{1}{g}\frac{\partial^2 \zeta}{\partial t^2} + \frac{\partial}{\partial s}\left(z + \frac{p}{\gamma}\right) = 0. \tag{4}$$

Integrating Eq. (4) about s from ζ to $\zeta - \ell$, we obtain

$$\int_\zeta^{\zeta-\ell} \frac{1}{g}\frac{\partial^2 \zeta}{\partial t^2}\, ds + \int_\zeta^{\zeta-\ell} \frac{\partial}{\partial s}\left(z + \frac{p}{\gamma}\right) ds = 0. \tag{5}$$

Since ζ is independent of s, we obtain

$$-\frac{\ell}{g}\frac{\partial^2 \zeta}{\partial t^2} + \left[z + \frac{p}{\gamma}\right]_\zeta^{\zeta-\ell} = 0. \tag{6}$$

The pressure p is zero (atmospheric pressure) at $s = \zeta$ and $s = \zeta - \ell$. The difference between the potential head at $s = \zeta - \ell$ and $s = \zeta$ is -2ζ. Since the number of the independent variables become one, the partial derivative is the total derivative. Thus, Eq. (6) is simplified as

$$\frac{\ell}{g} \frac{d^2\zeta}{dt^2} + 2\zeta = 0. \tag{7}$$

The general solution of Eq. (7) is

$$\zeta = C_1 \sin \sqrt{\frac{2g}{\ell}} t + C_2 \cos \sqrt{\frac{2g}{\ell}} t, \tag{8}$$

in which C_1 and C_2 = integral constants. This indicates that the water level oscillation continues permanently as long as energy loss happens.

Problem 5.6. When the energy loss is expressed by kv in Eq. (1), Eq. (1) is given by

$$\frac{1}{g} \frac{\partial v}{\partial t} + \frac{\partial}{\partial s} \left(\frac{v^2}{2g} + z + \frac{p}{\gamma} \right) = -kv, \tag{9}$$

in which k = a constant and $k < \sqrt{\frac{8}{g\ell}}$. Then, solve Eq. (9).

5.7. Variation of Parameters

As the application of the method of undetermined coefficients, let us consider the method of variation of parameters. This is more general and may be applied to linear differential equations with variable coefficients. The method of undetermined coefficients is easier to apply in the case of differential equations with constant coefficients. The method of variation of parameters may be illustrated by considering the following linear differential equation:

$$y'' + p(x)y' + q(x)y = r(x), \tag{5.92}$$

in which $'' = d/dx$. This method assumes that the general homogeneous solution is known *a priori*, that is

$$y_h(x) = c_1 y_1(x) + c_2 y_2(x), \tag{5.93}$$

in which $y_1(x)$ and $y_2(x)$ satisfy

$$y_i'' + p(x)y_i' + q(x)y_i = 0, \quad \text{for } i = 1, 2. \tag{5.94}$$

A particular integral to Eq. (5.92) is assumed to have the form of Eq. (5.93), except the arbitrary constant c_1 and c_2 are replaced by arbitrary functions $A(x)$ and $B(x)$:

$$y_p(x) = A(x)y_1(x) + B(x)y_2(x). \tag{5.95}$$

At this point we evaluate the first derivative of Eq. (5.95) as

$$y_p' = Ay_1' + By_2' + A'y_1 + B'y_2. \tag{5.96}$$

The last two terms in Eq. (5.96) are arbitrarily set to zero to be formed

$$A'y_1 + B'y_2 = 0. \tag{5.97}$$

Applying Eq. (5.97) to Eq. (5.96) and the differentiation of Eq. (5.96) yields

$$y_p'' = Ay_1'' + By_2'' + A'y_1' + B'y_2'. \tag{5.98}$$

Substituting Eqs. (5.95), (5.96), (5.97), and (5.98) into (5.92) gives

$$A'y_1' + B'y_2' = r. \tag{5.99}$$

Equations (5.97) and (5.99) constitute two simultaneous differential equations for A and B. Solving these two equations by Cramer's rule yields

$$A' = -\frac{ry_2}{|W|}, \quad B' = \frac{ry_1}{|W|}, \tag{5.100}$$

in which $|W|$ is the Wronskian of the homogeneous solution. It is expressed by

$$W = \begin{vmatrix} y_1 & y_2 \\ y_1' & y_2' \end{vmatrix}. \tag{5.101}$$

Integrating Eq. (5.100) and substituting into Eq. (5.95) yields the complete general solution as

$$y(x) = y_1 \left(-\int \frac{ry_2}{|W|} dx + c_1 \right) + y_2 \left(\int \frac{ry_1}{|W|} dx + c_2 \right). \tag{5.102}$$

As an example consider the equation

$$y'' + y = \sin x. \tag{5.103}$$

For the equation

$$y_1 = \sin x, \quad y_2 = \cos x, \quad |W| = -1, \quad r = \sin x. \tag{5.104}$$

Substituting these values into Eq. (5.102) and performing the indicated integrations yields the general solution

$$y = c_1 \sin x + c_2 \cos x - \left(\frac{1}{2} \sin^3 x - \frac{x}{2} \cos x + \frac{1}{4} \sin 2x \cos x \right). \tag{5.105}$$

For the second example let us consider the problem of obtaining a solution to the equation

$$x^2 y'' - 3xy' - 4y = x^3. \tag{5.106}$$

For the equation

$$x^2 = \frac{1}{x}, \quad y_2 = -\frac{x^3}{7}, \quad |W| = -\frac{1}{2}x, \quad r = x^3. \tag{5.107}$$

Substituting into Eq. (5.102), we have

$$y(x) = \frac{1}{x} \left(-\frac{5}{42} x^6 + c_1 \right) - \frac{x^3}{7} \left(-x^2 + c_2 \right). \tag{5.108}$$

Example 5.7. Solve the following equation:

$$x^2 y'' + xy' - y = x. \tag{1}$$

Solution. For this equation

$$y_1 = x, \quad y_2 = \frac{1}{x}, \quad |W| = -\frac{2}{x}, \quad r = x. \tag{2}$$

Substituting into Eq. (5.102), we get

$$y = x \left(\frac{x^2}{4} + c_1 \right) + \frac{1}{x} \left(-\frac{x^4}{8} + c_2 \right). \tag{3}$$

Problem 5.7. Solve the following equation:

$$x^2 y'' + y = x^3. \tag{4}$$

5.8. Ordinary Differential Equations and Integral Equations

We have been studying linear ordinary differential equations. These are suitable to analyze micro phenomena. On the other hand, an integral equation which includes integral operation is appropriate to study macro phenomena. We call the integral equation when it contains unknown dependent variables inside the integral operation. This was first studied by Abel who is a mathematician genius in Norway. We do not need to define an initial or boundary conditions in the integral equation. It is used in the wing theory in aerodynamics, wave mechanics in fluid, diffusion phenomena, and rainfall-runoff analysis in hydrology. For example, if a relationship between rainfall and runoff is assumed to be linear, the following equation is derived:

$$\hat{Q}(t) = \int_0^t h(t - \tau) R(\tau) d\tau, \qquad (5.109)$$

in which $\hat{Q}(t)$ = predicted runoff, $h(t)$ = unit hydrograph, and $R(\tau)$ = rainfall. To statistically minimize the difference between observation and computation in Eq. (5.109), the unit hydrograph $h(t)$ is determined as

$$E\{[\hat{Q}(t) - Q_{obs}]^2\} \Rightarrow \min, \qquad (5.110)$$

in which Q_{obs} = observed runoff and $E\{\cdot\}$ = operation of expectation. Equation (5.110) leads to the Wiener–Hopf integral equation. The further development of Wiener's theory is the Kalman filter (Mizumura, 2008b) in control theory. The number of the kind of integral equations is two. One is the Volterra type. The first kind of the Volterra type is

$$\int_a^t K(t, \xi) x(\xi) d\xi = f(t), \qquad (5.111)$$

in which t = an independent variable, a = an arbitrary constant, ξ = a dummy variable, $x(t)$ = a dependent variable, $f(t)$ = an arbitrary function, and $K(t, \xi)$ = a kernel. The second kind of the Volterra type is

$$x(t) - \int_a^t K(t, \xi) x(\xi) d\xi = f(t). \qquad (5.112)$$

Next is the Fredholm type. The first kind of the Fredholm type is

$$\int_a^b K(t,\xi)x(\xi)d\xi = f(t).$$
(5.113)

$a, b =$ arbitrary constants. The second kind of the Fredholm type is

$$x(t) - \lambda \int_a^b K(t,\xi)x(\xi)d\xi = f(t),$$
(5.114)

in which $\lambda =$ an eigen value. These integral equations of the Volterra and Fredholm types often appear in engineering. These are usually numerically solved whether these are linear or nonlinear.

Example 5.8. Solve the following two integral equations:

$$\int_0^t \xi x(\xi)d\xi = (t-1)e^t + 1,$$
(1)

$$x(t) - \frac{1}{4}\int_0^{\pi/2} t\xi x(\xi)d\xi = \sin t - \frac{t}{4}.$$
(2)

Solution. Differentiating Eq. (1) about t, we obtain

$$tx(t) = e^t + (t-1)e^t.$$
(3)

Then,

$$x(t) = e^t.$$
(4)

This is a solution. Comparing both sides of Eq. (2), $\sin t$ in the right side corresponds to $x(t)$ in the left side. The second term in the left side does not produce $\sin t$ in the left side. Thus, as a solution we assume $x(t) = \sin t$. Then, we get

$$\int_0^{\pi/2} t\xi \sin \xi \, d\xi = t.$$
(5)

The function $x(t) = \sin t$ satisfies Eq. (2) and it is a solution, because

$$\int_0^{\pi/2} \xi \sin \xi \, d\xi = 1.$$
(6)

Problem 5.8. Solve the following integral equation:

$$x(t) - \int_0^t \xi^2 x(\xi)d\xi = (t^2 - 1)e^t + 1.$$
(7)

Chapter 6

Complex Functions and Complex Integrals

A beautiful theory has been formed to calculate the resistance of a body in perfect fluid by the theory of complex functions. Especially, the theory of wings has been developed in aerodynamics to design aircrafts. The theory of residues was created to calculate integration and inverse Laplace transformation, although it is not possible to do so in integrations of real variables.

6.1. Complex Functions

A complex variable which is defined by $z = x + iy$, in which x and y = real numbers and $i = \sqrt{-1}$. The plane that consists of the x and iy axes is called a complex or Gauss plane. The constant i is called an imaginary unit. Reals x and y are called a real and imaginary part of z, respectively. They are written as

$$x = \Re(z), \quad y = \Im(z). \tag{6.1}$$

In the polar coordinate system $z = re^{i\theta}$, in which r is the absolute value of z and defined as $|z| = \sqrt{x^2 + y^2}$. An angle θ is an argument of z and defined as $\arg z = \arctan(y/x) = \tan^{-1}(y/x)$, in which $x \neq 0$. In the case $-\frac{\pi}{2} \leq \arg z \leq \frac{\pi}{2}$, the argument θ is defined by

$$\theta = Tan^{-1}(y/x), \quad ArcTan(y/x), \tag{6.2}$$

The symbol $arg\ z$ is called a principal value. The conjugate complex variable $\bar{z}$ of z is defined by

$$\bar{z} = x - iy. \tag{6.3}$$

The Euler equation is given between the exponential and trigonometric function as follows:

$$e^{i\theta} = \cos\theta + i\sin\theta. \tag{6.4}$$

When z is the complex variable,

$$w = f(z) \tag{6.5}$$

is called a complex function. If ϕ and ψ are the real and imaginary parts of w, respectively, the complex function w is expressed as

$$w = \phi + i\psi, \tag{6.6}$$

in which ϕ and ψ are the functions of x and y. The exponential, logarithmic, trigonometric, and hypergeometric functions are defined in the complex variable as the real function. Especially, the following relations are derived:

$$\sin z = \frac{e^{iz} - e^{-iz}}{2i}, \tag{6.7}$$

$$\cos z = \frac{e^{iz} + e^{-iz}}{2}, \tag{6.8}$$

$$\sinh z = \frac{e^z - e^{-z}}{2}, \tag{6.9}$$

$$\cosh z = \frac{e^z + e^{-z}}{2}. \tag{6.10}$$

Equations (6.7), (6.8), (6.9), and (6.10) show

$$\sinh iz = i\sin z, \quad \cosh iz = \cos z, \tag{6.11}$$

$$\sin iz = i\sinh z, \quad \cos iz = \cosh z. \tag{6.12}$$

When the function $f(z)$ is differentiable at a point in a region, $f(z)$ is called to be analytic or regular. The derivative is expressed by $f'(z)$. Then, it is defined by

$$f'(z) = \lim_{\Delta z \to 0} \frac{f(z + \Delta z) - f(z)}{\Delta z}. \tag{6.13}$$

Define $f(z) = \phi + i\psi$ and $\Delta z = \Delta x + i\Delta y$. Then, in the case of $\Delta y = 0$, from Eq. (6.13) we have

$$f'(z) = \lim_{\Delta x \to 0} \frac{f(z + \Delta x) - f(z)}{\Delta x} = \frac{\partial \phi}{\partial x} + i\frac{\partial \psi}{\partial x}. \qquad (6.14)$$

In the case of $\Delta x = 0$, we have

$$f'(z) = \lim_{i\Delta y \to 0} \frac{f(z + i\Delta y) - f(z)}{i\Delta y} = \frac{1}{i}\left(\frac{\partial \phi}{\partial y} + i\frac{\partial \psi}{\partial y}\right). \qquad (6.15)$$

Since the derivative $f'(z)$ is independent of the selection of Δz, Eq. (6.14) is equal to Eq. (6.15). Thus, the following Cauchy–Riemann equations are derived from Eqs. (6.14) and (6.15):

$$\frac{\partial \phi}{\partial x} = \frac{\partial \psi}{\partial y}, \qquad (6.16)$$

$$\frac{\partial \phi}{\partial y} = -\frac{\partial \psi}{\partial x}. \qquad (6.17)$$

The Cauchy–Riemann equations are a necessary and satisfactory condition in order that the complex function $f(z)$ has its derivative of finite values. Then, the function $f(z)$ is called analytic or regular. The point that the derivative of $f(z)$ is not finite is called a singular point. The Cauchy–Riemann equations indicate that the curve of $\phi = $ constant is orthogonal to the curve of $\psi = $ constant. When the complex function $f(z)$ or w is expressed by the complex variable z except the singular points, Eqs. (6.16) and (6.17) are derived. In fluid mechanics, ϕ and ψ show the velocity potential and stream function, respectively. From the Cauchy–Riemann equations we obtain

$$\nabla^2 \phi = 0, \qquad (6.18)$$

$$\nabla^2 \psi = 0, \qquad (6.19)$$

in which ∇^2 is the Laplacian,

$$\nabla^2 = \frac{\partial^2}{\partial x^2} + \frac{\partial^2}{\partial y^2}.$$

Two curves $\phi = $ constant and $\psi = $ constant are equi-potential and streamline, respectively. Solving Eqs. (6.18) or (6.19), we can obtain a flow pattern in fluid mechanics.

Example 6.1. Show that

$$\sin(x + iy) = \sin x \cosh y + i \sinh y \cos x. \tag{1}$$

Solution. From the formula of the trigonometric functions, we obtain

$$\sin(x + iy) = \sin x \cos iy + \cos x \sin iy. \tag{2}$$

From Eq. (6.4), we obtain

$$\cos iy = \frac{e^{-y} + e^y}{2} = \cosh y. \tag{3}$$

From Eq. (6.4), we obtain

$$\sin iy = \frac{e^{-y} - e^y}{2i} = i\frac{e^y - e^{-y}}{2} = i \sinh y. \tag{4}$$

Thus, we obtain

$$\sin(x + iy) = \sin x \cosh y + i \sinh y \cos x. \tag{5}$$

Problem 6.1. Show that $\phi = x^2 - y^2$ and $\psi = 2xy$ satisfy the Cauchy–Riemann equations. Then, show that the complex function w does not satisfy the Cauchy–Riemann equations when the complex function w is not expressed by z only.

6.2. Complex Integral

The integral of the function $f(z)$ from a point A to a point B along a curve C on a complex plane is defined by

$$\int_A^B f(z)dz = \int_C f(z)dz. \tag{6.20}$$

The complex plane is also called a z plane and expressed by a z-pl. When the function $f(z)$ is analytic inside and continuous on a closed curve C, the following equation is derived (Ahlfors, 1966).

$$\int_C f(z)dz = 0. \tag{6.21}$$

This is called a Cauchy theorem. When the direction of the integral is counter-clockwise, the direction of the closed curve C is defined to be positive. When $f(z) = \phi + i\psi$ and $dz = dx + idy$, by the Green theorem in Chap. 4, Eq. (6.21) is rewritten as

$$\int_C (\phi dx - \psi dy) + i \int_C (\psi dx + \phi dy) = - \iint_S \left(\frac{\partial \psi}{\partial x} + \frac{\partial \phi}{\partial y} \right) dx dy$$

$$+ i \iint_S \left(\frac{\partial \phi}{\partial x} - \frac{\partial \psi}{\partial y} \right) dx dy,$$

$$(6.22)$$

in which a region S includes the curve C and the inside the curve C. When the Cauchy–Riemann equations are applied to Eq. (6.22), Eq. (6.22) becomes zero. The Cauchy theorem on the contrary, the function $f(z)$ is analytic in the curve C and continuous on the curve C if Eq. (6.21) is used. This is called a Morera theorem (Ahlfors, 1966). When the function $f(z)$ is analytic on and inside the curve C, let us consider a circle C' of which radius is r inside the curve C as shown in Fig. 6.1. The integration of $f(z)$ along a path $BCAA'(-C')B'B$ indicates the following equation from the Cauchy theorem:

$$\int_{BCAA'(-C')B'B} f(z)dz = 0. \tag{6.23}$$

The direction of the integral of $(-C')$ is clockwise. Equation (6.23) is rewritten as

$$\int_C f(z)dz + \int_A^{A'} f(z)dz + \int_{-C'} f(z)dz + \int_{B'}^B f(z)dz = 0. \tag{6.24}$$

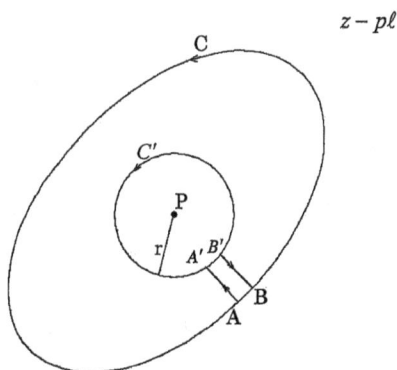

Figure 6.1. Cauchy theorem.

As the two points A and A' approach the two points B and B', respectively, the integrals of $f(z)$ along a path AA' and BB' become the same. Thus, Eq. (6.24) becomes

$$\int_C f(z)dz + \int_{-C'} f(z)dz = 0. \qquad (6.25)$$

Therefore, the integral along the curve C is equal to that along the curve C'. This indicates that the integral along an arbitrary curve C is equal to the integral along a circle of which radius is r. That is

$$\int_C f(z)dz = \int_{C'} f(z)dz. \qquad (6.26)$$

Next, consider a residue theorem. When the following equation is derived at a point A inside the closed curve C,

$$\lim_{z \to a} (z - a)f(z) = R, \qquad (6.27)$$

in which $a = $ a complex coordinate of point A in Fig. 6.2. The value R is called the residue of the function $f(z)$ at the point A. Then, R is a finite complex number. Thus, we have

$$\int_C f(z)dz = 2\pi i R. \qquad (6.28)$$

Applying Eq. (6.27) to $\frac{f(z)}{z-a}$, we obtain the following Cauchy integral formula (Ahlfors, 1966):

$$f(a) = \frac{1}{2\pi i} \int_C \frac{f(z)}{z - a} dz. \qquad (6.29)$$

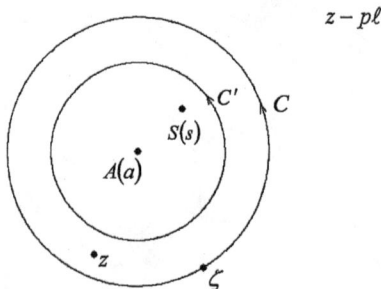

Figure 6.2.   Laurent series.

There is a circle C of which center is A and coordinate is a in Fig. 6.2. Assume that the function $f(z)$ is analytic on and inside the circle C. If ζ is on the circle C and z is inside the circle C, $f(z)$ is expressed by the Cauchy integral formula as:

$$f(z) = \frac{1}{2\pi i} \int_C \frac{f(\zeta)}{\zeta - z} d\zeta. \tag{6.30}$$

From Fig. 6.2 we have an inequality

$$|z - a| < |\zeta - a|$$

in which a indicates the complex number to show the coordinate of the point A. Thus, $\dfrac{1}{\zeta - z}$ is expanded in the series as follows:

$$\frac{1}{\zeta - z} = \frac{1}{\zeta - a}\left(1 - \frac{z - a}{\zeta - a}\right)^{-1} = \frac{1}{\zeta - a} + \frac{z - a}{(\zeta - a)^2}$$
$$+ \frac{(z - a)^2}{(\zeta - a)^3} + \cdots. \tag{6.31}$$

Substituting Eq. (6.31) into Eq. (6.30), we obtain

$$f(z) = f(a) + f'(a) \cdot (z - a) + \frac{f''(a)}{2!} \cdot (z - a)^2 + \cdots. \tag{6.32}$$

This is the Taylor series. If the function $f(z)$ has a singular point S inside the curve C in Fig. 6.2, there is a circle C' which includes the singular point S. The singular point S is defined such that the function $f(z)$ is not finite at the point S. The complex number s indicates the coordinate of the point S. Then, the function $f(z)$ is analytic in the annulus region between the circle C and C'. From the Cauchy integral formula, we obtain

$$f(z) = \frac{1}{2\pi i} \int_C \frac{f(\zeta)}{\zeta - z} d\zeta - \frac{1}{2\pi i} \int_{C'} \frac{f(\zeta)}{\zeta - z} d\zeta. \tag{6.33}$$

The directions of the integrals along the circles C and C' are counter clockwise. When ζ is on the circle C, the inequality $|\zeta - a| > |z - a|$ is formed as shown in Fig. 6.2. Thus, we obtain in the series as follows:

$$\frac{1}{\zeta - z} = \frac{1}{\zeta - a}\left(1 - \frac{z - a}{\zeta - a}\right)^{-1} = \frac{1}{\zeta - a} + \frac{z - a}{(\zeta - a)^2}$$

$$+ \frac{(z - a)^2}{(\zeta - a)^3} + \cdots . \qquad (6.34)$$

When ζ is on the circle C', $|\zeta - a| < |z - a|$ is used. Thus,

$$-\frac{1}{\zeta - z} = \frac{1}{z - a}\left(1 - \frac{\zeta - a}{z - a}\right)^{-1} = \frac{1}{z - a} + \frac{\zeta - a}{(z - a)^2}$$

$$+ \frac{(\zeta - a)^2}{(z - a)^3} + \cdots . \qquad (6.35)$$

Substituting Eqs. (6.34) and (6.35) into Eq. (6.33), the computation like the Taylor series expansion shows

$$f(z) = c_0 + c_1(z - a) + c_2(z - a)^2 + \cdots + \frac{c_{-1}}{z - a} + \frac{c_{-2}}{(z - a)^2} + \cdots ,$$

$$(6.36)$$

in which

$$c_n = \frac{1}{2\pi i}\int_C \frac{f(\zeta)}{(\zeta - a)^{n+1}}d\zeta, \quad (n \geq 0), \qquad (6.37)$$

$$c_{-n} = \frac{1}{2\pi i}\int_{C'} f(\zeta)(\zeta - a)^{n-1}d\zeta, \quad (n > 0). \qquad (6.38)$$

Equation (6.36) is called a Laurent series. Then the function $f(z)$ is expanded in the Laurent series in powers of $z - a$. This expansion will contain powers of $z - a$ with negative exponents, for otherwise $z - a$ would not be a singular point. Hence, there are two possibilities. The expansion Eq. (6.36) has only a finite number of powers of $z - a$ with negative exponents. In this case $f(z)$ is said to have a pole at $z = a$. If m is the largest of the negative exponents and if the function $(z - a)^m f(z)$ behaves regularly and is not zero at the point a, m is a positive integer and called the order of the pole $z = a$. The complex function $f(z)$ is said to have a pole of the m-th order at the point A. The sum of the terms with negative exponents in Eq. (6.36) is called the principal part of the function $f(z)$ at $z = a$. The other possibility is that the Laurent series of the function $f(z)$ about the point $z = a$ will have an infinite number of negative powers of $z - a$ and is of the form

$$f(z) = \sum_{n=-\infty}^{+\infty} c_n (z-a)^n. \tag{6.39}$$

In this case $f(z)$ is said to have an essential singularity at $z = a$.

Residues at Simple Poles of $f(z)$. Frequently it is required to evaluate residues of the function $f(z)$. If $z = a$ is a simple pole of $f(z)$, then we have

$$c_{-1} = Res\,[f(z)]|_{z=a}. \tag{6.40}$$

Residues at a Multiple Pole of $f(z)$. If the function $f(z)$ has a multiple pole at $z = a$ of order m, then the Laurent series of $f(z)$ is

$$w(z) = \frac{c_{-m}}{(z-a)^m} + \frac{c_{-m+1}}{(z-a)^{m-1}} + \cdots + \frac{c_{-1}}{z-a} + c_0 + c_1(z-a)$$
$$+ c_2(z-a)^2 + \cdots. \tag{6.41}$$

The residue at $z = a$ is c_{-1}. To obtain it we multiply Eq. (6.41) by $(z-a)^m$ and have

$$(z-a)^m f(z) = c_{-m} + c_{-m+1}(z-a) + \cdots + c_{-1}(z-a)^{m-1}$$
$$+ c_0(z-a)^m + c_1(z-a)^{m-1} + \cdots. \tag{6.42}$$

If we differentiate both sides of Eq. (6.42) $m-1$ times about z and place $z = a$, we obtain

$$\frac{d^{m-1}}{dz^{m-1}}[(z-a)^m f(z)]_{z=a} = c_{-1}(m-1)!. \tag{6.43}$$

Hence the residue c_{-1} at the multiple pole is

$$c_{-1} = \frac{1}{(m-1)!}\frac{d^{m-1}}{dz^{m-1}}[(z-a)^m f(z)]_{z=a}. \tag{6.44}$$

A function $f(z)$ which has finite number of pole on the z plane is called a meromorphic function.

Example 6.2. Expand the following $f(z)$ by the Laurent series at $z = 0$:

$$f(z) = \frac{1}{(z-1)(z-2)}. \tag{1}$$

Solution. Since the singular points are $z = 1$ and $z = 2$, draw circles of which centers are $z = 0$ and pass $z = 1$ and $z = 2$. The Laurent series in the annulus $(1 < |z| < 2)$ is as follows: Because $\left|\frac{z}{2}\right| < 1$ and $\left|\frac{1}{z}\right| < 1$ hold, we have

$$f(z) = \frac{1}{z-2} - \frac{1}{z-1}$$

$$= -\frac{1}{2\left(1-\frac{z}{2}\right)} - \frac{1}{z\left(1-\frac{1}{z}\right)}$$

$$= -\frac{1}{2}\left[1 + \frac{z}{2} + \left(\frac{z}{2}\right)^2 + \cdots\right] - \frac{1}{z}\left[1 + \frac{1}{z} + \frac{1}{z^2} + \cdots\right]$$

$$= \cdots - \frac{1}{z^n} - \cdots - \frac{1}{z^2} - \frac{1}{z} - \frac{1}{2} - \frac{z}{4} - \frac{z^2}{8} - \cdots . \tag{2}$$

Problem 6.2. Obtain the Laurent series of $f(z)$ at $z = 0$.

$$f(z) = \frac{1}{z(z-1)}. \tag{3}$$

6.3. Application of Complex Integral

When the degree of a polynomial $P(z)$ is smaller than that of a polynomial $Q(z)$ by more than 2, define the residues $R_1, R_2, \ldots, R_m$ at the singular points of $P(z)/Q(z)$ and the residues are distributed on the upper z plane. This indicates that the imaginary part of the singular point is positive. Then, the following equation is derived:

$$\int_{-\infty}^{\infty} \frac{P(x)}{Q(x)}\,dx = 2\pi i(R_1 + R_2 + \cdots R_m). \tag{6.45}$$

Selecting integral pathes appropriately, we derive the following equations:

$$\int_0^\infty \frac{\sin x}{x}\,dx = \frac{\pi}{2}, \tag{6.46}$$

$$\int_0^\infty \frac{\cos x}{1+x^2}\,dx = \frac{\pi}{2e}, \tag{6.47}$$

$$\int_0^\infty \sin x^2\,dx = \frac{1}{2}\sqrt{\frac{\pi}{2}}. \tag{6.48}$$

As an example, consider the following integrals:

(A) *Definite integral 1.*

Calculate the following integral:

$$\int_{-\infty}^{\infty} \frac{x^2 dx}{(x^2 + a^2)(x^2 + b^2)} = \frac{\pi}{a + b} \quad (a, b \text{ are positive real numbers}).$$

(6.49)

To prove Eq. (6.49), consider the complex function

$$f(z) = \frac{z^2}{(z^2 + a^2)(z^2 + b^2)}.$$

(6.50)

This has poles at $z = \pm ai$ and $\pm bi$. Integrate Eq. (6.50) along the path as shown in Fig. 6.3. Then, we have

$$\int_{-R}^{R} f(z)dz + \int_{C} f(z)dz = 2\pi i(R_1 + R_2),$$

(6.51)

in which R_1 and R_2 are residues at $z = ai$ and bi, respectively. These residues correspond to the singular points on the upper z plane. The differentiator dz in the second term in Eq. (6.51) is transformed to $dz = Rie^{i\theta}d\theta$. Thus, Eq. (6.51) is rewritten as

$$\int_{-R}^{R} f(x)dx + \int_{0}^{\pi} f(Re^{i\theta})Rie^{i\theta}d\theta = 2\pi i(R_1 + R_2).$$

(6.52)

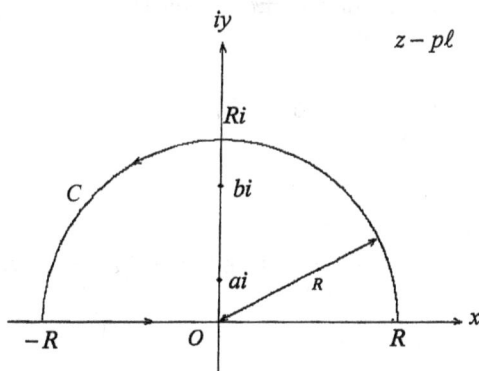

Figure 6.3.   Complex integral.

As $R \to \infty$ in Eq. (6.52), the second term becomes zero and Eq. (6.52) is equal to Eq. (6.49). The residue R_1 at $z = ai$ is

$$R_1 = Res|_{z=ai} = \lim_{z \to ai} (z - ai) \frac{z^2}{(z^2 + a^2)(z^2 + b^2)} = \frac{a}{2i(a^2 - b^2)}.$$
(6.53)

The residue R_2 at $z = bi$ is

$$R_2 = Res|_{z=bi} = \lim_{z \to bi} (z - bi) \frac{z^2}{(z^2 + a^2)(z^2 + b^2)} = \frac{b}{2i(b^2 - a^2)}.$$
(6.54)

Therefore,

$$2\pi i(R_1 + R_2) = \frac{\pi}{a + b}.$$
(6.55)

This shows Eq. (6.49).

(B) *Definite integral 2.*

$$\int_{-\infty}^{\infty} \frac{\cos mx}{1 + x^2} dx = \pi e^{-m} \quad (m : \text{ a positive integer}).$$
(6.56)

To prove Eq. (6.56), consider the complex function

$$f(z) = \frac{e^{imz}}{1 + z^2}.$$
(6.57)

This function has poles at $z = \pm i$. Using the integral path as shown in Fig. 6.4, we obtain

$$\int_{-R}^{R} f(z)dz + \int_{C} f(z)dz = 2\pi i R_1,$$
(6.58)

in which R_1 is the residue at $z = i$. Equation (6.58) is rewritten as

$$\int_{-R}^{R} f(x)dx + \int_{0}^{\pi} f(Re^{i\theta}) Rie^{i\theta} d\theta = 2\pi i R_1.$$
(6.59)

As $R \to \infty$, the second term on the left side becomes 0. Then,

$$\int_{-\infty}^{\infty} \frac{\cos mx}{1 + x^2} dx = 2\pi i R_1.$$
(6.60)

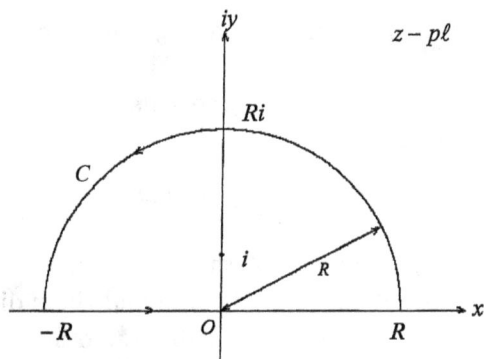

Figure 6.4.   Complex integral.

The residue R_1 is

$$R_1 = Res|_{z=i} = \lim_{z \to i}(z - i)\frac{e^{imz}}{1 + z^2} = -\frac{ie^{-m}}{2}. \qquad (6.61)$$

Thus, we obtain

$$\int_{-\infty}^{\infty} \frac{\cos mx}{1 + x^2}\,dx = \pi e^{-m}. \qquad (6.62)$$

Next, consider a Blasius equation and a Kutta–Joukowski theorem. The theory of wings in aerodynamics is based on these theorems. Figure 6.5 represents a two-dimensional irrotational flow past a body which is shown by the closed curve C'. When the pressure p is exerted

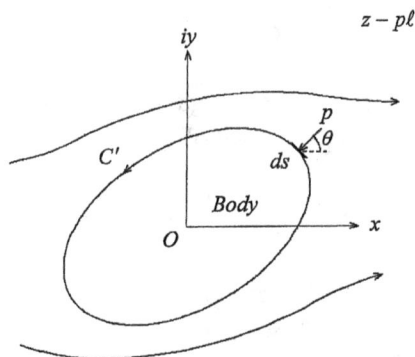

Figure 6.5.   Blasius equation.

on the body C', the forces in the x and y directions are given by

$$X = -\int_{C'} p \cos \theta \, ds, \qquad (6.63)$$

$$Y = -\int_{C'} p \sin \theta \, ds, \qquad (6.64)$$

in which $ds = $ a line segment of the closed curve C' and $\theta = $ an angle between a normal on the body (C') and the x direction. When $dx = -ds \sin \theta$ and $dy = ds \cos \theta$ are substituted into Eqs. (6.63) and (6.64), we obtain

$$X = -\int_{C'} p \, dy, \qquad (6.65)$$

$$Y = \int_{C'} p \, dx. \qquad (6.66)$$

Using the Bernoulli equation, we obtain the pressure p as Bernoulli equation,

$$p = -\frac{\rho}{2}(u^2 + v^2), \qquad (6.67)$$

in which u and $v = $ the velocities in the x and y directions, respectively. Substituting Eq. (6.67) into Eqs. (6.65) and (6.66), we obtain $X - iY$ as follows:

$$X - iY = -i \int_{C'} p(dx - idy) = i \overline{\int_{C'} p \, dz} = -\frac{\rho i}{2} \overline{\int_{C'} (u^2 + v^2) dz}, \tag{6.68}$$

in which "$-$" indicates the conjugate complex variable. Since the complex velocity $\frac{dw}{dz} = u - iv$, we obtain

$$u^2 + v^2 = \frac{dw}{dz} \cdot \left(\overline{\frac{dw}{dz}}\right). \qquad (6.69)$$

Since the closed curve C' corresponds to one of streamlines, $\psi = $ constant and $d\psi = 0$ on C', we get

$$\frac{dw}{dz} = \overline{\frac{dw}{dz}} \quad \text{on } C' \qquad (6.70)$$

and

$$u^2 + v^2 = \left(\frac{dw}{dz}\right)^2 \tag{6.71}$$

are formed. When Eq. (6.69) is used, Eq. (6.68) is

$$X - iY = \frac{\rho i}{2} \int_{C'} \left(\frac{dw}{dz}\right)^2 dz. \tag{6.72}$$

This is called the Blasius theorem or Blasius equation. The integral of the constant on the closed curve is zero. Next, consider the Kutta–Joukowski theorem. There is a steady uniform flow and the velocity is U at infinity. The flow is past a finite body C' with clockwise circulation Γ. The complex potential for the steady uniform flow is Uz and the circulation on the finite body is $-\frac{i\Gamma}{2\pi}\log z$. The other terms that indicate the influence of the finite body on the flow disappear as the distance from the finite body becomes large. Thus, we assume the complex potential w as

$$w = Uz - \frac{i\Gamma}{2\pi}\log z + \frac{c_1}{z} + \frac{c_2}{z^2} + \frac{c_3}{z^3} + \cdots . \tag{6.73}$$

The complex velocity is also

$$\frac{dw}{dz} = U - \frac{i\Gamma}{2\pi z} - \frac{c_1}{z^2} - \frac{2c_2}{z^3} - \frac{3c_3}{z^4} - \cdots . \tag{6.74}$$

Substituting Eq. (6.74) into (6.72), we obtain

$$X - iY = \frac{\rho i}{2} \int_{C'} \left(\frac{dw}{dz}\right)^2 dz = \frac{\rho i}{2} \int_{C'} \left(U^2 - \frac{iU\Gamma}{\pi z} + \frac{b_1}{z^2} + \cdots \right) dz, \tag{6.75}$$

in which

$$b_1 = -2Uc_1 - \frac{\Gamma^2}{4\pi^2}. \tag{6.76}$$

In the integral along the closed curve C', the residue at $z = 0$ is $-\rho U\Gamma/(2\pi)$. Thus, multiply $2\pi i$, we obtain

$$X - iY = -i\rho U\Gamma. \tag{6.77}$$

Equation (6.77) indicates $X = 0$ and $Y = \rho U\Gamma$. This means that the drag and lift forces are 0 and $\rho U\Gamma$, respectively. The lift force is a force exerted on the body normal to the flow direction. The lift force can float an air claft in air.

Example 6.3. When a circular cylinder of radius a is located in the uniform flow of U, the complex potential is given by

$$w = U\left(z + \frac{a^2}{z}\right). \tag{1}$$

Obtain the drag and lift forces on the circular cylinder.

Solution. The complex velocity is

$$\frac{dw}{dz} = U\left(1 - \frac{a^2}{z^2}\right). \tag{2}$$

Substituting Eq. (2) into Eq. (6.72), we obtain

$$X - iY = \frac{\rho i}{2}\int_C U^2\left(1 - \frac{2a^2}{z^2} + \frac{a^4}{z^4}\right)dz = 0. \tag{3}$$

Because the integrated function does not contain $1/z$. This shows that the drag and lift forces are both zero. This is the contradiction for real fluid such as water or air, because the body in flow is exerted upon by an influence of drag. This is called d'Alembert's contradiction. To improve this fact, the Navier–Stokes equations were derived by Navier and Stokes independently. In the 20th century, Prandtl developed the boundary-layer theory to solve the Navier–Stokes equations and developed fluid mechanics.

Problem 6.3. Integrate

$$\int_0^\infty \frac{dx}{1 + x^4} = \frac{\pi}{\sqrt{2}}. \tag{4}$$

Chapter 7

Conformal Mapping

When the motion of the potential flow is studied, the method of conformal mapping on the complex plane can often analytically solve it. When $w = f(z)$ is analytic, the curve for $\phi = $ constant is orthogonal to the curve for $\psi = $ constant, in which $w = \phi + i\psi$ and $z = x + iy$. This theory can apply to the potential motion, wave motion, and groundwater flow motion with the Darcy law. Since the functions ϕ and ψ satisfy the Laplace equation, ϕ and ψ express linear phenomena. Thus, a superposition principle is applicable.

7.1. Conformal Mapping by Elementary Functions

Let us consider the conformal mapping by elementary functions. We understand and become familiar with elementary mapping functions to explain simple fluid flow.

(A) Parallel flow

When the complex potential $w = \phi + i\psi$ is given by

$$w = Uz, \quad (U : \text{real number}), \tag{7.1}$$

the comparison of the real and imaginary parts of Eq. (7.1) shows the following equations:

$$\phi = Ux, \quad \psi = Uy. \tag{7.2}$$

These indicate

$$u = \frac{\partial \phi}{\partial x} = \frac{\partial \psi}{\partial y} = U, \quad v = \frac{\partial \phi}{\partial y} = -\frac{\partial \psi}{\partial x} = 0. \tag{7.3}$$

This is the uniform flow with the velocity U in the x direction. Next, consider the complex potential

$$w = (U_r + iU_i)z, \quad (U_r, U_i: \text{real numbers}). \tag{7.4}$$

Equation (7.4) indicates

$$\phi = U_r x - U_i y, \quad \psi = U_i x + U_r y. \tag{7.5}$$

Thus, we have

$$u = \frac{\partial \phi}{\partial x} = \frac{\partial \psi}{\partial y} = U_r, \quad v = \frac{\partial \phi}{\partial y} = -\frac{\partial \psi}{\partial x} = -U_i. \tag{7.6}$$

These show the uniform flow with the velocity components U_r in the x direction and $-U_i$ in the y direction.

(B) Flow past an angle

Consider the complex potential

$$w = z^n, \quad (n: \text{positive real number}). \tag{7.7}$$

Substituting $w = \phi + i\psi$ and $z = re^{i\theta}$ into Eq. (7.7), we obtain

$$\phi = r^n \cos n\theta, \quad \psi = r^n \sin n\theta. \tag{7.8}$$

The constant stream function gives the boundary of the flow field. Therefore, $\sin n\theta = 0$ shows the walls of which stream function is zero. The walls correspond to $\theta = 0$ and π/n. The complex velocity is

$$\frac{dw}{dz} = u - iv = nz^{n-1} = nr^{n-1}[\cos(n-1)\theta + i\sin(n-1)\theta]. \tag{7.9}$$

Equation (7.9) for $n > 1$ represents the flow past an angle which is less than π. The velocity is zero at $r = 0$ or $z = 0$. Equation (7.9) for $n < 1$ shows the flow past an angle which is larger than π. The velocity is $\pm\infty$ at $r = 0$ or $z = 0$ and the pressure is $-\infty$ at a vertex of the angle. Since the real fluid such as air or water cannot retain the pressure $-\infty$, the flow separates from the wall and the vortex is formed in the lee of the vertex of the angle. Equation (7.9) for $n = 1$ describes the uniform flow as the same as Eq. (7.1).

(C) Flow of source or sink

Consider the following complex potential of the logarithmic function:

$$w = A \log z, \quad (A: \text{real number}). \tag{7.10}$$

Substituting $z = re^{i\theta}$ into Eq. (7.10), we obtain

$$\phi = A \log r, \quad \psi = A\theta. \tag{7.11}$$

The streamline in Eq. (7.11) is the radial line from the origin indicates $\theta =$ constant. This streamline shows the flow pattern when the source or sink is located at the origin. Equi-potential lines correspond to concentric circles which are orthogonal to the streamlines. The radial velocity v_r and tangential velocity v_θ are, respectively, given by

$$v_r = \frac{\partial \phi}{\partial r} = \frac{A}{r}, \tag{7.12}$$

and

$$v_\theta = \frac{1}{r}\frac{\partial \phi}{\partial \theta} = 0. \tag{7.13}$$

This shows that the fluid goes in or comes out from the origin according to the sign of A. When the sign of A is positive and negative, the origin corresponds to the source and sink, respectively. The discharge is computed by the product of v_r and cross-sectional area $2\pi r$.

$$m = v_r \cdot 2\pi r = 2\pi A. \tag{7.14}$$

The value m is called the strength of the sources or sinks. When the source or sink are located at $z = z_1$ and $z = z_2$, the complex potential is shown by

$$w = A \log(z - z_1) + B \log(z - z_2), \tag{7.15}$$

in which $2\pi A$ and $2\pi B$ are the strengths of the sources or sinks.

(D) Flow past a doublet

Consider the following complex potential:

$$w = \frac{m}{z}, \quad (m: \text{real number}). \tag{7.16}$$

From Eq. (7.16) we obtain

$$\phi = \frac{mx}{x^2 + y^2}, \quad \psi = -\frac{my}{x^2 + y^2}. \tag{7.17}$$

An equi-potential line ($\phi = \phi_0$ = const) is given by

$$\left(x - \frac{m}{2\phi_0}\right)^2 + y^2 = \left(\frac{m}{2\phi_0}\right)^2. \tag{7.18}$$

A streamline ($\psi = \psi_0$ = const) is given by

$$x^2 + \left(y + \frac{m}{2\psi_0}\right)^2 = \left(\frac{m}{2\psi_0}\right)^2. \tag{7.19}$$

The equi-potential lines are a group of circles of which center is on the x axis and pass through the origin. The streamlines are a group of circles of which the center is on the y axis and pass through the origin. Both are orthogonal to each other. The strength of the doublet is $2\pi m$. The flow due to the doublet is simulated by the source and sink. As the source approaches the sink in the condition such that their strengths multiplied by their distance is kept constant, the flow pattern becomes a function by the doublet. When the source and sink are located at z and $z - \Delta z$, the complex potential of which strength $2\pi A$ is

$$w = A \log z - A \log(z - \Delta z). \tag{7.20}$$

With the Taylor series Eq. (7.20) is rearranged to

$$w = A \log \frac{z}{z - \Delta z} = A \log \left(1 + \frac{\Delta z}{z} + \left(\frac{\Delta z}{z}\right)^2 + \cdots\right) \cong A \frac{\Delta z}{z}. \tag{7.21}$$

If $A\Delta z = m'$ for $\Delta z \to 0$ and $A \to \infty$, we have

$$w = \frac{m'}{z}. \tag{7.22}$$

Thus, m' can be a complex number, because Δz is a complex number.

(E) Flow due to vortex

When A is an imaginary number in $-i\Gamma/2\pi$ Eq. (7.10), Eq. (7.10) is written by

$$w = -\frac{i\Gamma}{2\pi} \log z, \quad (\Gamma: \text{real number}). \tag{7.23}$$

Substituting $z = re^{i\theta}$ into Eq. (7.23), we obtain

$$\phi = \frac{\Gamma}{2\pi}\theta, \quad \psi = -\frac{\Gamma}{2\pi}\log r. \tag{7.24}$$

The streamline ($\psi = $ const) is shown by the curve of $r = $ const. These are concentric circles of which center is on the origin. The equi-potential line $\phi = $ const is given by $\theta = $ const and the radial lines from the origin. This indicates a vortex at the origin. The radial and tangential velocities are respectively

$$v_r = \frac{\partial \phi}{\partial r} = 0, \tag{7.25}$$

$$v_\theta = \frac{1}{r}\frac{\partial \phi}{\partial \theta} = \frac{\Gamma}{2\pi r}. \tag{7.26}$$

Equations (7.25) and (7.26) indicate that Eq. (7.23) forms a circular motion about the origin. The strength of the circulation Γ is defined by

$$\Gamma = 2\pi r \cdot v_\theta. \tag{7.27}$$

This is the same as the definition of the circulation as

$$\Gamma = \int_C v_\theta \, ds, \tag{7.28}$$

in which $C = $ the circle of which center is the origin and $s = $ the coordinate on the circle C. When the vortex is located on $z = z_1$, the complex potential is

$$w = -\frac{i\Gamma}{2\pi}\log(z - z_1), \tag{7.29}$$

in which $z_1 = $ a complex number.

(F) Flow past a cylinder

Consider the following complex potential:

$$w = U\left(z + \frac{a^2}{z}\right), \tag{7.30}$$

in which U and a are real numbers. Substituting $z = re^{i\theta}$ and $w = \phi + i\psi$ into Eq. (7.30) and from the real and imaginary parts of

Eq. (7.30), we obtain

$$\phi = U \left(r + \frac{a^2}{r} \right) \cos \theta \tag{7.31}$$

and

$$\psi = U \left(r - \frac{a^2}{r} \right) \sin \theta. \tag{7.32}$$

The radial and tangential velocities are respectively

$$v_r = \frac{\partial \phi}{\partial r} = U \left(1 - \frac{a^2}{r^2} \right) \cos \theta \tag{7.33}$$

and

$$v_\theta = \frac{1}{r} \frac{\partial \phi}{\partial \theta} = -U \left(r + \frac{a^2}{r} \right) \sin \theta. \tag{7.34}$$

At $r = a$, $v_r = 0$ and $\psi = 0$ are formed. Thus, $r = a$ shows the streamline or the boundary. The streamline $r = a$ indicates the surface of the cylinder. This corresponds to $z = ae^{i\theta}$. The maximum velocity on the surface of the cylinder is $v_\theta = -2U$ at $\theta = \pi/2$. As $z \to \pm\infty$, Eq. (7.30) approaches $w = Uz$ which is the complex potential of the uniform flow with the velocity U in the x direction. Therefore, Eq. (7.30) describes the complex potential of the flow past a cylinder of which radius is a at the origin in the uniform flow U.

Example 7.1. Obtain the complex potential when a vortex of which strength is Γ is located at $x = a$ and the y axis is the boundary as shown in Fig. 7.1.

Solution. When Eq. (7.23) is used and there is no wall, the complex potential w_1 is

$$w_1 = -\frac{i\Gamma}{2\pi} \log(z - a). \tag{1}$$

When the wall exists at $x = 0$, the velocity in the x direction is zero. If there is a vortex of which strength is the same and direction is

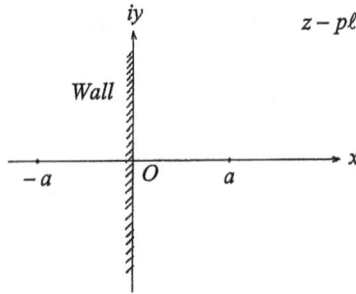

Figure 7.1. Example 7.1.

opposite to the vortex w_1 at $x = -a$, the velocity of the x component at $x = 0$ is zero. The complex potential w_2 is given by

$$w_2 = \frac{i\Gamma}{2\pi} \log(z + a). \qquad (2)$$

The resultant complex potential $w = w_1 + w_2$ is given by

$$w = -\frac{i\Gamma}{2\pi} \log(z - a) + \frac{i\Gamma}{2\pi} \log(z + a) = \frac{i\Gamma}{2\pi} \log \frac{z + a}{z - a}. \qquad (3)$$

The complex velocity is

$$\frac{dw}{dz} = \frac{\partial \phi}{\partial x} + i\frac{\partial \psi}{\partial x} = -\frac{i\Gamma a}{\pi} \cdot \frac{1}{z^2 - a^2}. \qquad (4)$$

At the wall or the y axis, $z = iy$ is formed. Thus, dw/dz becomes imaginary. From Eq. (4) we have

$$\left.\frac{\partial \phi}{\partial x}\right|_{x=0} = 0. \qquad (5)$$

Equation (5) shows that the velocity in the x direction at the wall is zero.

Problem 7.1. When a vortex is located at (a, ib) and surrounded by two orthogonal walls of the x axis $(x > 0)$ and the y axis $(y > 0)$ as shown in Fig. 7.2, obtain the complex potential. The constants a and b are positive real numbers.

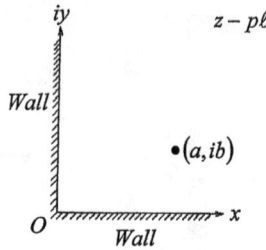

Figure 7.2.   Example 7.2.

7.2. Schwartz–Christoffel Transformation

We have studied analysis of flow fields using various elementary functions. But we need to compute the complex potential when an arbitrary body or boundary is in the flow field. Herein consider an analytic complex function to transform the inside of the polygon to the upper half complex plane as shown in Fig. 7.3. This is called a Schwartz–Chritoffel transformation (Ahlfors, 1966). When two simply connected regions D and D' are given on the z and ζ planes, respectively, there exists an analytic complex function $\zeta = f(z)$, such that $f(z)$ defines a one-to-one mapping of the region D onto D'. The simply connected region defines that when a closed curve shrinks in the simply connected region it reduces a point. This is not always true in a doubly connected region such as a doughnut. When

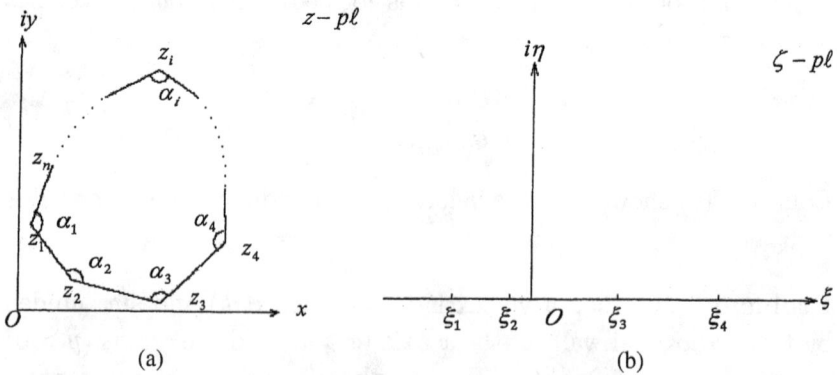

Figure 7.3.   Schwartz–Christoffel transform.

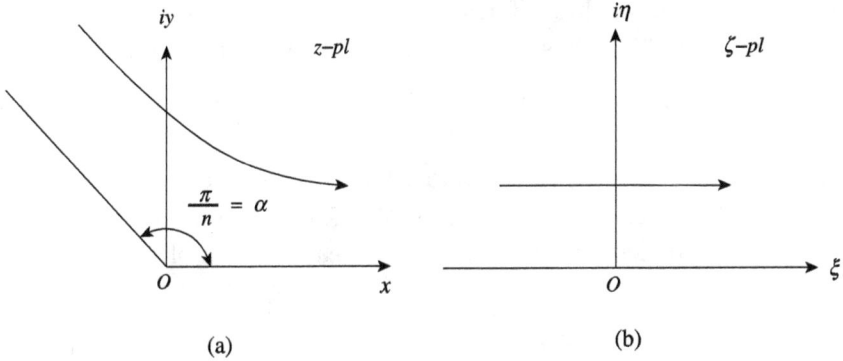

Figure 7.4.   Flow past an angle.

the polygon is assumed to be a wall or boundary in the flow for the Schwartz–Christoffel transformation, the side of the polygon forms the streamline. If the side of the polygon is mapped on the real axis on the ζ plane, the flow inside the polygon corresponds to the flow on the upper half ζ plane in Fig. 7.3. The mapping function which transforms the flow along the real axis on the ζ plane to the flow past an angle on the z plane in Fig. 7.4 is given by

$$\zeta = Az^n \tag{7.35}$$

or

$$z = A^{-1/n}\zeta^{1/n}, \tag{7.36}$$

in which $z = x + iy$ and $\zeta = \xi + i\eta$. The variables ξ and η are real numbers. This is the case for $z = 0$ corresponding to $\zeta = 0$. When $z = z_i$ corresponds to $\zeta = \xi_i$ which is a real number, the derivative of $z - z_i = A^{-1/n}(\zeta - \zeta_i)^{1/n}$ is

$$\frac{dz}{d\zeta} = A^{-1/n}\frac{1}{n}(\zeta - \xi_i)^{\frac{1}{n}-1}. \tag{7.37}$$

Since an angle of the vertex of the polygon is α_i, Eq. (7.37) is written by $1/n = \alpha_i/\pi$. Thus, we have

$$\frac{dz}{d\zeta} = \left(A^{-1/n}\frac{\alpha_i}{\pi}\right)(\zeta - \xi_i)^{\frac{\alpha_i}{\pi}-1}. \tag{7.38}$$

The similar relationship is formed at the other vertices of the polygon (Mizumura, 1997). Thus, we obtain the following function:

$$\frac{dz}{d\zeta} = A(\zeta - \xi_1)^{\frac{\alpha_1}{\pi}-1}(\zeta - \xi_2)^{\frac{\alpha_2}{\pi}-1}\cdots(\zeta - \xi_n)^{\frac{\alpha_n}{\pi}-1}, \qquad (7.39)$$

in which $\xi_1, \xi_2, \ldots, \xi_n$ are points which correspond to the vertices of the polygon on the real axis of the ζ plane and $\alpha_1, \alpha_2, \ldots, \alpha_n$ are internal angles which correspond to each vertex of the polygon. Thus,

$$\alpha_1 + \alpha_2 + \cdots + \alpha_n = (n-2)\pi \qquad (7.40)$$

is formed. The remarks on the usage of the Schwartz–Christoffel transformation are as follows:

(i) three points are arbitrarily selected on the ζ plane; and

(ii) the points at infinity on the ζ plane are independent of the Schwartz–Christoffel transformation.

As an example, if the limiting shape of the upper half ζ plane is a unit circle as shown in Fig. 7.5, let us consider the Schwartz–Christoffel transformation of an infinite strip onto the unit circle. From Eq. (7.39) we obtain

$$\frac{dz}{d\zeta} = A(\zeta - 1)^{-1}(\zeta + 1)^{-1}. \qquad (7.41)$$

Integrating Eq. (7.41), we get

$$z = \frac{A}{2}\log\frac{1-\zeta}{1+\zeta} + B, \qquad (7.42)$$

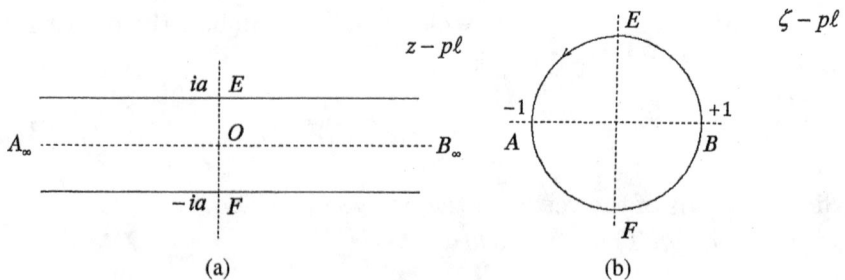

Figure 7.5.   Mapping into a circle.

in which B is an integral constant. The correspondence of $z = 0$ to $\zeta = 0$ indicates

$$B = 0. \tag{7.43}$$

Since $z = ai$ corresponds to $\zeta = i$, we have

$$A = -\frac{4a}{\pi}. \tag{7.44}$$

Thus, we obtain the mapping function as

$$z = \frac{4a}{\pi} \log \frac{1+\zeta}{1-\zeta} \quad \text{or} \quad \zeta = \tanh \frac{\pi z}{4a}. \tag{7.45}$$

Example 7.2. As shown in Fig. 7.6, obtain the Schwartz–Christoffel transformation of a semi-infinite strip on the upper half ζ plane.

Solution. The four vertices z_1, z_2, z_3, and z_4 of the semi-infinite strip on the z plane correspond to the four points ξ_1, ξ_2, ξ_3, and ξ_4 on the real axis of the ζ plane, respectively. Since the three points on the real axis on the ζ plane are arbitrarily selected, we have

$$\xi_1 = -\infty, \quad \xi_2 = -1, \quad \xi_3 = 1. \tag{1}$$

Since the point z_1 is the same as the point z_4 at infinity on the z plane, ξ_4 must be ∞. The terms corresponding to $\xi = \pm\infty$ are excluded from

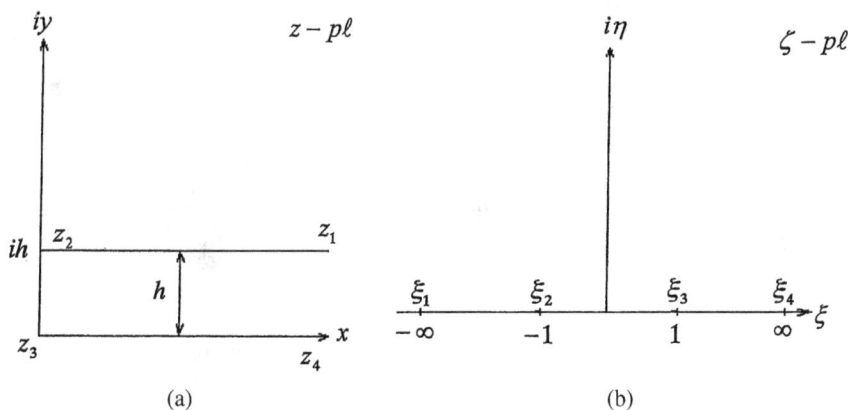

Figure 7.6.   Mapping of a semi infinite strip.

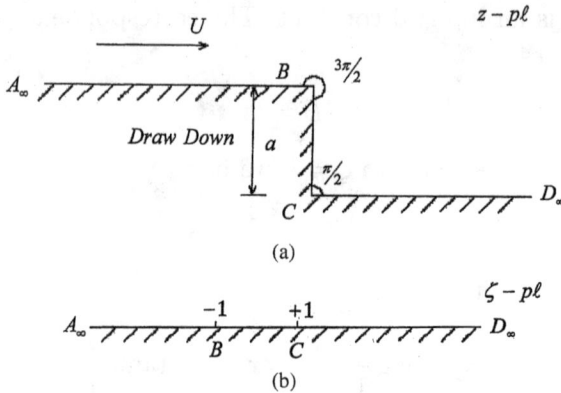

Figure 7.7.   Flow on a step.

the Schwartz–Christoffel transformation. Equation (7.39) becomes

$$\frac{dz}{d\zeta} = \frac{A}{\sqrt{\zeta^2 - 1}}. \tag{2}$$

Integrating Eq. (2) with the $\zeta = \cosh X$, we obtain

$$z = A \cosh^{-1} \zeta + C, \tag{3}$$

in which $C =$ an integral constant. The conditions that $\xi_2 = -1$ and $\xi_3 = 1$ at $z_2 = ih$ and $z_3 = 0$, respectively, give $C = 0$ and $A = h/\pi$. Thus, from Eq. (3) we obtain

$$z = \frac{h}{\pi} \cosh^{-1} \zeta, \tag{4}$$

or

$$\zeta = \cosh \frac{\pi z}{h}. \tag{5}$$

Problem 7.2. Map a flow on a step to the upper half ζ plane as shown in Fig. 7.7.

7.3. Applications to Hydraulic Engineering

Consider an infinite strip on the z plane as shown in Fig. 7.8 (Milne-Thomson, 1968). The width of the infinite strip is a. The six points

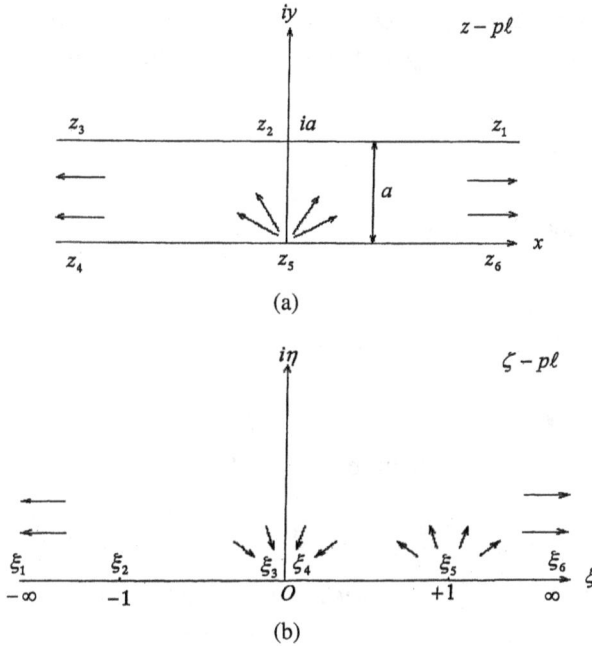

Figure 7.8.   Flow in an infinite strip.

z_1, z_2, z_3, z_4, z_5, and z_6 on the z plane correspond to ξ_1, ξ_2, ξ_3, ξ_4, ξ_5, and ξ_6 on the ζ plane. Three out of six points ξ_1, ξ_2, ξ_3, ξ_4, ξ_5, and ξ_6 on the ζ plane are arbitrarily selected. Since ξ_1 and ξ_6 are $-\infty$ and ∞ on the ζ plane, respectively, they are not considered in the Schwartz–Christoffel transformation. Since z_3 coincides with z_4 at infinity on the z plane, ξ_3 coincides with ξ_4 at the origin on the ζ plane. Thus, since the three points on the z plane are z_2, z_3, and z_5, the points ξ_2, ξ_3, and ξ_5 are given by

$$\xi_2 = -1, \quad \xi_3 = 0, \quad \xi_5 = 1. \tag{7.46}$$

The Schwartz–Christoffel transformation is given by

$$\frac{dz}{d\zeta} = \frac{A}{\zeta}. \tag{7.47}$$

Integrating Eq. (7.47), we obtain

$$z = A \log \zeta + C_1, \tag{7.48}$$

in which $C_1 =$ an integral constant. The correspondence between z_2 and ξ_2, between z_5 and ξ_5 are, respectively, shown by

$$ia = A\log(-1) + C_1, \tag{7.49}$$

$$0 = A\log(1) + C_1. \tag{7.50}$$

Equation (7.50) indicates

$$C_1 = 0. \tag{7.51}$$

Equation (7.49) and $e^{i\pi} = -1$ lead to

$$A = a/\pi. \tag{7.52}$$

Thus, the transformation function Eq. (7.48) becomes

$$z = \frac{a}{\pi}\log\zeta \quad \text{or} \quad \zeta = e^{\frac{\pi z}{a}}. \tag{7.53}$$

Then, consider the flow pattern when fluid flows into the z plane from the point z_5 and out from the points z_1, z_3, z_4, and z_6. The flow corresponding to that is formed by a source of the strength m at ξ_5, sinks of the strength $m/2$ at ξ_1 (or ξ_6) and ξ_3 (or ξ_4). Then, the complex potential is given by

$$w = \frac{m}{2\pi}\log(\zeta - 1) - \frac{m}{4\pi}\log\zeta. \tag{7.54}$$

When the discharge from ξ_5 is Q, we get $Q = m/2$. Thus, Eq. (7.54) is rewriten by

$$w = \frac{Q}{\pi}\left[\log(\zeta - 1) - \frac{1}{2}\log\zeta\right] = \frac{Q}{\pi}\log(\zeta^{1/2} - \zeta^{-1/2}). \tag{7.55}$$

Substituting Eq. (7.53) into Eq. (7.55), we obtain

$$w = \frac{Q}{\pi}\log\left(e^{\frac{\pi z}{2a}} - e^{-\frac{\pi z}{2a}}\right) = \frac{Q}{\pi}\log\left(\sinh\frac{\pi z}{2a}\right) - \frac{Q}{\pi}\log 2. \tag{7.56}$$

Taking the derivative of w about z, we obtain the complex velocity on the z plane as

$$\frac{dw}{dz} = \frac{Q}{2a}\coth\frac{\pi z}{2a}, \tag{7.57}$$

in which

$$\coth \frac{\pi z}{2a} = \frac{e^{\frac{\pi z}{a}} + 1}{e^{\frac{\pi z}{a}} - 1}. \tag{7.58}$$

Equation (7.57) is written as

$$u - iv = \frac{Q}{2a} \frac{e^{\frac{\pi z}{a}} + 1}{e^{\frac{\pi z}{a}} - 1}. \tag{7.59}$$

Since the complex velocity at $z = ia$ is zero, we have

$$u = v = 0. \tag{7.60}$$

We used the relationship $e^{\pi i} = -1$. Thus, $z = ia$ is the stagnation point on the z plane. The velocity at $z = \pm\infty$ is $Q/(2a)$.

Next, consider the flow on a step in a ditch as shown in Fig. 7.9. Since three points ξ_1, ξ_2, and ξ_4 on the real axis on the ζ plane are arbitrarily selected, we have

$$\xi_1 = -\infty, \quad \xi_2 = 0, \quad \xi_4 = 1. \tag{7.61}$$

Figure 7.9 indicates

$$\xi_3 = \xi_2, \quad \xi_6 = \infty. \tag{7.62}$$

The coordinate a at ξ_5 is unknown. The Schwartz–Christoffel transformation is derived as

$$\frac{dz}{d\zeta} = A\zeta^{-1}(\zeta - 1)^{1/2}(\zeta - a)^{-1/2}. \tag{7.63}$$

The flow in Fig. 7.9 corresponds to the flow that there is a source at the origin on the ζ plane. Thus,

$$w = \frac{m}{2\pi} \log \zeta. \tag{7.64}$$

The discharge through the ditch is $U_1 h_1$ and this is the discharge from the source. The strength of the source m is $2U_1 h_1$. Thus, we have the complex potential as

$$w = \frac{U_1 h_1}{\pi} \log \zeta. \tag{7.65}$$

(a)

(b)

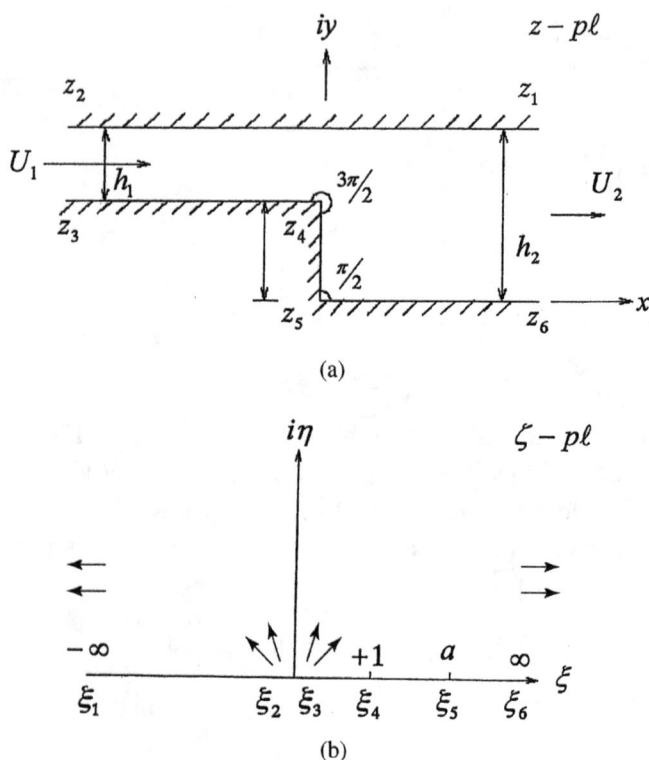

Figure 7.9.   Abrupt change in the breadth of a channel.

The complex velocity on the z plane is computed by Eqs. (7.65) and (7.63) as follows:

$$\frac{dw}{dz} = \frac{dw}{d\zeta} \cdot \frac{d\zeta}{dz} = \frac{U_1 h_1}{\pi \zeta} \cdot \frac{\zeta}{A} \sqrt{\frac{\zeta - a}{\zeta - 1}} = \frac{U_1 h_1}{\pi A} \sqrt{\frac{\zeta - a}{\zeta - 1}}. \qquad (7.66)$$

At $z = \infty$ ($\zeta = \infty$) we obtain from Eq. (7.66) as

$$U_2 = \frac{U_1 h_1}{\pi A} \sqrt{\frac{\zeta - a}{\zeta - 1}}\Bigg|_{\zeta \to \infty}. \qquad (7.67)$$

Thus, we get

$$A = \frac{U_1 h_1}{U_2 \pi}. \qquad (7.68)$$

From the continuity equation, we obtain

$$U_1 h_1 = U_2 h_2. \tag{7.69}$$

From Eqs. (7.68) and (7.69), we have

$$A = \frac{h_2}{\pi}. \tag{7.70}$$

Since from Eq. (7.66) $dw/dz = U_1$ at $\zeta = 0$, we have

$$U_1 = \frac{U_1 h_1 \sqrt{a}}{h_2}. \tag{7.71}$$

Equation (7.71) determines the value of a as

$$a = \left(\frac{h_2}{h_1}\right)^2. \tag{7.72}$$

To integrate Eq. (7.63), we define a new variable t as follows:

$$t = \sqrt{\frac{\zeta - a}{\zeta - 1}}. \tag{7.73}$$

With Eq. (7.73), Eq. (7.63) is rearranged as

$$\frac{d\zeta}{\zeta} = \left(\frac{2t}{1 - t^2} - \frac{2t}{a - t^2}\right) dt. \tag{7.74}$$

The usage of Eqs. (7.63), (7.73), and (7.74) lead to

$$\frac{dz}{dt} = A\left(\frac{2}{1 - t^2} - \frac{2}{a - t^2}\right). \tag{7.75}$$

Integrating Eq. (7.75), we obtain

$$z = \frac{h_2}{\pi}\left(\log \frac{1 + t}{1 - t} - \frac{1}{\sqrt{a}} \log \frac{\sqrt{a} + t}{\sqrt{a} - t}\right) + C_1. \tag{7.76}$$

Since $z_5 = 0$ corresponds to $\xi_5 = a$, $t = 0$. Thus, we get $C_1 = 0$ in Eq. (7.76).

Next, consider the open channel flow at the confluence by Modi *et al.* (1981) as shown in Fig. 7.10. The cross-sectional shape of the three open channels are rectangular. Two open channels on the

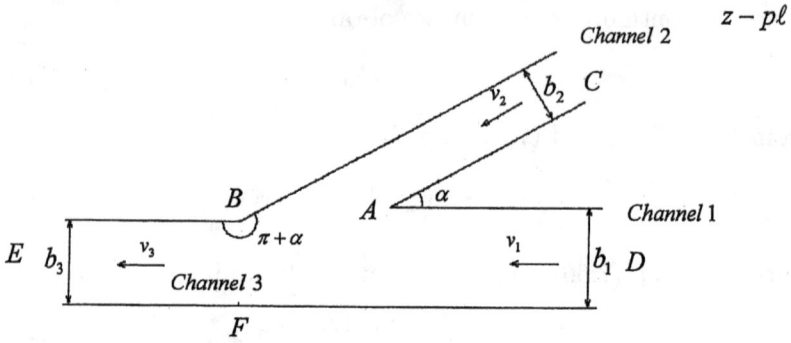

(a)

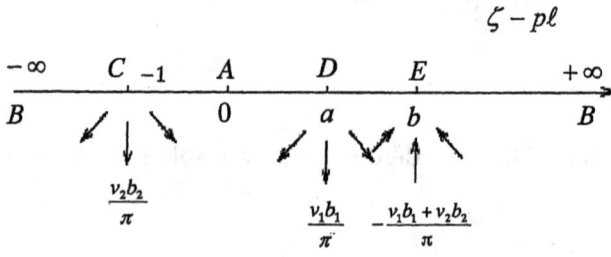

(b)

Figure 7.10.   Flow at the confluence of the open channels.

horizontal join at an angle at the confluence. Let us calculate a discharge relation such that the point A is a stagnation point. In Fig. 7.10, transform the z plane onto the lower half ζ plane. The Schwartz–Chritoffel transformation is given by

$$\frac{dz}{d\zeta} = \frac{K\zeta^{1-\alpha/\pi}}{(\zeta+1)(\zeta-a)(\zeta-b)}. \qquad (7.77)$$

Equation (7.77) is rearranged in the form of partial fractions by

$$\frac{dz}{d\zeta} = K\zeta^{-\alpha/\pi}\left[-\frac{1}{(1+a)(1+b)}\cdot\frac{1}{(1+\zeta)} + \frac{a}{(1+a)(b-a)}\cdot\frac{1}{(a-\zeta)}\right.$$
$$\left. -\frac{b}{(1+b)(b-a)}\cdot\frac{1}{(b-\zeta)}\right], \qquad (7.78)$$

in which K = a complex constant and a and b = unknowns. Integrating Eq. (7.75) from the point A to B in Fig. 7.10, we obtain

$$b_2 \csc \alpha + (b_1 - b_3) \cot \alpha + i(b_1 - b_3)$$

$$= K\pi \left[\frac{1}{(1+a)(1+b)} + \frac{a^{1-\alpha/\pi} \cos \alpha}{(1+a)(b-a)} - \frac{b^{1-\alpha/\pi} \cos \alpha}{(1+b)(b-a)} \right] \csc \alpha,$$

$$(7.79)$$

in which csc = $cosec$ and b_1, b_2 and b_3 = channel widths. Since K is the complex number, we may write

$$K = K_1 + iK_2. \tag{7.80}$$

Substituting Eq. (7.80) into Eq. (7.79), we obtain the following two equations of the real and imaginary parts:

$$b_2 + (b_1 - b_3) \cos \alpha = K_1 \pi \left[\frac{1}{(1+a)(1+b)} + \frac{a^{1-\alpha/\pi} \cos \alpha}{(1+a)(b-a)} \right.$$

$$\left. - \frac{b^{1-\alpha/\pi} \cos \alpha}{(1+b)(b-a)} \right] \tag{7.81}$$

and

$$b_1 - b_3 = K_2 \pi \left[\frac{1}{(1+a)(1+b)} + \frac{a^{1-\alpha/\pi} \cos \alpha}{(1+a)(b-a)} \right.$$

$$\left. - \frac{b^{1-\alpha/\pi} \cos \alpha}{(1+b)(b-a)} \right] \csc \alpha. \tag{7.82}$$

Integrating Eq. (7.78) from $\zeta = a - \epsilon$ to $a + \epsilon$ on the real axis and approaching ϵ to 0, we obtain

$$\int_0^{b_1 i} ds = b_1 i$$

$$= K \int_{a-\epsilon}^{a+\epsilon} \xi^{-\alpha/\pi} \left[-\frac{1}{(1+a)(1+b)} \cdot \frac{1}{(1+\zeta)} \right.$$

$$\left. + \frac{a}{(1+a)(b-a)} \cdot \frac{1}{(a-\zeta)} - \frac{b}{(1+b)(b-a)} \cdot \frac{1}{(b-\zeta)} \right] d\xi$$

$$= K\pi \frac{ia^{1-\alpha\pi}}{(1+a)(b-a)}. \tag{7.83}$$

Equation (7.83) shows that K is a real number. This indicates

$$b_1 = K_1 \pi \frac{a^{1-\alpha/\pi}}{(1+a)(b-a)}. \tag{7.84}$$

Since the integration is done on the real axis ξ, the complex constant K is real and equal to K_1 ($K_2 = 0$). In the same way, integrating Eq. (7.78) from $\zeta = b - \epsilon$ to $b + \epsilon$ on the real axis and approaching ϵ to 0, we obtain

$$b_3 = K_1 \pi \frac{b^{1-\alpha/\pi}}{(1+b)(b-a)}. \tag{7.85}$$

Eliminating K_1 from Eqs. (7.84) and (7.85), we obtain

$$\frac{a^{1-\alpha/\pi}}{b_1(1+a)} = \frac{b^{1-\alpha/\pi}}{b_3(1+b)}. \tag{7.86}$$

Substituting Eqs. (7.84) and (7.85) into Eq. (7.81), we obtain

$$b_2 = \frac{K_1 \pi}{(1+a)(1+b)}. \tag{7.87}$$

Dividing Eq. (7.87) by Eq. (7.84), we get

$$\frac{b_2}{b_1} = \frac{b-a}{(1+b)a^{1-\alpha/\pi}}. \tag{7.88}$$

Therefore, the unknowns a, b, K_1, and K_2 are expressed by the knowns b_1, b_2, b_3, and α. Next, consider the flow field on the ζ plane. The discharges from the channel 1, 2, and 3 are $b_1 v_1$, $b_2 v_2$, and $b_3 v_3$, respectively. The symbols v_1, v_2, and v_3 are velocity in the channel 1, 2, and 3, respectively. The complex potential on the ζ plane is

$$w = -\frac{v_1 b_1 + v_2 b_2}{\pi} \log(\zeta - b) + \frac{v_1 b_1}{\pi} \log(\zeta - a) + \frac{v_2 b_2}{\pi} \log(\zeta + 1). \tag{7.89}$$

Differentiating Eq. (7.89) about ζ, we obtain

$$\frac{dw}{d\zeta} = -\frac{v_1 b_1 + v_2 b_2}{\pi} \cdot \frac{1}{\zeta - b} + \frac{v_1 b_1}{\pi} \cdot \frac{1}{\zeta - a} + \frac{v_2 b_2}{\pi} \cdot \frac{1}{\zeta + 1}. \tag{7.90}$$

Thus, the complex velocity is given by

$$
\frac{dw}{dz} = \frac{dw}{d\zeta} \cdot \frac{d\zeta}{dz}
$$

$$
= \frac{v_1 b_1 + v_2 b_2}{\pi} \cdot \frac{(\zeta + 1)(\zeta - a)}{K\zeta^{1-\alpha/\pi}} - \frac{v_1 b_1}{\pi} \cdot \frac{(\zeta + 1)(\zeta - b)}{K\zeta^{1-\alpha/\pi}}
$$

$$
- \frac{v_2 b_2}{\pi} \cdot \frac{(\zeta - b)(\zeta - a)}{K\zeta^{1-\alpha/\pi}}. \tag{7.91}
$$

Since $dw/dz = 0$ at the stagnation point, we have

$$
\frac{v_1 b_1 + v_2 b_2}{\pi} \cdot \frac{(\zeta + 1)(\zeta - a)}{K\zeta^{1-\alpha/\pi}} - \frac{v_1 b_1}{\pi} \cdot \frac{(\zeta + 1)(\zeta - b)}{K\zeta^{1-\alpha/\pi}}
$$

$$
- \frac{v_2 b_2}{\pi} \cdot \frac{(\zeta - b)(\zeta - a)}{K\zeta^{1-\alpha/\pi}} = 0. \tag{7.92}
$$

Solving Eq. (7.92) about ζ, we have

$$
\zeta = \frac{v_2 b_2 a(1 + b) - v_1 b_1 (b - a)}{v_1 b_1 (b - a) + v_2 b_2 (1 + b)}. \tag{7.93}
$$

If n_q is a discharge ratio of the channel 1 and 2, we get

$$
n_q = \frac{v_2 b_2}{v_1 b_1 + v_2 b_2}. \tag{7.94}
$$

Thus, Eq. (7.93) is given by

$$
\zeta = \frac{n_q b(1 + a) - (b - a)}{(b - a) + n_q(1 + a)}. \tag{7.95}
$$

This shows the coordinate of the stagnation point. If the point A is the stagnation point, $\zeta = 0$. Thus,

$$
n_q = \frac{b - a}{b(1 + a)}. \tag{7.96}
$$

When the flow separates at the confluence, a free streamline is formed. This theory is also applicable for obtaining the shape of a free streamline. When the flow of the perfect fluid separates, the velocity is ∞ and the pressure is $-\infty$ from the Bernoulli equation. But real fluid cannot retain the velocity of ∞ and the pressure of $-\infty$. Thus, we assume the free streamline in which the pressure and the velocity are constant in a wake. The flow from a breach of the open channel

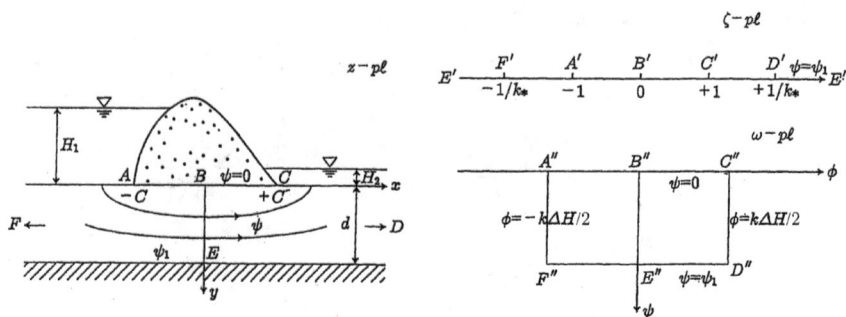

Figure 7.11. Seepage flow under a dam.

flow is investigated by Mizumura *et al.* (2003b). They theoretically and experimentally studied this problem using the Euler equations of motion. They found that the discharge ratio of the outside flow to the main channel flow is inversely proportional to the Froude number of the main channel flow when the flow is supercritical. This is explained in Chap. 12.

The next example is the seepage flow under a dam in Fig. 7.11 (Harr, 1962). The seepage flow is assumed to be governed by the Darcy law and the velocity q in the s direction is proportional to the hydraulic gradient in the s direction. That is, the velocity is given by

$$q = -k\frac{dh}{ds}, \tag{7.97}$$

in which k is the hydraulic conductivity and h is the piezometric head. When $-kh$ is equal to the potential ϕ, the groundwater flow is assumed to have a potential ϕ. As shown in Sec. 6.1, the stream function ψ is determined such that they satisfy the Cauchy–Riemann equations. The functions ϕ and ψ correspond to the w plane in Fig. 7.11. The ζ plane maps the z plane and the w plane. The z and w planes are the coordinate system of the left hand. Thus, clockwise direction is defined to be positive in the coordinate system of the left hand. The clockwise direction is defined to be negative in the coordinate system of the right hand. The difference of the water levels in the upstream and downstream is defined $\Delta H = H_1 - H_2$. The correspondence between the w and z planes is as follows:

z plane	w plane
CD (equi-potential)	$C''D''$ ($\phi = k\Delta H/2$)
FA (equi-potential)	$F''A''$ ($\phi = -k\Delta H/2$)
AC (streamline)	$A''C''$ ($\psi = 0$)
DF (streamline)	$D''F''$ ($\psi = \psi_1$)

Consider the mapping relationship between the z and ζ planes. The symmetry of the flow about the z plane shows the numerals on the ζ plane as shown in Fig. 7.11. An unknown parameter k_* is larger than 1, as shown in Fig. 7.11. The Schwartz–Christoffel transformation is

$$\frac{dz}{d\zeta} = C_1 \left(\zeta + \frac{1}{k_*}\right)^{0/\pi - 1} (\zeta + 1)^{\pi/\pi - 1} \left(\zeta - \frac{1}{k_*}\right)^{0/\pi - 1}$$

$$= C_1 \left(\zeta^2 - \frac{1}{k_*^2}\right)^{-1}. \tag{7.98}$$

Integrating Eq. (7.98), we obtain,

$$z = C_1 \int \frac{d\zeta}{\zeta^2 - (1/k_*)^2} = C_1 k_* \tanh^{-1}(k_*\zeta) + C_2, \tag{7.99}$$

in which $C_2 =$ an integral constant. The correspondence between the z and ζ planes is as follows:

$$\text{Point } A \quad z = -c \Longleftrightarrow \zeta = -1$$

$$\text{Point } C \quad z = +c \Longleftrightarrow \zeta = +1$$

$$\text{Point } E \quad z = id \Longleftrightarrow \zeta = \infty$$

Thus, we have

$$-c = -C_1 k_* \tanh^{-1} k_* + C_2, \tag{7.100}$$

$$c = C_1 k_* \tanh^{-1} k_* + C_2, \tag{7.101}$$

$$id = C_1 k_* \tanh^{-1} \infty + C_2 \tag{7.102}$$

and the relationship $\tanh iy = i \tan y$ (Chap. 6) derives

$$C_1 = \frac{2d}{\pi k_*}, \quad C_2 = 0, \quad k_* = \tanh \frac{c\pi}{2d}. \tag{7.103}$$

Therefore, the Schwartz–Christoffel transformation between the ζ and z planes is

$$\zeta = \frac{\tanh{(\pi z/2d)}}{\tanh{(\pi c/2d)}}. \tag{7.104}$$

The Schwartz–Christoffel transformation between the w and ζ planes is given by

$$w = C_3 \int_0^\zeta \frac{d\zeta}{\sqrt{(\zeta^2 - 1)(\zeta^2 - 1/k_*^2)}} + C_4, \tag{7.105}$$

in which C_4 is an integral constant. The correspondence between the w and z planes is as follows:

$$\text{Point } A \quad w = -k\Delta H/2 \Leftrightarrow \zeta = -1$$

$$\text{Point } C \quad w = k\Delta H/2 \Leftrightarrow \zeta = +1.$$

Substituting the above correspondence into Eq. (7.105), we have

$$-\frac{k\Delta H}{2} = C_3 \int_0^{-1} \frac{d\zeta}{\sqrt{(\zeta^2 - 1)(\zeta^2 - 1/k_*^2)}} + C_4$$

$$= -C_3 K(k_*) + C_4, \tag{7.106}$$

$$\frac{k\Delta H}{2} = C_3 \int_0^{+1} \frac{d\zeta}{\sqrt{(\zeta^2 - 1)(\zeta^2 - 1/k_*^2)}} + C_4 = C_3 K(k_*) + C_4, \tag{7.107}$$

in which $K(k_*)$ is the elliptic integral of the first kind of which modulus is k_*. Equations (7.106) and (7.107) indicate

$$C_4 = 0 \quad \text{and} \quad C_3 = \frac{k\Delta H}{2K(k_*)}. \tag{7.108}$$

Thus, the complex potential w is given by

$$w = \frac{k\Delta H}{2K(k_*)} \int_0^\zeta \frac{d\zeta}{\sqrt{(1 - \zeta^2)(1 - k_*^2\zeta^2)}}. \tag{7.109}$$

Next, consider the seepage discharge under the dam. Since the point D'' in the w plane corresponds to $1/k_*$ on the ζ plane in Eq. (7.105), we have

$$w = \phi_D + i\psi_D = \frac{k\Delta H}{2K(k_*)} \int_0^{1/k_*} \frac{d\zeta}{\sqrt{(1 - \zeta^2)(1 - k_*^2\zeta^2)}}$$

$$= \frac{k\Delta H}{2K(k_*)}(K + iK'), \tag{7.110}$$

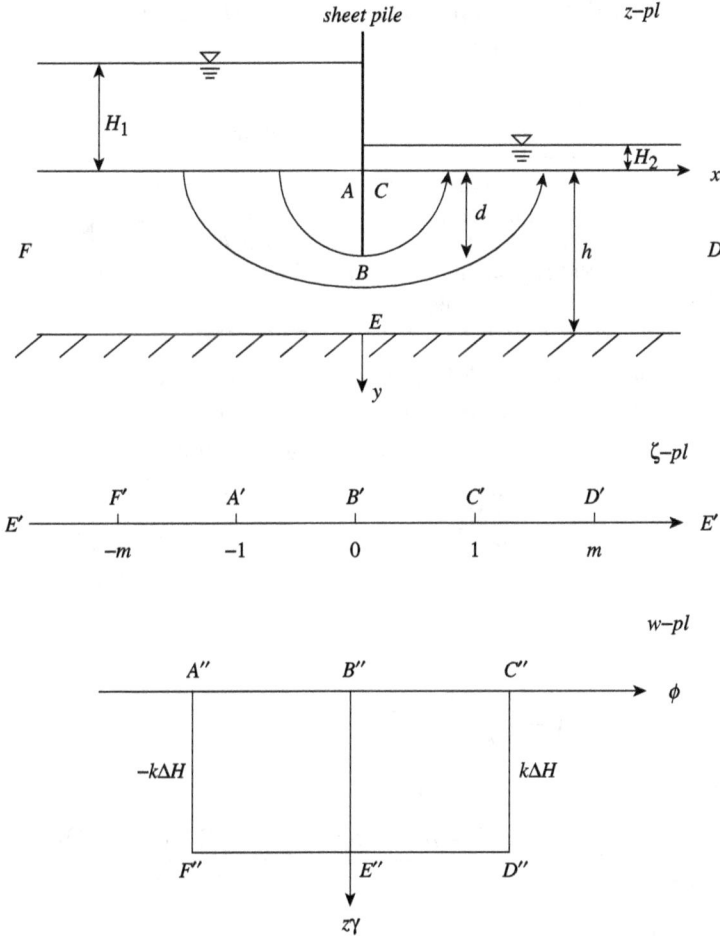

Figure 7.12. Seepage flow under a sheet pile.

in which K' is the elliptic integral of the first kind with modulus $\sqrt{1 - k_*^2}$ (Chap. 9). Since the seepage discharge corresponds to ψ_D, we get

$$Q = \psi_D = \frac{k \Delta H}{2K(k_*)} K'. \tag{7.111}$$

Another problem is given in Fig. 7.12. Instead of the dam in Fig. 7.11, we consider a sheet pile as shown in Fig. 7.12. The ζ and w planes are almost the same as Fig. 7.11. The difference of the water levels in the upstream and downstream is $\Delta H = H_1 - H_2$. The correspondence between the z and w planes is as follows:

z plane	w plane
CD (equi-potential)	$C''D''$ $(\phi = k\Delta H/2)$
FA (equi-potential)	$F''A''$ $(\phi = -k\Delta H/2)$
AB (streamline)	$A''B''$ $(\psi = 0)$
BC (streamline)	$B''C''$ $(\psi = 0)$
DF (streamline)	$D''F''$ $(\psi = \psi_1)$

Consider the mapping relationship between the z and ζ planes. The symmetry of the flow about the y axis of the z plane shows the numerals on the ζ plane as shown in Fig. 7.12. A real number m is unknown and greater than 1. The Schwartz–Christoffel transformation is

$$\frac{dz}{d\zeta} = C_1(\zeta + m)^{-1}(\zeta + 1)^{-1/2}\zeta(\zeta - 1)^{-1/2}(\zeta - m)^{-1}$$

$$= \frac{C_1\zeta}{\zeta^2 - m^2}\frac{1}{\sqrt{\zeta^2 - 1}}$$

$$= \frac{C_1}{2}\left(\frac{1}{\zeta - m} + \frac{1}{\zeta + m}\right)\frac{1}{\sqrt{\zeta^2 - 1}}. \tag{7.112}$$

Integrating Eq. (7.112), we obtain

$$z = \frac{C_1}{\sqrt{m^2 - 1}}\left[\log\left|\frac{2m(\zeta - m) + 2(m^2 - 1) - 2\sqrt{(m^2 - 1)(\zeta^2 - 1)}}{\zeta - m}\right|\right.$$

$$\left. + \log\left|\frac{-2m(\zeta + m) + 2(m^2 - 1) - 2\sqrt{(m^2 - 1)(\zeta^2 - 1)}}{\zeta + m}\right|\right] + C_2, \tag{7.113}$$

in which C_2 = an integral constant. The correspondence between the z and ζ planes is as follows:

$$\text{Point } A \quad z = 0 \Leftrightarrow \zeta = -1$$

$$\text{Point } C \quad z = 0 \Leftrightarrow \zeta = +1$$

$$\text{Point } B \quad z = id \Leftrightarrow \zeta = 0$$

$$\text{Point } E \quad z = ih \Leftrightarrow \zeta = -\pm\infty.$$

The correspondence of four points derives the following equations:

$$0 = \frac{C_1}{\sqrt{m^2 - 1}} \log \left| \frac{-2m(1+m) + 2(m^2 - 1)}{-(1+m)} \right.$$

$$\left. \cdot \frac{-2m(m-1) + 2(m^2 - 1)}{m - 1} \right| + C_2. \tag{7.114}$$

The equations at the points A and C are the same. They are

$$id = \frac{C_1}{\sqrt{m^2 - 1}} \log \left| \frac{-2m^2 + 2(m^2 - 1) - 2i\sqrt{m^2 - 1}}{-m} \right.$$

$$\left. \cdot \frac{-2m^2 + 2(m^2 - 1) - 2i\sqrt{m^2 - 1}}{m} \right| + C_2, \tag{7.115}$$

$$ih = \frac{C_1}{\sqrt{m^2 - 1}} \log \left| (2m - 2\sqrt{m^2 - 1})(-2m - 2\sqrt{m^2 - 1}) \right| + C_2. \tag{7.116}$$

Equations (7.114), (7.115), and (7.116) numerically derive C_1, C_2, and m. The Schwartz–Christoffel transformation between the w and ζ planes is

$$w = C_3 \int \frac{d\zeta}{\sqrt{(\zeta^2 - 1)(\zeta^2 - 1/k_*^2)}} + C_4, \tag{7.117}$$

in which $k_* = 1/m$ and $C_4 =$ an integral constant. The correspondence between the points A and C on the w and z planes is

$$\text{Point } A \quad w = -k\Delta H/2 \Leftrightarrow \zeta = -1$$

$$\text{Point } C \quad w = k\Delta H/2 \Leftrightarrow \zeta = +1.$$

Substituting the correspondence into Eq. (7.117), we obtain

$$-\frac{k\Delta H}{2} = C_3 \int_0^{-1} \frac{d\zeta}{\sqrt{(1 - \zeta^2)(1 - k_*^2 \zeta^2)}} + C_4$$

$$= -C_3 K(k_*) + C_4, \tag{7.118}$$

$$\frac{k\Delta H}{2} = C_3 \int_0^1 \frac{d\zeta}{\sqrt{(1 - \zeta^2)(1 - k_*^2 \zeta^2)}} + C_4 = C_3 K(k_*) + C_4, \tag{7.119}$$

in which $K(k_*)$ is the elliptic integral of the first kind of which modulus is k_*. Equations (7.118) and (7.119) indicate

$$C_4 = 0 \quad \text{and} \quad C_3 = \frac{k\Delta H}{2K(k_*)}. \tag{7.120}$$

Thus, from Eq. (7.117) w is given by

$$w = \frac{k\Delta H}{2K(k_*)} \int_0^\zeta \frac{d\zeta}{\sqrt{(1 - \zeta^2)(1 - k_*^2\zeta^2)}}. \tag{7.121}$$

Consider the seepage discharge under the sheet pile. Since the point D'' on the w plane corresponds to m on the ζ plane in Eq. (7.121),

$$w = \phi + \psi = \frac{k\Delta H}{2K(k_*)} \int_0^{1/k_*} \frac{d\zeta}{\sqrt{(1 - \zeta^2)(1 - k_*^2\zeta^2)}}$$

$$= \frac{k\Delta H}{2K(k_*)}(K + iK'), \tag{7.122}$$

in which K' is the elliptic integral of the first kind with the modulus $\sqrt{1 - k_*^2}$. Since the seepage discharge corresponds to ψ_D,

$$Q = \psi_D = \frac{k\Delta H}{2K(k_*)}K'. \tag{7.123}$$

This is the same form as Eq. (7.111). The value of k_* is different from the seepage flow under the dam. The difference is the numerical value of k_*.

Example 7.3. When the longshore current flows past a groyne as shown in Fig. 7.13, obtain the Schwartz–Christoffel transformation function.

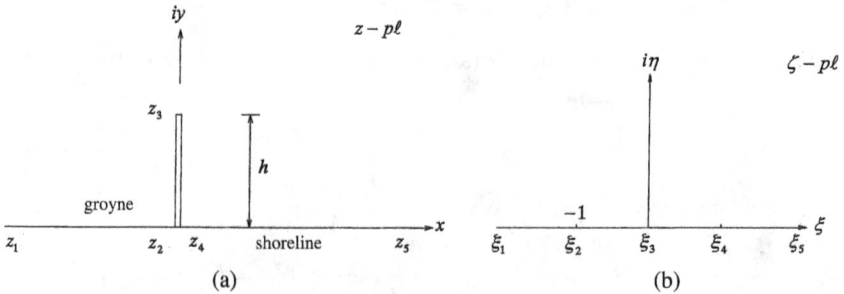

Figure 7.13.   Flow past a groyne.

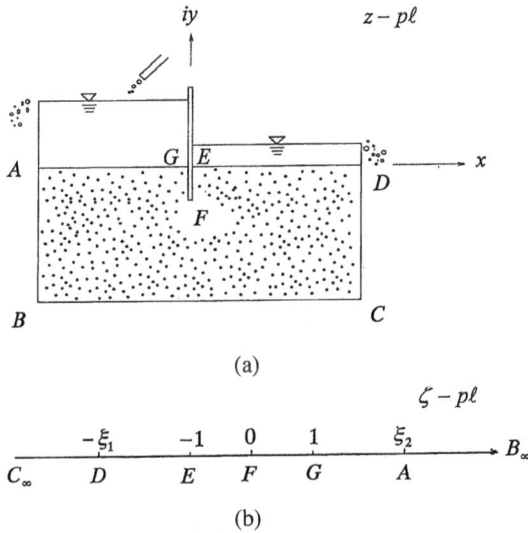

(a)

(b)

Figure 7.14.   Problem 7.3.

Solution. Assume that z_1, z_2, z_3, z_4, and z_5 on the z plane correspond to $\xi_1 = -\infty$, $\xi_2 = -1$, $\xi_3 = 0$, $\xi_4 = 1$, and $\xi_5 = \infty$ on the ζ plane, we obtain

$$\frac{dz}{d\zeta} = A(\zeta + 1)^{-1/2}\zeta(\zeta - 1)^{-1/2} = \frac{A\zeta}{\sqrt{\zeta^2 - 1}}. \tag{1}$$

Integrating Eq. (1), we obtain

$$z = A\sqrt{\zeta^2 - 1} + C_1, \tag{2}$$

in which C_1 = an integral constant, $z = 0$ at $\zeta = \pm 1$ shows $C_1 = 0$. $z = ih$ at $\zeta = 0$ indicates $A = h$. Thus, the mapping function is

$$z = h\sqrt{\zeta^2 - 1}. \tag{3}$$

Problem 7.3. There is the seepage flow as shown in Fig. 7.14. Show that the Schwartz–Christoffel transformation from the z plane to the ζ plane is given by

$$\frac{dz}{d\zeta} = A\zeta \frac{1}{\sqrt{(\zeta + \xi_1)(\zeta - \xi_2)}} \frac{1}{\sqrt{\zeta^2 - 1}}. \tag{4}$$

Chapter 8

Partial Differential Equation
of the First-Order

The continuity equations in hydraulic engineering are often expressed by the partial differential equations of the first-order. The partial differential equations of the first-order are analytically solved in the nonlinear cases if they are integrable. Let us study these lucky cases.

8.1. General

When a dependent variable z is a differentiable function of independent variables x and y, define the partial derivatives as

$$\frac{\partial z}{\partial x} = p, \quad \frac{\partial z}{\partial y} = q, \tag{8.1}$$

then, the following relationship

$$f(x, y, z, p, q) = 0 \tag{8.2}$$

is called the partial differential equations of the first-order. This forms a group of curved surfaces in space. The ordinary and partial differential equations of the first-order contain one constant and arbitrary function, respectively. As an example, the following partial differential equation of the first-order (Sneddon, 1957):

$$p + q = 0 \tag{8.3}$$

has a solution

$$z = F(x - y), \tag{8.4}$$

in which $F(\cdot)$ is an arbitrary function. Thus,

$$z = (x - y)^2, \quad z = \exp[A(x - y)^3], \quad z = \sin[B(x - y)] \tag{8.5}$$

121

are the solutions of Eq. (8.3). The constants A and B are arbitrary. Equation (8.5) forms an integral surface of which family of straight lines are parallel to an equation $x = y$ on the $z = 0$ plane. When $zx = a$ at $y = 0$ in Eq. (8.4), Eq. (8.4) becomes $z = F(x)$. Since $z = a/x = F(x)$,

$$z = \frac{a}{x - y} \tag{8.6}$$

is obtained. Equation (8.6) is the solution of Eq. (8.3). When $z^2 + y^2 = b^2$ at $x = a$ is formed, $b^2 - y^2 = F^2(a - y)$ with the usage of $z = F(a - y)$. Substituting $y = a - Y$, we obtain

$$F^2(Y) = b^2 - (a - Y)^2. \tag{8.7}$$

The comparison of $z^2 = F^2(x - y)$ in Eq. (8.4) with Eq. (8.7) shows

$$z^2 + (x - y - a)^2 = b^2, \tag{8.8}$$

in which a and b are integral constants. This is the general solution. When the partial differential equation of the first-order is given by

$$f(x, y, z)p + g(x, y, z)q = h(x, y, z), \tag{8.9}$$

in which $f(x, y, z)$, $g(x, y, z)$, and $h(x, y, z)$ are arbitrary differentiable functions. This is called a linear partial differential equation of the first-order. When the solution of the integral surface is S, consider the normal vector $\overrightarrow{PN}$ and a vector $\overrightarrow{PQ}$ of which components in the x, y, and z directions are $f(x, y, z)$, $g(x, y, z)$, and $h(x, y, z)$, respectively, as shown in Fig. 8.1. The direction cosines of the vector $\overrightarrow{PN}$ in the x, y, and z directions are p, q, and -1, respectively. Since

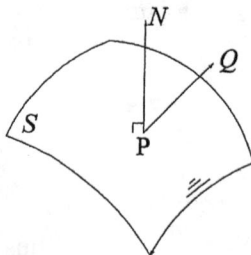

Figure 8.1.   Normal $\overrightarrow{PN}$ and parallel vector $\overrightarrow{PQ}$.

Eq. (8.9) is

$$f(x, y, z)p + g(x, y, z)q - h(x, y, z) = 0, \qquad (8.10)$$

the vector $\overrightarrow{PQ}$ is orthogonal to $\overrightarrow{PN}$. Since $\overrightarrow{PN}$ is tangent to an integral surface, the differential equation of the integral surface is

$$\frac{dx}{f(x, y, z)} = \frac{dy}{g(x, y, z)} = \frac{dz}{h(x, y, z)}. \qquad (8.11)$$

Equation (8.11) is called a characteristic differential equation of Eq. (8.9). If you compare Eq. (8.9) with

$$p\,dx + q\,dy = dz, \qquad (8.12)$$

the geometrical description is easily understood. As an example, consider the following equation

$$(y^2 + z^2 - x^2)p - 2xyq + 2xz = 0. \qquad (8.13)$$

The characteristic differential equation is

$$\frac{dx}{y^2 + z^2 - x^2} = \frac{dy}{-2xy} = \frac{dz}{-2xz}. \qquad (8.14)$$

The second and third terms in Eq. (8.14) show

$$z = ay, \qquad (8.15)$$

in which a = an integral constant. Substituting Eq. (8.15) into the first and second terms in Eq. (8.14), we obtain

$$\frac{dx}{y^2(1 + a^2) - x^2} = \frac{dy}{-2xy}. \qquad (8.16)$$

Defining

$$x^2 = X, \quad \text{and} \quad y^2 = Y, \qquad (8.17)$$

we obtain the homogeneous differential equation. Then, substituting $\xi = X/Y$ into the homogeneous differential equation, we have the solution by the separation of variables. The solution is

$$x^2 + y^2(1 + a^2) = by, \qquad (8.18)$$

in which b = an integral constant. Thus, rewriting Eqs. (8.15) and (8.18), we get

$$z/y = a \qquad (8.19)$$

and

$$(x^2 + y^2 + z^2)/y = b. \qquad (8.20)$$

The general solution of Eq. (8.13) is derived from Eqs. (8.19) and (8.20) as follows:

$$F\left(\frac{z}{y}, \frac{x^2 + y^2 + z^2}{y}\right) = 0, \qquad (8.21)$$

in which $F(\cdot)$ = an arbitrary function. When there are n independent variables, the partial differential equation is expressed by

$$f_1 \frac{\partial z}{\partial x_1} + f_2 \frac{\partial z}{\partial x_2} + \cdots + f_n \frac{\partial z}{\partial x_n} = f, \qquad (8.22)$$

in which $f_1, f_2, \ldots, f_n$, and f are the functions of $x_1, x_2, \ldots, x_n$, and z. When we solve Eq. (8.22), the characteristic differential equations are written as

$$\frac{dx_1}{f_1} = \frac{dx_2}{f_2} = \cdots = \frac{dx_n}{f_n} = \frac{dz}{f}. \qquad (8.23)$$

The solutions for $i = 1, 2, \ldots, n$ are given by

$$u_i(x_1, x_2, \ldots, x_n, z) = C_i \quad \text{for } i = 1, 2, \ldots, n, \qquad (8.24)$$

in which u_i = a solution of equations and C_i = an integral constant. The general solution of Eq. (8.22) is

$$F(u_1, u_2, \ldots, u_n) = 0, \qquad (8.25)$$

in which $F(\cdot)$ = an arbitrary function. Equation (8.23) is the characteristic differential equation of Eq. (8.22). An equation is assumed to contain n parameters when the number of independent variables is n. When this equation satisfies the partial differential equation, this equation is called a complete solution. This complete solution is considered to be a curved surface with n parameters being eliminated from the equation of the envelope surface of this complete solution. Then, the resultant equation is called a singular solution.

The solution which satisfies an arbitrary function is called a general solution. The solution which does not correspond to the complete solution, the singular solution, and the general solution is called a particular solution. As an example, the following differential equation

$$p - q = 2\sqrt{z - 1} \qquad (8.26)$$

has a particular solution such as $z = 1$.

Example 8.1. Assuming that a and b are constants, solve the following equation:

$$a\frac{\partial z}{\partial x} + b\frac{\partial z}{\partial y} = 1. \qquad (1)$$

Solution. The characteristic differential equation is

$$\frac{dx}{a} = \frac{dy}{b} = \frac{dz}{1}. \qquad (2)$$

The solutions are $x - az = C_1$ and $y - bz = C_2$, in which C_1 and C_2 are arbitrary constants. Introducing an arbitarary function $F(\cdot)$, we obtain

$$y - bz = F(x - az). \qquad (3)$$

This is a general solution. The arbitrary function can be determined by the boundary condition.

Problem 8.1. Show that the general solution of the following equation:

$$x^2\frac{\partial z}{\partial x} + y^2\frac{\partial z}{\partial y} = (x + y)z \qquad (4)$$

is

$$z = xy \cdot f\left(\frac{x - y}{z}\right) \qquad (5)$$

or

$$z = xy \cdot g\left(\frac{x - y}{xy}\right) \qquad (6)$$

in which $f(\cdot)$ and $g(\cdot)$ are arbitrary functions.

8.2. Charpit's Method

To solve a general partial differential equation of the first-order

$$f(x, y, z, p, q) = 0, \tag{8.27}$$

consider another equation of x, y, z, p, q, and a parameter a as follows:

$$g(x, y, z, p, q, a) = 0. \tag{8.28}$$

Then, p and q are derived from Eqs. (8.27) and (8.28). These are substituted in the equation of the complete differentials as

$$p\,dx + q\,dy - dz = 0. \tag{8.29}$$

When $\frac{\partial(f,g)}{\partial(p,q)} \neq 0$, the integrability of Eq. (8.29) leads to

$$\frac{\partial q}{\partial x} - \frac{\partial p}{\partial y} + p\frac{\partial q}{\partial z} - q\frac{\partial p}{\partial z} = 0, \tag{8.30}$$

in which

$$\frac{\partial(f, g)}{\partial(p, q)} = \begin{vmatrix} \dfrac{\partial f}{\partial p} & \dfrac{\partial g}{\partial p} \\ \dfrac{\partial f}{\partial q} & \dfrac{\partial g}{\partial q} \end{vmatrix}. \tag{8.31}$$

Deriving $\partial q/\partial x, \partial p/\partial y, \partial q/\partial z, \partial p/\partial z$ from Eqs. (8.27) and (8.28) and substituting them into Eq. (8.30), we get

$$P\frac{\partial g}{\partial x} + Q\frac{\partial g}{\partial y} + (pP + qQ)\frac{\partial g}{\partial z} - (X + pZ)\frac{\partial g}{\partial p} - (Y + qZ)\frac{\partial g}{\partial q} = 0, \tag{8.32}$$

in which $P = \partial f/\partial p$, $Q = \partial f/\partial q$, $X = \partial f/\partial x$, $Y = \partial f/\partial y$, and $Z = \partial f/\partial z$. The characteristic differential equation of Eq. (8.32) is

$$\frac{dx}{P} = \frac{dy}{Q} = \frac{dz}{pP + qQ} = -\frac{dp}{X + pZ} = -\frac{dq}{Y + qZ}. \tag{8.33}$$

The solution of Eq. (8.33) forms the general solution. This is called Charpit's method.

Example 8.2. Solve the following differential equation:

$$pxy + pq + qy = yz. \tag{1}$$

Solution. Applying Eq. (8.33) to Eq. (1), we obtain

$$\frac{dx}{xy+q} = \frac{dy}{p+y} = \frac{dz}{yz+pq} = \frac{dp}{0} = -\frac{dq}{px+q-z-qy}. \tag{2}$$

One of the solutions of Eq. (2) is

$$p = a: \quad a = \text{constant}. \tag{3}$$

Substituting Eq. (3) into Eq. (1), we get

$$q = \frac{y(z-ax)}{a+y}. \tag{4}$$

The relationship of the complete diffferentiation derives

$$dz = \frac{\partial z}{\partial x}dx + \frac{\partial z}{\partial y}dy. \tag{5}$$

That is

$$dz = pdx + qdy = adx + \frac{y(z-ax)}{a+y}dy. \tag{6}$$

Integrating Eq. (6), we have

$$\int \frac{d(z-ax)}{z-ax} = \int \frac{y}{a+y}dy. \tag{7}$$

This reduces to the following complete solution:

$$(z-ax)(a+y)^a = be^y, \tag{8}$$

in which a and b are arbitrary constants.

Problem 8.2. Show that the complete solution of the following partial differential equation:

$$p^2x + q^2y = z \tag{9}$$

is

$$[(1+a)z]^{1/2} = (a/x)^{1/2} + y^{1/2} + b, \tag{10}$$

in which a and b are arbitrary constants.

8.3. Application to Hydrology

Rainfall and runoff analysis on a steep slope is done by the kinematic wave method. The kinematic wave method treats the overland flow on the slope when the rainfall is considered. The kinematic wave method is the first step for analysis of flood routing. Since the overland flow plays an important role in rainfall and runoff analysis, many hydrologists have conducted many studies (Singh, 1996). The continuity equation of the overland flow is

$$\frac{\partial h}{\partial t} + \frac{\partial uh}{\partial x} = r(x,t), \tag{8.34}$$

in which h = water depth, t = time, u = velocity, x = flow direction along the slope, and $r(x,t)$ = rainfall on the slope. When the slope is constant and the rainfall is time-varying, Eq. (8.34) is

$$\frac{\partial h}{\partial t} + \frac{\partial uh}{\partial x} = r(t), \tag{8.35}$$

in which $r(t) = r(x,t)$. Applying Manning's formula to uh as the equation of motion, the following approximation:

$$uh \cong ah^2 + bh + c \tag{8.36}$$

is used to obtain a closed form solution. The parameters a, b, and c are calculated by the least squares method. As a result, Eq. (8.35) becomes the following nonlinear partial differential equation:

$$\frac{\partial h}{\partial t} + (2ah + b)\frac{\partial h}{\partial x} = r(t). \tag{8.37}$$

The characteristic differential equation of Eq. (8.37) is

$$\frac{dt}{1} = \frac{dx}{2ah + b} = \frac{dh}{r(t)} = d\sigma, \tag{8.38}$$

in which σ = a dummy variable. The first and third terms in Eq. (8.38) lead to

$$h = \int_\tau^t r(\sigma)d\sigma + h_\tau, \tag{8.39}$$

in which τ = time that a characteristic starts from the t axis as explained in Fig. 8.2 and h_τ = water depth at $t = \tau$. From the first

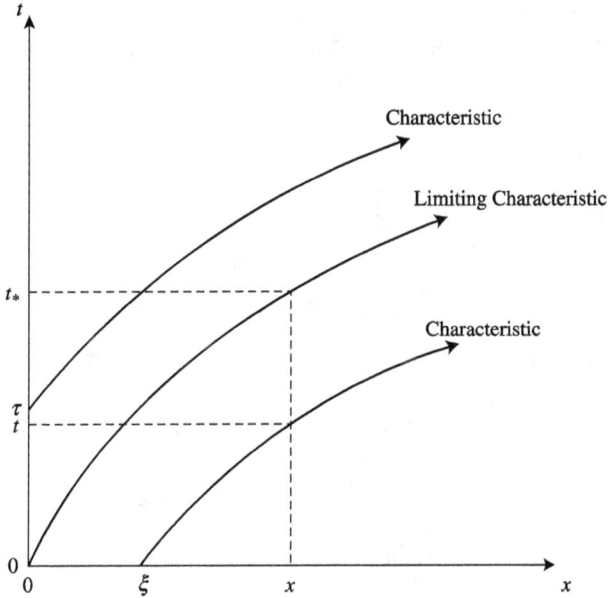

Figure 8.2.   Characteristics.

and second terms of Eq. (8.38), we get

$$x = \int_{\tau}^{t} (2ah + b)d\sigma + \xi,\tag{8.40}$$

in which ξ = the x coordinate that a characteristic starts at the x axis as explained in Fig. 8.2. Substituting Eq. (8.39) into Eq. (8.40), we have

$$x - \xi = \int_{\tau}^{t} \int_{\tau}^{\eta} 2ar(\sigma)d\sigma d\eta + (2ah_{\tau} + b)(t - \tau),\tag{8.41}$$

in which η = a dummy variable. If $f(\cdot)$ is the initial water depth, the general solution of Eq. (8.37) is derived from Eqs. (8.39) and (8.41),

$$x - h_{\tau} - \int_{\tau}^{t} r(\sigma)d\sigma = f\left\{ x - \xi - \int_{\tau}^{t} \int_{\tau}^{\eta} 2ar(\sigma)d\sigma d\eta - 2a(t - \tau)\right.$$

$$\left. \times \left[h - \int_{\tau}^{t} r(\sigma)d\sigma\right] - b(t - \tau)\right\}.\tag{8.42}$$

We do not need ξ, because the initial water depth gives ξ. When the rainfall $r(x, t)$ is the function of x in Eq. (8.34), Eq. (8.34) is written by

$$\frac{\partial h}{\partial t} + \frac{\partial uh}{\partial x} = r(x). \tag{8.43}$$

With Eq. (8.36) the characteristic differential equation of Eq. (8.43) is

$$\frac{dt}{1} = \frac{dx}{2ah + b} = \frac{dh}{r(x)} = d\sigma. \tag{8.44}$$

After the limiting characteristic reaches, the solution is

$$ah^2 + bh = \int_0^x r(\sigma)d\sigma, \tag{8.45}$$

in which $\sigma = $ a dummy variable. Thus, the water depth is written as

$$h = \frac{\sqrt{b^2 + 4a \int_0^x r(\sigma)d\sigma} - b}{2a}. \tag{8.46}$$

Before the limiting characteristic reaches, the water depth is

$$h = \frac{\sqrt{b^2 + 4a \int_\xi^x r(\sigma)d\sigma} - b}{2a}, \tag{8.47}$$

in which $\xi = $ a point a characteristic intersects the x axis. From the first and second terms of Eq. (8.44) and Eq. (8.47) we have

$$t = \int_\xi^x \frac{d\eta}{\sqrt{b^2 + 4a \int_\xi^x r(\sigma)d\sigma}}, \tag{8.48}$$

in which $\eta = $ a dummy variable. The variable ξ is obtained by the given x and t from Eq. (8.48). Defining

$$f(\xi) = t - \int_\xi^x \frac{d\eta}{\sqrt{b^2 + 4a \int_\xi^x r(\sigma)d\sigma}}, \tag{8.49}$$

we have from Fig. 8.2,

$$f(0) = t - t_* < 0, \tag{8.50}$$

$$f(x) = t > 0. \tag{8.51}$$

Taking the derivative of Eq. (8.49) about ξ, we obtain

$$\frac{df(\xi)}{d\xi} = \frac{1}{\sqrt{b^2 + 4a\int_\xi^x r(\sigma)d\sigma}} > 0. \qquad (8.52)$$

Thus, the function $f(\xi)$ has a root between 0 and x. This is solved by the secant method in Chap. 15.

Example 8.3. When the initial water depth is given by

$$h = \frac{\sqrt{b^2 + 4ar_o x} - b}{2a}, \qquad (1)$$

compute $f(\cdot)$ in Eq. (8.42). A symbol r_o is rainfall at $t = 0$ and the water depth $h = 0$ at $x = 0$.

Solution. The water depth at $x = 0$ is

$$h = 0. \qquad (2)$$

Substituting $t = \tau = 0$ into Eq. (8.42), we obtain the water depth at $x = \xi$:

$$h_\tau = \frac{\sqrt{b^2 + 4ar_o\xi} - b}{2a}. \qquad (3)$$

The function $f(\cdot)$ is from Eq. (8.42),

$$f(x - \xi) = \frac{\sqrt{b^2 + 4ar_o x} - \sqrt{b^2 + 4ar_o\xi}}{2a}. \qquad (4)$$

Thus, the arbitrary function becomes

$$f(x) = \frac{\sqrt{b^2 + 4ar_o(x + \xi)} - \sqrt{b^2 + 4ar_o\xi}}{2a}. \qquad (5)$$

Problem 8.3. Show that the solution of Eq. (8.42) is given by

$$h = -W_2 + \left\{ W_2^2 - [W_2 - (t - \tau)r_o]^2 + \frac{b^2 + 4ar_o W_1}{4a^2} \right\}^{1/2}, \qquad (6)$$

in which

$$W_1 = x - \int_\tau^t \int_\tau^\eta 2ar(\sigma)d\sigma d\eta + (t - \tau)\left[\int_\tau^t 2ar(\sigma)d\sigma - b \right], \qquad (7)$$

$$W_2 = \frac{b}{2a} - \int_\tau^t r(\sigma)d\sigma + (t - \tau)r_o. \qquad (8)$$

Chapter 9

Special Functions

When the solution of differential equations is not expressed by elementary functions, a function to describe the solution is called a special function. As an example, when the Laplace equation is transformed from the Cartesian coordinate system to the cylindrical coordinate system, one of two ordinary differential equations derived from the separation of variables is a Bessel differential equation and its solution is Bessel functions. When the Laplace equation is transformed from the Cartesian coordinate system to the spherical coordinate system, one of two ordinary differential equations from the separation of variables is a Legendre differential equation and its solution is Legendre functions. We study important special functions in mathematical physics.

9.1. Gamma, Beta, and Error Functions

Euler defined Gamma function as follows:

$$\Gamma(n) = \int_0^\infty x^{n-1} e^{-x} dx, \quad n > 0. \tag{9.1}$$

This integral of Eq. (9.1) converges for $n > 0$ and Eq. (9.1) is the function of n. For $n = 1$ we have

$$\Gamma(1) = \int_0^\infty e^{-x} dx = 1. \tag{9.2}$$

Using the partial integral, we have

$$\Gamma(n+1) = \int_0^\infty x^n e^{-x} dx = [-x^n e^{-x}]_0^\infty + n \int_0^\infty x^{n-1} e^{-x} dx = n\Gamma(n). \tag{9.3}$$

The function $\Gamma(n)$ is defined except $n = 0, -1, -2, \ldots$. From Eqs. (9.2) and (9.3) we obtain

$$\Gamma(1) = 1, \quad \Gamma(2) = 1 \times \Gamma(1) = 1, \quad \Gamma(3) = 2 \times \Gamma(2) = 2 \times 1,$$

$$\Gamma(4) = 3 \times \Gamma(3) = 3 \times 2 \times 1, \ldots, \quad \Gamma(n+1) = n!, \ldots, \quad (9.4)$$

in which n is a positive integer. Defining $\Gamma(1) = 0! = 1$, we get a Gauss Π function as follows:

$$\Pi(n) = \Gamma(n+1) = n!. \qquad (9.5)$$

Substituting $x = y^2$ into Eq. (9.1), we get

$$\Gamma(n) = 2 \int_0^\infty y^{2n-1} e^{-y^2} \, dy. \qquad (9.6)$$

Referring to Example 3.31 and substituting $n = 1/2$ in Eq. (9.6), we have

$$\Gamma(1/2) = 2 \int_0^\infty e^{-y^2} \, dy = \sqrt{\pi}. \qquad (9.7)$$

From Eq. (9.3) we obtain

$$\Gamma(-1/2) = \frac{\Gamma(1/2)}{-1/2} = -2\sqrt{\pi}, \qquad (9.8)$$

$$\Gamma(-3/2) = \frac{\Gamma(-1/2)}{-3/2} = \frac{4\sqrt{\pi}}{3}. \qquad (9.9)$$

In the same way, we obtain

$$\Gamma(-5/2) = \frac{\Gamma(-3/2)}{-5/2} = -\frac{8\sqrt{\pi}}{15}, \quad \Gamma(-7/2) = \frac{\Gamma(-5/2)}{-7/2} = \frac{16\sqrt{\pi}}{105}. \qquad (9.10)$$

The Gamma function can be computed except $n = 0, -1, -2, -3, \ldots$. The Gamma function can be applicable to the Laplace transform in Chap. 11 and shown in Fig. 9.1.

A Beta function $\beta(m, n)$ is defined by

$$\beta(m, n) = \int_0^1 x^{m-1}(1 - x)^{n-1} dx \quad (m > 0, \ n > 0). \qquad (9.11)$$

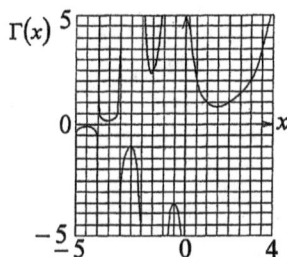

Figure 9.1. Gamma function.

Substituting

$$x = 1 - y \tag{9.12}$$

into Eq. (9.11), we have

$$\beta(m, n) = \int_0^1 (1 - y)^{m-1} y^{n-1} dy = \beta(n, m). \tag{9.13}$$

Substituting $x = \sin^2 \phi$ into Eq. (9.11), we get

$$\beta(m, n) = 2 \int_0^{\pi/2} (\sin \phi)^{2m-1} (\cos \phi)^{2n-1} d\phi. \tag{9.14}$$

Substituting $x = y/(1 + y)$ into Eq. (9.11), we obtain

$$\beta(m, n) = \int_0^\infty \frac{y^{n-1}}{(1 + y)^{m+n}} dy. \tag{9.15}$$

According to Eq. (9.6), we have

$$\Gamma(m) = 2 \int_0^\infty y^{2m-1} e^{-y^2} dy, \tag{9.16}$$

$$\Gamma(n) = 2 \int_0^\infty y^{2n-1} e^{-y^2} dy. \tag{9.17}$$

The product of Eqs. (9.16) and (9.17) gives

$$\Gamma(m)\Gamma(n) = 4 \int_0^\infty x^{2m-1} e^{-x^2} dx \int_0^\infty y^{2n-1} e^{-y^2} dy$$

$$= 4 \int_0^\infty \int_0^\infty x^{2m-1} y^{2n-1} e^{-(x^2+y^2)} dx dy. \tag{9.18}$$

Substituting $x = r\cos\theta$ and $y = r\sin\theta$ into Eq. (9.18) and using the Jacobian $dxdy = rdrd\theta$, we have

$$\Gamma(m)\Gamma(n) = 4 \int_0^{\pi/2} (\cos\theta)^{2m-1}(\sin\theta)^{2n-1}d\theta \int_0^\infty r^{2(m+n)-1}e^{-r^2}dr.$$

(9.19)

From Eq. (9.14), we obtain

$$\beta(n,m) = 2 \int_0^{\pi/2} (\cos\theta)^{2m-1}(\sin\theta)^{2n-1}d\theta.$$

(9.20)

From Eq. (9.6), we get

$$\Gamma(m+n) = 2 \int_0^\infty r^{2(m+n)-1}e^{-r^2}dr.$$

(9.21)

Substituting Eqs. (9.14) and (9.21) into Eq. (9.19), we get

$$\Gamma(m)\Gamma(n) = \beta(m,n)\Gamma(m+n)$$

(9.22)

or

$$\beta(m,n) = \frac{\Gamma(m)\Gamma(n)}{\Gamma(m+n)}.$$

(9.23)

Using Eqs. (9.20) and (9.23), we get

$$\int_0^{\pi/2} (\cos\theta)^{2m-1}(\sin\theta)^{2n-1}d\theta = \frac{\Gamma(m)\Gamma(n)}{2\Gamma(m+n)}, \quad \text{for} \quad m > 0, \quad n > 0.$$

(9.24)

Equation (9.15) is written by

$$\int_0^\infty \frac{y^{n-1}}{(1+y)^{n+m}}dy = \frac{\Gamma(m)\Gamma(n)}{\Gamma(m+n)}, \quad \text{for} \quad m > 0, \quad n > 0.$$

(9.25)

When $m = 1 - n$ in Eq. (9.25), we have

$$\int_0^\infty \frac{y^{n-1}}{1+y}dy = \frac{\Gamma(1-n)\Gamma(n)}{\Gamma(1)}.$$

(9.26)

The application of the complex integral shows

$$\int_0^\infty \frac{y^{n-1}}{1+y}dy = \frac{\pi}{\sin n\pi}, \quad \text{for} \quad 0 < n < 1.$$

(9.27)

Since $\Gamma(1) = 1$, Eqs. (9.26) and (9.27) lead to

$$\Gamma(n)\Gamma(1-n) = \frac{\pi}{\sin n\pi}. \tag{9.28}$$

Next, consider an error function. This is defined by

$$erf(x) = \frac{2}{\sqrt{\pi}} \int_0^x e^{-n^2} dn. \tag{9.29}$$

This function plays an important role in the theory of probability and statistics. This function is also the fundamental solution of the partial differential equations for diffusion or heat conduction type. The error function has the following properties:

$$erf(-x) = -erf(x), \tag{9.30}$$

$$erf(0) = 0, \tag{9.31}$$

$$erf(\infty) = \frac{2}{\sqrt{\pi}} \int_0^\infty e^{-n^2} dn = 1. \tag{9.32}$$

Example 9.1. Show that

$$\int_0^x e^{-n^2} dn = \frac{\sqrt{\pi}}{2} - \frac{e^{-x^2}}{2x} \left(1 - \frac{1}{2x^2} + \frac{1 \times 3}{2^2 x^4} \right)$$

$$+ \frac{1 \times 3 \times 5}{2^3} \int_0^\infty \frac{e^{-n^2}}{n^6} dn. \tag{1}$$

Solution.

$$\int_0^x e^{-n^2} dn = \int_0^\infty e^{-n^2} dn - \int_x^\infty e^{-n^2} dn$$

$$= \frac{\sqrt{\pi}}{2} - \int_{x^2}^\infty \frac{e^{-y}}{2\sqrt{y}} dy \quad (\text{define } y = n^2)$$

$$= \frac{\sqrt{\pi}}{2} - \frac{e^{-x^2}}{2x} + \int_{x^2}^\infty \frac{e^{-y}}{4} y^{-3/2} dy = \frac{\sqrt{\pi}}{2} - \frac{e^{-x^2}}{2x}$$

$$- \left[\frac{e^{-y}}{4} y^{-3/2} \right]_{x^2}^\infty + \frac{3}{8} \int_{x^2}^\infty e^{-y} y^{-5/2} dy. \tag{2}$$

The iterations of Eq. (2) derives Eq. (1). The integral $\int_0^x e^{-n^2} dn$ is usually obtained from the mathematical table.

Problem 9.1. Calculate the following integral using the partial integral:

$$\int_0^\infty e^{-x^4}\, dx. \tag{3}$$

9.2. Bessel Function

Applications of the separation of variables to the Laplace equation in the cylindrical coordinate system produces the Bessel differential equation. The solution of the Bessel Differential equation is not expressed by the elementary functions, but it is defined by the sum of convergent power series as the special function. This is called the Bessel function or the cylindrical function. The differential equation which the Bessel function satisfies is given by

$$x^2\frac{d^2y}{dx^2} + x\frac{dy}{dx} + (x^2 - n^2)y = 0, \tag{9.33}$$

in which $n =$ an integer. Since Eq. (9.33) is the linear differential equation of the second-order, it contains two independent solutions. The standard form of the solution is

$$y = C_1 J_n(x) + C_2 Y_n(x), \tag{9.34}$$

in which C_1 and C_2 are integral constants. Symbols $J_n(x)$ and $Y_n(x)$ are the Bessel function of the first kind of order n and the Bessel function of the second kind of order n, respectively. These functions are similar to the sine or cosine functions with recession property. Assume the solution of Eq. (9.33) as follows:

$$y = \sum_{s=0}^\infty C_s x^{r+s}. \tag{9.35}$$

Substituting Eq. (9.35) into Eq. (9.33), we obtain

$$\sum_{s=0}^\infty [(r+s)^2 + x^2 - n^2] C_s x^{r+s} = 0. \tag{9.36}$$

If we now equate the coefficients of the various powers of x, x^r, x^{r+1}, x^{r+2}, etc., to zero in Eq. (9.36), we have the following set of equations:

$$C_s[(r+s)^2 - n^2] + C_{s-2} = 0. \tag{9.37}$$

This is valid for $s = 0, 1, 2, \ldots$ in view of the fact that

$$C_{-1} = C_{-2} = 0. \tag{9.38}$$

Thus, the leading coefficient in the expansion in Eq. (9.35) is C_0. Letting $s = 0$ in Eq. (9.37), we obtain

$$C_0(r^2 - n^2) = 0. \tag{9.39}$$

This equation is known as the indicial equation and since $C_0 \neq 0$, it follows that

$$r = \pm n. \tag{9.40}$$

For $s = 1$, we have

$$C_1[(r+1)^2 - n^2] = 0, \tag{9.41}$$

it follows that

$$C_1 = 0. \tag{9.42}$$

Equations (9.37) and (9.42) shows $C_s = 0$ for $s = 1, 3, 5, \ldots$. For $s = 2, 4, 6, \ldots$, Eq. (9.37) gives

$$C_s = -\frac{C_{s-2}}{s(2n+s)}, \quad \text{for } s = 2, 4, 6, \ldots. \tag{9.43}$$

The coefficients $C_2, C_4, C_6, \ldots$ are expressed by C_0. Thus, Eq. (9.35) becomes

$$y = C_0 \left[x^n - \frac{x^{n+2}}{2^2(n+1)} + \frac{x^{n+4}}{2^4(n+1)(n+2)2!} + \cdots \right.$$
$$\left. + \frac{(-1)^s x^{n+2s}}{2^{2s}(n+1)\cdots(n+s)s!} + \cdots \right]. \tag{9.44}$$

The coefficients are infinite when n is a negative integer. Excluding this case, we standardize the solution by taking

$$C_0 = \frac{1}{2^n \Gamma(n+1)} = \frac{1}{2^n \Pi(n)} = \frac{1}{2^n n!}, \tag{9.45}$$

we obtain

$$J_n(x) = \sum_{s=0}^{\infty} \frac{(-1)^s}{\Pi(s)\Pi(s+n)} \left(\frac{x}{2}\right)^{2s+n}. \tag{9.46}$$

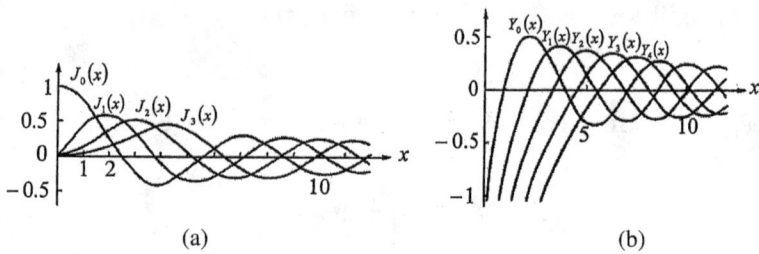

Figure 9.2.   (a) Bessel function of the first kind. (b) Bessel function of the second kind.

This series converges for any finite value of x and represents a function $J_n(x)$ that is known as the Bessel function of the first kind of order n as shown in Fig. 9.2(a). When n is not an integer, the second solution is obtained by replacing n by $-n$ in Eq. (9.46) in accordance with Eq. (9.40). It is therefore written by

$$J_{-n}(x) = \sum_{s=0}^{\infty} \frac{(-1)^s}{\Pi(s)\Pi(s-n)} \left(\frac{x}{2}\right)^{2s-n}. \tag{9.47}$$

Since $J_n(x)$ is independent of $J_{-n}(x)$, the general solution is expressed in the form

$$y = AJ_n(x) + BJ_{-n}(x), \tag{9.48}$$

in which A and B are arbitrary constants provided that n is not an integer. When n is an integer, we have

$$\frac{1}{\Pi(s-n)} = 0, \quad \text{for } s < n. \tag{9.49}$$

Eq. (9.47) is expressed by

$$J_{-n}(x) = \sum_{s=n}^{\infty} \frac{(-1)^s}{\Pi(s)\Pi(s-n)} \left(\frac{x}{2}\right)^{2s-n}. \tag{9.50}$$

Substituting $r = s - n$ or $s = r + n$ into Eq. (9.50), we obtain

$$J_{-n}(x) = \sum_{r=0}^{\infty} \frac{(-1)^{r+n}}{\Pi(r+n)\Pi(r)} \left(\frac{x}{2}\right)^{2r+n}$$

$$= (-1)^n J_n(x), \quad \text{for } n = 0, 1, 2, \ldots \tag{9.51}$$

When n is an integer, the solution is

$$y = AJ_n(x) + B(-1)^n J_n(x) = [A + B(-1)^n]J_n(x) = CJ_n(x),$$
$$(9.52)$$

in which $C = A + B(-1)^n$. Because Eq. (9.52) is not the general solution. As Eq. (9.52) shows, $J_{-n}(x)$ is not independent of $J_n(x)$, we must find another solution. Another solution is found in the following way. Consider the function

$$Y_n(x) = \frac{1}{\sin n\pi}[J_n(x)\cos n\pi - J_{-n}(x)]. \qquad (9.53)$$

If n is not an integer, the function $Y_n(x)$ is dependent on $J_n(x)$, and since it is a linear combination of $J_n(x)$ and $J_{-n}(x)$, it is a solution of the Bessel differential equation of order n. If n is an integer, from Eq. (9.51) we have

$$Y_n(x) = \frac{0}{0}. \qquad (9.54)$$

Here, when n is an integer, we define $Y_n(x)$ to be

$$Y_n(x) = \lim_{r \to n} \frac{J_r(x)\cos r\pi - J_{-r}(x)}{\sin r\pi}. \qquad (9.55)$$

With this definition of $Y_n(x)$ we have, on carrying out the limiting process,

$$\frac{\pi}{2}Y_0(x) = J_0(x)\left(\log\frac{x}{2} + \gamma\right) + \left(\frac{x}{2}\right)^2 - \frac{(1 + \frac{1}{2})(x/2)^4}{(2!)^2}$$
$$+ \left(1 + \frac{1}{2} + \frac{1}{3}\right)\frac{(x/2)^6}{(3!)^2} - \cdots, \qquad (9.56)$$

in which γ is an Euler number $0.5772157\cdots$. For $n \neq 0$ we have

$$\pi Y_n(x) = 2J_n(x)\left(\log\frac{x}{2} + \gamma\right) - \sum_{r=0}^{\infty}(-1)^r\frac{(x/2)^{n+2r}}{r!(n+r)!}$$
$$\times \left(\sum_{m=1}^{n+r}m^{-1} + \sum_{m=1}^{r}m^{-1}\right) - \sum_{r=0}^{n-1}\left(\frac{x}{2}\right)^{-n+2r}\frac{(n-r-1)!}{r!}. \qquad (9.57)$$

As $x \to 0$, $Y_n(x) \to -\infty$. Thus, when n is an integer, the general solution is

$$y = C_1 J_n(x) + C_2 Y_n(x). \tag{9.58}$$

$Y_n(x)$ is the Bessel function of the second kind of order n as shown in Fig. 9.2(b) and also called Neumann's function and written as $N_n(x)$. Hankel function is also defined by $J_n(x) \pm iY_n(x)$ and called the Bessel function of the third kind of order n.

According to Eq. (9.46), we get

$$xJ_n'(x) = nJ_n(x) - xJ_{n+1}(x), \tag{9.59}$$

$$2J_n'(x) = J_{n-1}(x) - J_{n+1}(x), \tag{9.60}$$

in which "$'$" $= d/dx$. Substituting $n = 0$ in Eq. (9.59), we get

$$J_0'(x) = -J_1(x). \tag{9.61}$$

The following differential equations

$$\frac{d^2y}{dx^2} + bxy = 0, \tag{9.62}$$

$$\frac{d^2y}{dx^2} + bx^2y = 0, \tag{9.63}$$

have the solution of the Bessel function. In Eqs. (9.62) and (9.63), b is a constant. Except the Bessel function of the first and second kinds, there are modified Bessel functions such as $I_n(x)$ and $K_n(x)$. The general type of the Bessel function is given by

$$\frac{d^2y}{dx^2} + \frac{1 - 2\alpha}{x}\frac{dy}{dx} + \left[(\beta\gamma x^{\gamma-1})^2 + \frac{\alpha^2 - n^2\gamma^2}{x^2}\right]y = 0. \tag{9.64}$$

The solution of Eq. (9.64) is

$$y = x^\alpha Z_n(\beta x^\gamma), \tag{9.65}$$

in which α, β, and γ are constants. When n in Eq. (9.65) is a nonintegral, Z_n indicates the Bessel functions. When the sign of the term x^2 is negative in Eq. (9.33), the solutions are modified Bessel functions. Mizumura and Yamamoto (1991) obtained the reflection and the transmission coefficients through a sloping offshore breakwater of concrete blocks using the modified Bessel functions.

Example 9.2. The water depth of a bay of rectangular cross section is constant h_0 and the width of the bay is given by $b_0 x/a$. When long waves of an angular frequency σ propagate into the bay, the deviation of the water surface level from the still water level in the bay is given by η. According to the theory of long waves of infinitesimally small amplitude, η satisfies the following equation:

$$\frac{d^2\eta}{dx^2} + \frac{1}{x}\frac{d\eta}{dx} + \frac{\sigma^2}{gh_0}\eta = 0. \tag{1}$$

Solve η in Eq. (1) and x is the coordinate system from the onshore to the offshore in the bay.

Solution. We multiply x^2 to Eq. (1). Then, we get

$$x^2\frac{d^2\eta}{dx^2} + x\frac{d\eta}{dx} + x^2\frac{\sigma^2}{gh_0}\eta = 0, \tag{2}$$

in which

$$x^2\frac{\sigma^2}{gh_0} = X^2 \quad \text{or} \quad X = \frac{\sigma}{\sqrt{gh_0}}x. \tag{3}$$

Subsituting Eq. (3) into Eq. (2), we get

$$X^2\frac{d^2\eta}{dX^2} + X\frac{d\eta}{dX} + X^2\eta = 0. \tag{4}$$

Substituting $n = 0$ into Eq. (9.33) and using Eq. (9.58), we obtain

$$\eta = C_1 J_0(X) + C_2 Y_0(X) = C_1 J_0\left(\frac{\sigma}{\sqrt{gh_0}}x\right) + C_2 Y_0\left(\frac{\sigma}{\sqrt{gh_0}}x\right), \tag{5}$$

in which C_1 and C_2 are integral constants. The water depth is finite at $x = 0$. Thus, since $Y_0(0) = -\infty$, C_2 in Eq. (5) is zero. The function $J_0(x)$ in Fig. 9.2(a) indicates that the amplitude of the waves increases as x becomes small.

Problem 9.2. In Example 9.2, when the water depth is constant h_0 and the width of the bay is given by $b_0 x^m/a$ ($m \neq 1$), the displacement of the water surface level from the still water level η is expressed by Eq. (6).

$$\frac{d^2\eta}{dx^2} + \frac{m}{x}\frac{d\eta}{dx} + \frac{\sigma^2}{gh_0}\eta = 0. \tag{6}$$

Show that the solution of Eq. (6) is

$$\eta = x^\nu \left[C_1 J_\nu \left(\frac{\sigma}{\sqrt{gh_0}} x \right) + C_2 Y_\nu \left(\frac{\sigma}{\sqrt{gh_0}} x \right) \right], \tag{7}$$

in which $\nu = (1 - m)/2$.

9.3. Legendre Function

The Legendre differential equation appears when the Laplace equation is expressed by the spherical coordinate system and solved by the method of the separation of variables. The solution of the Legendre differential equation is also called the Legendre function or spherical function. The Legendre differential equation is expressed by

$$(1 - x^2)\frac{d^2 y}{dx^2} - 2x\frac{dy}{dx} + n(n + 1)y = 0, \tag{9.66}$$

in which n is the order of the Legendre differential equation, and n is equal to 0 or a positive integer. Assume the solution of powers of x as follows:

$$y = \sum_{r=0}^{\infty} a_r x^{m+r}. \tag{9.67}$$

Substituting Eq. (9.67) into Eq. (9.66) and equating the coefficients of the various powers of x to zero in Eq. (9.66), we obtain

$$(m+r)(m+r-1)a_r + (n-m-r+2)(n+m+r-1)a_{r-2} = 0, \tag{9.68}$$

in which $a_{-1} = a_{-2} = 0$. The condition of $r = 0$ in Eq. (9.68) shows

$$m(m - 1)a_0 = 0. \tag{9.69}$$

The condition of $r = 1$ in Eq. (9.68) shows

$$(m + 1)ma_1 = 0. \tag{9.70}$$

In order that a_0 and a_1 are arbitrary, $m = 0$ is selected in Eqs. (9.69) and Eq. (9.70). Equation (9.68) with $m = 0$ shows

$$a_r = -\frac{(n - r + 2)(n + r - 1)}{r(r - 1)} a_{r-2}. \tag{9.71}$$

Substituting Eq. (9.71) into Eq. (9.67), we obtain

$$y = a_0 \left[1 - \frac{n(n+1)}{2!}x^2 + \frac{n(n-2)(n+1)(n+3)}{4!}x^4 - \cdots \right]$$

$$+ a_1 \left[x - \frac{(n-1)(n+2)}{3!}x^3 \right.$$

$$\left. + \frac{(n-1)(n-3)(n+2)(n+4)}{5!}x^5 - \cdots \right]. \tag{9.72}$$

Equation (9.72) for $n = 0$ gives

$$y = a_0 \cdot 1. \tag{9.73}$$

Equation (9.72) for $n = 1$ derives

$$y = a_1 \cdot x. \tag{9.74}$$

For $n = 2$ we have

$$y = a_0(1 - 3x^2). \tag{9.75}$$

For $n = 3$ we get

$$y = a_1 \left(x - \frac{5}{3}x^3 \right). \tag{9.76}$$

Now if we give the arbitrary coefficients a_0 or a_1, as the case may be, such a numerical value that the polynomial becomes equal to unity when x is unity, we obtain the following system of polynomials:

$$P_0(x) = 1, \quad P_1(x) = x, \quad P_2(x) = \frac{1}{2}(3x^2 - 1),$$

$$P_3(x) = \frac{1}{2}(5x^3 - 3x), \ldots. \tag{9.77}$$

These are called Legendre polynomials or Legendre functions. These satisfy Eq. (9.66). The Legendre polynomials are generally given by Rodrigues:

$$P_n(x) = \frac{1}{2^n n!} \frac{d^n}{dx^n}(x^2 - 1)^n. \tag{9.78}$$

These are also called the Legendre function of the first kind, as shown in Fig. 9.3. The Legendre function of the second kind is also described

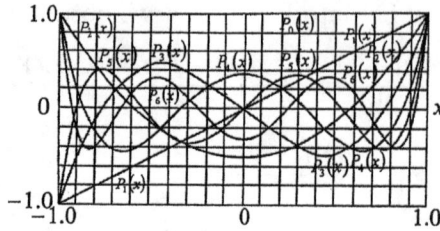

Figure 9.3.   Legendre function of the first kind.

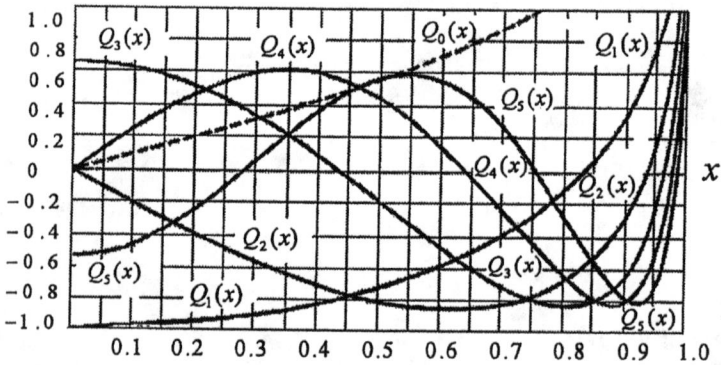

Figure 9.4.   Legendre function of the second kind.

in Fig. 9.4. These are given by

$$Q_n(x) = \sum_{r=0}^{\infty} \frac{2^n (n+r)!(n+2r)!}{r!(2n+2r+1)!} x^{-n-2r-1}. \tag{9.79}$$

The Legendre functions $P_n(x)$ and $Q_n(x)$ are the special type of the hypergeometric function. Therefore, the general solution of Eq. (9.66) is

$$y = AP_n(x) + BQ_n(x), \tag{9.80}$$

in which A and B are arbitrary constants. $Q_n(x)$ is derived by the method that is beyond the scope of this discussion.

Example 9.3. A sphere of which radius a moves in a still fluid with the velocity U. The origin of the spherical coordinate system (r, θ, ω) is defined at the center of the sphere (Lamb, 1945). Since the flow condition is independent of the moving direction of the sphere, the governing equation is independent of ω. The velocity potential ϕ is written as

$$\frac{\partial}{\partial r}\left(r^2\frac{\partial\phi}{\partial r}\right) + \frac{1}{\sin\theta}\frac{\partial}{\partial\theta}\left(\sin\theta\frac{\partial\phi}{\partial\theta}\right) = 0. \tag{1}$$

Solve Eq. (1) and obtain the velocity potential ϕ.

Solution. Assume $\phi = R(r)\Theta(\theta)$. Substituting the velocity potential ϕ into Eq. (1), we have

$$\frac{1}{R}\frac{d}{dr}\left(r^2\frac{dR}{dr}\right) = -\frac{1}{\Theta\sin\theta}\frac{d}{d\theta}\left(\sin\theta\frac{d\Theta}{d\theta}\right). \tag{2}$$

Equating Eq. (2) by $n(n+1)$, we obtain

$$\frac{d}{dr}\left(r^2\frac{dR}{dr}\right) - n(n+1)R = 0 \tag{3}$$

and

$$\frac{1}{\sin\theta}\frac{d}{d\theta}\left(\sin\theta\frac{d\Theta}{d\theta}\right) + n(n+1)\Theta = 0. \tag{4}$$

Solving Eq. (3), we get R as

$$R = Ar^n + Br^{-n-1}. \tag{5}$$

Using $\mu = \cos\theta$ in Eq. (4), we obtain Θ as

$$\frac{d}{d\mu}\left[(1-\mu^2)\frac{d\Theta}{d\mu}\right] + n(n+1)\Theta = 0. \tag{6}$$

Equation (6) is the Legendre differential equation. The general solution of Eq. (6) is

$$\Theta(\mu) = CP_n(\mu) + DQ_n(\mu), \tag{7}$$

in which A, B, C, and D are arbitrary constants. These are determined by the boundary conditions.

Problem 9.3. Show that

$$\int_{-1}^{1} xP_n(x)P_{n-1}(x)dx = \frac{2n}{4n^2 - 1}. \tag{8}$$

9.4. Elliptic Function

When there are two polynomials, a rational function is derived by
a ratio of one polynomial to another polynomial. When $R(x, y)$ is
a rational function and $\phi(x)$ is a parabolic function, the following
integral

$$\int R(x, \sqrt{\phi(x)})\, dx \tag{9.81}$$

is generally integrable. But when the degree of $\phi(x)$ is more than two,
Eq. (9.81) is not expressed by elementary functions. The integral of
Eq. (9.81) is historically studied by Jacobi *et al.* Since this integral
appears in the computation of the length of the surrounding curve
of an ellipse, this is called an elliptic integral (Hancock, 1958). In
hydraulic engineering, the elliptic integral appears in the integral of
the Schwartz–Christoffel transformation in Chap. 7 or a KdV partial
differential equation in Chap. 16. According to Jacobi, the elliptic
integral of the first kind is defined by

$$u = \int_0^x \frac{dx}{\sqrt{(1-x^2)(1-k^2x^2)}}, \quad \text{for} \quad 0 \le k \le 1. \tag{9.82}$$

The inverse function of Eq. (9.82) is given by

$$x = sn(u, k) = sn\, u \quad \text{or} \quad u = sn^{-1}(x, k), \tag{9.83}$$

in which k is a modulus of an sn function. The function $sn\, u$ is defined
between $-K$ and K when x changes from -1 to 1. The function $sn\, u$
is defined outside the original range of $-K$ and K as a periodic
function. When $k = 0$, Eq. (9.82) is easily integrable as

$$u = \int_0^x \frac{dx}{\sqrt{1-x^2}} = \sin^{-1} x. \tag{9.84}$$

Thus, we get

$$sn(u, 0) = \sin u. \tag{9.85}$$

For $k = 1$ we have

$$u = \int_0^x \frac{dx}{1 - x^2} = \frac{1}{2} \log \frac{1 + x}{1 - x}. \tag{9.86}$$

Thus, the following equation is derived:

$$sn(u, 1) = \frac{e^{2u} - 1}{e^{2u} + 1} = \tanh u. \tag{9.87}$$

The function cn is defined in the range $-K \leq u \leq K$ as

$$cn\, u = \sqrt{1 - sn^2\, u}. \tag{9.88}$$

In the same way, the function dn is defined as

$$dn\, u = \sqrt{1 - k^2\, sn^2\, u}. \tag{9.89}$$

If we have

$$sn\, K = 1, \tag{9.90}$$

we have

$$cn\, K = 0, \tag{9.91}$$

$$dn\, K = \sqrt{1 - k^2}. \tag{9.92}$$

An addition theorem of the function sn (Hancock, 1958) is given by

$$sn(u + v) = \frac{sn\, u \, cn\, v \, dn\, v + sn\, v \, cn\, u \, dn\, u}{1 - k^2\, sn^2\, u \, sn^2\, v}. \tag{9.93}$$

Using Eqs. (9.93), (9.90), (9.91), (9.92), and $v = K$, we get

$$sn(u + K) = \frac{cn\, u}{dn\, u}. \tag{9.94}$$

If sn is defined between $-K$ and K, the range of sn is expanded to the range $[K, 3K]$ from Eq. (9.94). An addition theorem of cn gives

$$cn(u + K) = -\sqrt{1 - k^2} \frac{sn\, u}{dn\, u}. \tag{9.95}$$

In the same way, an addition theorem of dn shows

$$dn(u + K) = \sqrt{1 - k^2}\, dn\, u. \tag{9.96}$$

Substituting $u + K$ into u in Eq. (9.94) and using Eqs. (9.95) and (9.96), we obtain

$$sn(u + 2K) = -sn\, u. \tag{9.97}$$

Substituting $-u$ into u in Eq. (9.97), we have

$$sn(2K - u) = -sn(-u) = sn\, u. \tag{9.98}$$

We used that sn is an odd function. Substituting $u = 0$, K, and $2K$ in Eq. (9.97), we get

$$sn\, 2K = 0, \quad sn\, 3K = -1, \quad sn\, 4K = 0. \tag{9.99}$$

Differentiating Eq. (9.82) about x, we obtain

$$\frac{du}{dx} = \frac{1}{\sqrt{(1 - x^2)(1 - k^2 x^2)}}. \tag{9.100}$$

Since $x = sn\, u$, the following equation is derived:

$$\frac{d}{du} sn\, u = \frac{dx}{du} = \sqrt{(1 - x^2)(1 - k^2 x^2)} = \sqrt{(1 - sn^2\, u)(1 - k^2\, sn^2\, u)}. \tag{9.101}$$

These functions sn, cn, and dn are represented in Fig. 9.5. The periods of sn and cn are $4K$ and that of dn is $2K$. The elliptic integral of the second kind is defined by

$$\int_0^x \sqrt{\frac{1 - k^2 x^2}{1 - x^2}}\, dx. \tag{9.102}$$

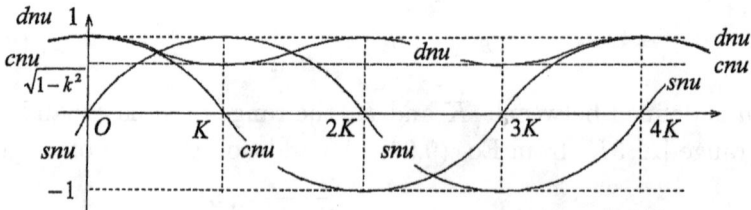

Figure 9.5.   Elliptic functions.

When $x = \sin\theta$ and $0 < k^2 < 1$, the elliptic functions of the first and second kinds are given by

$$\int_0^x \frac{dx}{\sqrt{(1 - x^2)(1 - k^2 x^2)}} = \int_0^\theta \frac{d\theta}{\sqrt{1 - k^2 \sin^2\theta}}, \qquad (9.103)$$

$$\int_0^x \sqrt{\frac{1 - k^2 x^2}{1 - x^2}}\, dx = \int_0^\theta \sqrt{1 - k^2 \sin^2\theta}\, d\theta. \qquad (9.104)$$

Example 9.4. Transform the following equation:

$$I = \int_x^a \frac{dx}{\sqrt{-(x - a)(x - b)(x - c)}} \qquad (a > b > c) \qquad (1)$$

to

$$I = \frac{2}{\sqrt{a - c}} \int_0^\theta \frac{d\theta}{\sqrt{1 - k^2 \sin^2\theta}}, \qquad (2)$$

in which

$$k = \sqrt{\frac{a - b}{a - c}}, \qquad \sin\theta = \sqrt{\frac{a - x}{a - b}}. \qquad (3)$$

Solution. The second equation of Eq. (3) shows

$$x = a - (a - b)\sin^2\theta, \qquad dx = -2(a - b)\sin\theta\cos\theta\, d\theta \qquad (4)$$

and

$$x - c = a - c - (a - b)\sin^2\theta,$$
$$x - b = (a - b)(1 - \sin^2\theta). \qquad (5)$$

Thus, Eq. (1) is

$$I = \int_\theta^0 \frac{-2(a - b)\sin\theta\cos\theta\, d\theta}{(a - b)\sin\theta\cos\theta\sqrt{a - c - (a - b)\sin^2\theta}}$$

$$= \frac{2}{\sqrt{a - c}} \int_0^\theta \frac{d\theta}{\sqrt{1 - k^2 \sin^2\theta}}. \qquad (6)$$

Problem 9.4. The equation of an ellipse is in the Cartesian coordinate given by

$$\frac{x^2}{a^2} + \frac{y^2}{b^2} = 1, \quad (a > b > 0). \tag{7}$$

Show that the length of the curve of the ellipse from $(0, b)$ to (x, y) is obtained by L as follows:

$$L = a \int_0^u \sqrt{\frac{1 - k^2 u^2}{1 - u^2}} du, \tag{8}$$

in which $u = x/a$ and $k^2 = (a^2 - b^2)/a^2$.

9.5. Orthogonality Principle

The Bessel function satisfies the following equations:

$$\int_0^1 x J_n^2(\alpha_k x) dx = \frac{1}{2} J_{n+1}^2(\alpha_k), \tag{9.105}$$

$$\int_0^1 J_n(ax) J_n(bx) dx = 0, \tag{9.106}$$

in which α_k, a, and b are the successive positive roots of $J_n(\alpha) = 0$. An arbitrary function $F(x)$ is expanded in the interval from $x = 0$ to $x = 1$ in a series of the form as

$$F(x) = \sum_{s=1}^{\infty} C_s J_n(\alpha_s x). \tag{9.107}$$

To obtain the coefficient C_k of this expansion, we multiply both members of Eq. (9.107) by $x J_n(\alpha_k x) dx$ and integrate from $x = 0$ to $x = 1$. We have by virtue of Eq. (9.106)

$$\int_0^1 x J_n(\alpha_k x) F(x) dx = C_k \int_0^1 x J_n^2(\alpha_k x) dx. \tag{9.108}$$

Hence the typical coefficient of the series expansion Eq. (9.107) is given by

$$C_s = \frac{2}{J_{n+1}^2(\alpha_s)} \int_0^1 x J_n(\alpha_s x) F(x) dx. \tag{9.109}$$

The Legendre function of Eq. (9.78) satisfies

$$\int_{-1}^{1} [P_n(x)]^2 dx = \frac{1}{2n+1}, \quad \text{for } n = 0, 1, 2, \ldots, \quad (9.110)$$

$$\int_{-1}^{1} P_n(x) P_m(x) dx = 0, \quad \text{if } m \neq n. \quad (9.111)$$

If $F(x)$ and its derivative are continuous in the interval $(-1, 1)$, $F(x)$ may be expanded in a series of the form

$$F(x) = \sum_{n=0}^{\infty} a_n P_n(x). \quad (9.112)$$

To obtain the coefficient a_m, we multiply both sides of Eq. (9.112) by $P_m(x)$ and integrate over the interval $(-1, 1)$. We then obtain

$$\int_{-1}^{1} F(x) P_m(x) dx = a_m \int_{-1}^{1} [P_m(x)]^2 dx = \frac{2a_m}{2m+1}. \quad (9.113)$$

The coefficient of the expansion Eq. (9.112) is given by

$$a_n = \frac{2n+1}{2} \int_{-1}^{1} F(x) P_n(x) dx. \quad (9.114)$$

The expansions Eqs. (9.107) and (9.112) are similar to an expansion of an arbitrary function into a Fourier series in Chap. 10. Equations (9.106) and (9.111) are called the orthogonality principle of the Bessel function and the Legendre function, respectively.

Example 9.5. Show Eq. (9.106).

Solution. Substituting $y = u/\sqrt{x}$ into Eq. (9.33), we get

$$\frac{d^2u}{dx^2} + \left(1 - \frac{n^2 - 1/4}{x^2}\right) u = 0. \quad (1)$$

In Eq. (1) we place ax instead of x, we get

$$\frac{d^2u}{dx^2} + \left(a^2 - \frac{n^2 - 1/4}{x^2}\right) u = 0. \quad (2)$$

The solution of Eq. (2) is

$$u = \sqrt{x} J_n(ax). \quad (3)$$

In the same manner of the replacement bx,

$$v = \sqrt{x} J_n(bx) \qquad (4)$$

satisfies the equation

$$\frac{d^2 v}{dx^2} + \left(b^2 - \frac{n^2 - 1/4}{x^2}\right) v = 0. \qquad (5)$$

If we multiply Eq. (2) by v and Eq. (5) by u and subtract the second product from the first, we obtain

$$(b^2 - a^2)uv = u''v - v''u, \qquad (6)$$

in which "$'$" indicates d/dx. Let us integrate both members of Eq. (6) from 0 to x. We thus obtain

$$(b^2 - a^2) \int_0^x uv\,dx = \int_0^x (u''v - v''u)\,dx$$

$$= \int_0^x \frac{d}{dx}(vu' - uv')\,dx = (vu' - uv')|_0^x. \qquad (7)$$

That is

$$(b^2 - a^2) \int_0^x x J_n(ax) J_n(bx)\,dx$$

$$= x[a J_n(bx) J_n'(ax) - b J_n(ax) J_n'(bx)]. \qquad (8)$$

Substituting $x = 1$ and assuming that a and b are distinct positive zeros of $J_n(\alpha)$, we have

$$\int_0^1 x J_n(ax) J_n(bx)\,dx = 0. \qquad (9)$$

Problem 9.5. Show

$$\int_{-1}^1 [P_n(x)]^2\,dx = \frac{2}{2n+1}, \quad n = 0, 1, 2, \ldots \qquad (10)$$

Chapter 10

Fourier Series and Integral

We, engineers use the method of the separation of variables when we solve linear partial differential equations. If we select sin or cos functions to satisfy the linear partial differential equations, it is very convenient that the sin or cos functions form a complete system. Fourier found that an arbitrary function could be expressed by sin or cos functions. Then, the series is called a Fourier series. This property is very useful to approximate the solution of the linear partial differential equations. The Fourier series is applicable to nonlinear problems using a variational method in Chap. 18. Finally, we consider Fourier transform to solve linear differential equations.

10.1. Fourier Series

The simplest periodic process that occurs in nature is mathematically described by the sin or cos functions. If the process repeats itself f times a second, the function representing the simple oscillation may be either

$$A \sin 2\pi f t \quad \text{or} \quad A \cos 2\pi f t, \tag{10.1}$$

in which t is time measured in seconds from a suitable starting point, A is called the amplitude, and f is the frequency of the vibration. Besides the frequency f, it is customary to speak of the angular frequency of the vibration $\omega = 2\pi f$. A phenomenon described by a simple sin or cos functions are called a simple harmonic vibration. We may simplify the computations involving these functions considerably by using imaginary exponentials instead of sin and cos functions. We have the Euler equation as

$$e^{i\omega t} = \cos \omega t + i \sin \omega t. \tag{10.2}$$

If we let

$$z = Ae^{i\omega t} : A \text{ is real,} \tag{10.3}$$

then, z represents a complex number whose representative point in the complex plane describes a circle of radius A with angular velocity ω. When $z = x + iy$, the projections on the real and imaginary axes are, respectively

$$x = \Re z, \quad y = \Im z, \tag{10.4}$$

in which $\Re$ and $\Im$ mean the real and imaginary parts, respectively. The conjugate complex variable z is

$$\bar{z} = Ae^{-i\omega t}, \tag{10.5}$$

then, we get

$$z\bar{z} = |z|^2 = A^2. \tag{10.6}$$

Let us consider an arbitrary process that is repeated every T sec and represented by $F(t)$. Since by hypothesis, the process repeats itself every T sec, we have

$$F(t) = F(t + T), \tag{10.7}$$

in which $F(t)$ is single-valued and finite and has a finite number of discontinuities and a finite number of maxima and minima in the interval of one oscillation, T. The method of determining the coefficients will now be given. It is most convenient to start from the complex representation and write

$$F(t) = a_0 + a_1 e^{i\omega t} + a_2 e^{2i\omega t} + \cdots + a_n e^{ni\omega t} + \cdots + a_{-1} e^{-i\omega t}$$
$$+ a_{-2} e^{-2i\omega t} + \cdots + a_{-n} e^{-ni\omega t} + \cdots$$
$$= \sum_{n=-\infty}^{\infty} a_n e^{ni\omega t}, \tag{10.8}$$

in which the angular frequency is

$$\omega = 2\pi/T. \tag{10.9}$$

Since the function $F(t)$ on the left side is real, the imaginary part on the right side is zero. To obtain a_n in Eq. (10.8), integrating

Eq. (10.8) over one period T, we have

$$\int_0^{2\pi/\omega} F(t)dt = \int_0^{2\pi/\omega} \left(\sum_{n=-\infty}^{\infty} a_n e^{ni\omega t} \right) dt$$

$$= \sum_{n=-\infty}^{\infty} a_n \int_0^{2\pi/\omega} e^{ni\omega t} dt. \qquad (10.10)$$

For $n \neq 0$,

$$\int_0^{2\pi/\omega} e^{ni\omega t} dt = \frac{1}{ni\omega} (e^{2ni\pi} - 1) = 0. \qquad (10.11)$$

For $n = 0$,

$$\int_0^{2\pi/\omega} dt = \frac{2\pi}{\omega} = T. \qquad (10.12)$$

Thus, Eq. (10.10) is

$$\int_0^T F(t)dt = a_0 T. \qquad (10.13)$$

Equation (10.13) leads to

$$a_0 = \frac{1}{T} \int_0^T F(t)dt = \overline{F(t)} \qquad (10.14)$$

in which "$-$" indicates an average. The coefficient a_0 is equal to the average of $F(t)$. To calculate the coefficients a_n for $n \neq 0$, let us integrate the product of Eq. (10.8) and $e^{-ni\omega t}$ from 0 to T. Then, we have

$$\int_0^T F(t)e^{-ni\omega t} dt = a_n T \qquad (10.15)$$

or

$$a_n = \frac{1}{T} \int_0^T F(t)e^{-ni\omega t} dt. \qquad (10.16)$$

In the same way, we get

$$a_{-n} = \frac{1}{T} \int_0^T F(t)e^{ni\omega t} dt. \qquad (10.17)$$

Thus, the following equation is derived:

$$a_{-n} = \bar{a}_n, \qquad (10.18)$$

in which $\bar{a}_n$ is the conjugate complex variable of a_n. Then, Eq. (10.8) is rewritten as

$$F(t) = a_0 + \sum_{n=1}^{\infty}(a_n e^{ni\omega t} + a_{-n}e^{-ni\omega t}). \qquad (10.19)$$

This is called a complex Fourier series. On the right side of Eq. (10.19), the following relationship is obtained:

$$a_n e^{ni\omega t} + a_{-n}e^{-ni\omega t} = (a_n + a_{-n})\cos n\omega t + i(a_n - a_{-n})\sin n\omega t. \qquad (10.20)$$

If we now let

$$A_n = a_n + a_{-n}, \quad B_n = i(a_n - a_{-n}), \quad \frac{A_0}{2} = a_0. \qquad (10.21)$$

The coefficients A_n, B_n, and A_0 are real. We obtain

$$F(t) = \frac{A_0}{2} + \sum_{n=1}^{\infty} A_n \cos n\omega t + \sum_{n=1}^{\infty} B_n \sin n\omega t. \qquad (10.22)$$

This is the Fourier series in real expression. The coefficients A_n and B_n are computed by $F(t)$ as

$$A_0 = \frac{1}{T}\int_0^T F(t)dt, \qquad (10.23)$$

$$A_n = \frac{2}{T}\int_0^T F(t)\cos n\omega t \, dt, \quad \text{for } n = 1, 2, 3, \ldots, \qquad (10.24)$$

$$B_n = \frac{2}{T}\int_0^T F(t)\sin n\omega t \, dt, \quad \text{for } n = 1, 2, 3, \ldots. \qquad (10.25)$$

Equations (10.23), (10.24), and (10.25) are also calculated by the least squares method. The integral of squared difference between $F(t)$ and Eq. (10.22), S is defined by

$$S = \int_0^T \left[F(t) - \frac{A_0}{2} - \sum_{n=1}^{\infty} A_n \cos n\omega t - \sum_{n=1}^{\infty} B_n \sin n\omega t\right]^2 dt. \qquad (10.26)$$

Differentiating Eq. (10.26) about A_0, we let 0 as

$$\frac{\partial S}{\partial A_0} = 0. \tag{10.27}$$

In the same way,

$$\frac{\partial S}{\partial A_n} = 0, \quad \text{for } n = 1, 2, 3, \ldots, \tag{10.28}$$

$$\frac{\partial S}{\partial B_n} = 0, \quad \text{for } n = 1, 2, 3, \ldots. \tag{10.29}$$

Equations (10.27), (10.28), and (10.29) lead to Eqs. (10.23), (10.24), and (10.25), respectively. As an example, consider the following function:

$$F(t) = A, \quad \text{for } 0 < t < T/2, \tag{10.30}$$

$$F(t) = -A, \quad \text{for } T/2 < t < T. \tag{10.31}$$

The coefficients of the complex Fourier series are given by Eq. (10.16) as

$$a_n = \frac{1}{T} \int_0^T F(t) e^{-ni\omega t} dt = 0, \quad \text{for } n = 0 \text{ or even}, \tag{10.32}$$

$$a_n = \frac{2A}{ni\pi}, \quad \text{for } n = \text{odd}. \tag{10.33}$$

Thus, the complex Fourier series is

$$F(t) = \frac{2A}{i\pi} \sum_{n=-\infty}^{\infty} \frac{e^{ni\omega t}}{n}, \quad \text{for } n = \text{odd}. \tag{10.34}$$

The coefficients of the Fourier series are given by Eqs. (10.24) and (10.25) as

$$A_n = a_n + a_{-n} = 0, \quad \text{for } n = 1, 2, 3, \ldots, \tag{10.35}$$

$$B_n = i(a_n - a_{-n}) = \frac{4A}{n\pi}, \quad \text{for } n = 1, 2, 3, \ldots. \tag{10.36}$$

The coefficient A_0 is zero. Thus, the Fourier series is written as

$$F(t) = \frac{4A}{\pi} \sum_{n=1}^{\infty} \frac{\sin n\omega t}{n}, \quad \text{for } n = 1, 2, \ldots. \tag{10.37}$$

Equation (10.34) is the same as Eq. (10.37).

Using the Fourier series, Mizumura and Yamasaka (2002) calculated flow pattern in the embayment of a river and solved the kinematic wave equation (Mizumura, 2006b) and the Boussinesq equation
(Mizumura, 2002, 2009a).

Example 10.1. A function $F(t)$ for $-\pi < t < \pi$ is expressed by

$$F(t) = 0, \quad -\pi < t \leq 0 \tag{1}$$

$$F(t) = t, \quad 0 \leq t < \pi. \tag{2}$$

Then, obtain A_0, A_n, and B_n.

Solution. Since the period is 2π, the coefficients are

$$A_0 = \frac{1}{2\pi} \int_{-\pi}^{\pi} F(t)dt = \frac{\pi}{4}, \tag{3}$$

$$A_n = \frac{1}{\pi} \int_{-\pi}^{\pi} F(t) \cos nt \, dt = \frac{(-1)^n - 1}{n^2 \pi}, \tag{4}$$

$$B_n = \frac{1}{\pi} \int_{-\pi}^{\pi} F(t) \sin nt \, dt = -\frac{(-1)^n}{n}, \tag{5}$$

in which $n = 1, 2, \ldots$. Thus, we have

$$F(t) = \frac{\pi}{4} + \sum_{n=1}^{\infty} \left[\frac{(-1)^n - 1}{n^2 \pi} \cos nt - \frac{(-1)^n}{n} \sin nt \right]. \tag{6}$$

Problem 10.1. Express the function of the period T and amplitude
a by the Fourier series as shown in Fig. 10.1.

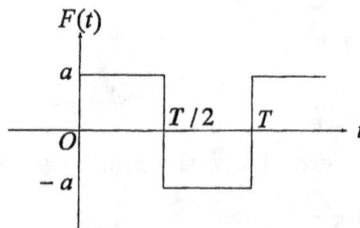

Figure 10.1.   Problem 10.1.

10.2. Fourier Integral

Let us consider the Fourier integral (Churchill, 1941) as the limiting form of the Fourier series. The complex Fourier series is written as

$$F(t) = \sum_{n=-\infty}^{\infty} a_n e^{ni\omega t}, \qquad (10.38)$$

in which the period $T = 2\pi/\omega$. The coefficients are

$$a_n = \frac{1}{T} \int_0^T F(u)e^{-ni\omega u} du. \qquad (10.39)$$

According to the periodicity of the function $F(t)$, we have

$$a_n = \frac{1}{T} \int_{-T/2}^{T/2} F(u)e^{-ni\omega u} du. \qquad (10.40)$$

Substituting Eq. (10.40) into Eq. (10.38), we have

$$F(t) = \sum_{n=-\infty}^{\infty} \frac{1}{T} \int_{-T/2}^{T/2} F(u)e^{\frac{2ni\pi}{T}(t-u)} du. \qquad (10.41)$$

Substituting $1/T = \Delta s$ in Eq. (10.41), we obtain

$$F(t) = \sum_{n=-\infty}^{\infty} \Delta s \int_{-T/2}^{T/2} F(u)e^{2ni\pi(t-u)\Delta s} du. \qquad (10.42)$$

Under the condition of $n\Delta s = s$, when $\Delta s \to ds$ and $T \to \infty$ are formed, we get

$$F(t) = \int_{-\infty}^{\infty} ds \int_{-\infty}^{\infty} F(u)e^{2i\pi s(t-u)} du$$

$$= \int_{-\infty}^{\infty} e^{2i\pi st} ds \int_{-\infty}^{\infty} F(u)e^{-2i\pi su} du. \qquad (10.43)$$

The second form of the identity Eq. (10.43) shows that the function $F(t)$ may be expressed by a continuous series of harmonics, one for each value of s. The conditions to satisfy the above are as follows:

(i) $F(t)$ is a single-valued function.
(ii) The mean value at a point t_0 of discontinuity to the function will be given by

$$F(t_0) = \frac{1}{2}[F(t_0 + 0) + F(t_0 - 0)]. \tag{10.44}$$

(iii) The integral $\int_{-\infty}^{\infty} |F(t)| dt$ must exist.

When we expand a function into the Fourier series in a certain range, the function is defined by the series outside this range in a periodic manner. However, by the Fourier integral, we obtain analytical expression for discontinuous functions that represent the function throughout the infinite range $-\infty < t < \infty$. As an example, we consider

$$F(t) = \begin{cases} A, & \text{for } -a < t < a, \\ 0, & \text{for } t < -a \quad \text{or} \quad t > a. \end{cases} \tag{10.45}$$

A function $F(t) = t$ is simple, but it is not applicable because this function does not satisfy condition (iii). Substituting Eq. (10.45) into Eq. (10.43), we have

$$F(t) = \frac{A}{2\pi i} \int_{-\infty}^{\infty} \frac{e^{iv(t+a)} - e^{iv(t-a)}}{v} dv, \tag{10.46}$$

in which $v = 2\pi s$. Next, we consider a Fourier transform and an inverse Fourier transform. Applying the Euler equation to Eq. (10.43), we obtain

$$F(t) = 2 \int_0^{\infty} ds \int_{-\infty}^{\infty} F(u) \cos 2\pi s(t - u) du. \tag{10.47}$$

Because the function $F(t)$ is real. Substituting $2\pi s = \omega$ into Eq. (10.43), we get

$$F(t) = \frac{1}{2\pi} \int_{-\infty}^{\infty} e^{i\omega t} d\omega \int_{-\infty}^{\infty} F(u) e^{-i\omega u} du. \tag{10.48}$$

If we define

$$g(\omega) = \frac{1}{2\pi} \int_{-\infty}^{\infty} F(u)e^{-i\omega u}du, \tag{10.49}$$

Eq. (10.48) is

$$F(t) = \int_{-\infty}^{\infty} g(\omega)e^{i\omega t}d\omega. \tag{10.50}$$

Equation (10.49) is called a Fourier transform of $F(t)$. Equation (10.50) is called an inverse Fourier transform. The another definition of the Fourier transform is also given by

$$g(\omega) = \frac{1}{\sqrt{2\pi}} \int_{-\infty}^{\infty} F(u)e^{-i\omega u}du \tag{10.51}$$

and the inverse Fourier transform is

$$F(t) = \frac{1}{\sqrt{2\pi}} \int_{-\infty}^{\infty} g(\omega)e^{i\omega t}d\omega. \tag{10.52}$$

We must be careful of the definition of the Fourier transform.

Example 10.2. Obtain the Fourier transform of $F(t) = e^{-|t|}$.

Solution. From Eq. (10.49) we have

$$g(\omega) = \frac{1}{2\pi} \int_{-\infty}^{\infty} e^{-|u|}e^{-i\omega u}du$$

$$= \frac{1}{2\pi}\left[\int_{-\infty}^{0} e^{u-i\omega u}du + \int_{0}^{\infty} e^{-u-i\omega u}du\right]$$

$$= \frac{1}{2\pi}\left[\frac{e^{u(1-i\omega)}}{1-i\omega}\right]_{-\infty}^{0} + \frac{1}{2\pi}\left[-\frac{e^{-u(1+i\omega)}}{1+i\omega}\right]_{0}^{\infty}$$

$$= \frac{1}{2\pi}\left[\frac{1}{1-i\omega} + \frac{1}{1+i\omega}\right] = \frac{1}{\pi(1+\omega^2)}. \tag{1}$$

Problem 10.2. Compute the Fourier transform of a function as shown in Fig. 10.2.

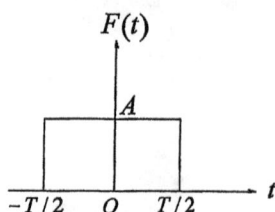

Figure 10.2.   Problem 10.2.

10.3. Fourier Transform and Its Applications

A Fourier transform is defined as follows:

$$g(\omega) = F\{F(u)\} = \frac{1}{2\pi} \int_{-\infty}^{\infty} F(u)e^{-i\omega u} d\omega. \tag{10.53}$$

Using the Euler equation in Eq. (10.53), we have

$$g(\omega) = \frac{1}{2\pi} \int_{-\infty}^{\infty} F(u)(\cos \omega u - i \sin \omega u) d\omega. \tag{10.54}$$

When $F(u)$ is an odd function, we obtain

$$g(\omega) = \frac{-i}{\pi} \int_{0}^{\infty} F(u) \sin \omega u \, d\omega. \tag{10.55}$$

If $i\pi g(\omega) = g_s(\omega)$ is defined in Eq. (10.55), we get

$$\int_{0}^{\infty} F(u) \sin \omega u \, du = g_s(\omega) = F_s\{F(u)\}. \tag{10.56}$$

This is called a Fourier sine transform. An inverse Fourier sine transform of Eq. (10.56) is given by

$$F_s^{-1}\{g_s(\omega)\} = \frac{2}{\pi} \int_{0}^{\infty} g_s(\omega) \sin \omega u \, d\omega = F(u). \tag{10.57}$$

In the same way, if a $F(u)$ is an even function, a Fourier cosine transform is defined by

$$\int_{0}^{\infty} F(u) \cos \omega u \, du = g_c(\omega) = F_c\{F(u)\}. \tag{10.58}$$

An inverse Fourier cosine transform of Eq. (10.58) is

$$F_c^{-1}\{g_c(\omega)\} = \frac{2}{\pi} \int_0^\infty g_c(\omega) \cos \omega u \, d\omega = F(u). \tag{10.59}$$

Apply the Fourier transform to the following partial differential equation of diffusion or heat conduction type (Chap. 14):

$$\frac{\partial v}{\partial t} = H^2 \frac{\partial^2 v}{\partial x^2}, \quad \text{for } x \geq 0. \tag{10.60}$$

Equation (10.60) also indicates the water infiltration into the ground. The boundary conditions are

$$v = v_0, \quad \text{for } x = 0, \quad t > 0, \tag{10.61}$$

$$v = 0, \quad \text{for } x = 0, \quad t = 0, \tag{10.62}$$

in which v = water content, t = time, x = vertically downward coordinate system from the ground surface, v_0 = water infiltration at the ground surface, and H^2 = moisture diffusivity. Applying the Fourier sine transform to $v(x,t)$, we have

$$\bar{v}(\omega, t) - F_s\{v(x, t)\} = \int_0^\infty v(x, t) \sin \omega x \, dx. \tag{10.63}$$

Applying the Fourier sine transform to Eq. (10.60), we get

$$\int_0^\infty \frac{\partial v}{\partial t} \sin \omega x \, dx = H^2 \int_0^\infty \frac{\partial^2 v}{\partial x^2} \sin \omega x \, dx. \tag{10.64}$$

The term on the left side of Eq. (10.64) is

$$\int_0^\infty \frac{\partial v}{\partial t} \sin \omega x \, dx = \frac{\partial}{\partial t} \int_0^\infty v \sin \omega x \, dx = \frac{\partial \bar{v}}{\partial t} = \frac{d\bar{v}}{dt}. \tag{10.65}$$

The physical condition of water content at $x = \infty$ indicates

$$\frac{\partial v}{\partial x}\bigg|_{x=\infty} = 0, \quad v|_{x=\infty} = 0, \tag{10.66}$$

We rewrite the term on the right side of Eq. (10.64) as

$$H^2 \int_0^\infty \frac{\partial^2 v}{\partial x^2} \sin \omega x \, dx = H^2 \left[\frac{\partial v}{\partial x} \sin \omega x \right]_0^\infty - H^2 \int_0^\infty \frac{\partial v}{\partial x} \omega \cos \omega x \, dx$$

$$= -H^2 [v\omega \cos \omega x]_0^\infty - H^2 \int_0^\infty v\omega^2 \sin \omega x \, dx$$

$$= H^2 v_0 \omega - H^2 \omega^2 \bar{v}. \tag{10.67}$$

Thus, Eq. (10.64) is written as

$$\frac{d\bar{v}}{dt} + H^2 \omega^2 \bar{v} = H^2 \omega v_0. \tag{10.68}$$

Equation (10.68) is the ordinary differential equation of the first-order. Referring to Eq. (5.12), we obtain

$$\bar{v} = \frac{v_0}{\omega} (1 - e^{-H^2 \omega^2 t}). \tag{10.69}$$

The inverse Fourier sine transform of Eq. (10.69) is given by

$$v(x,t) = F_s^{-1}\{\bar{v}(\omega,t)\} = \frac{2v_0}{\pi} \int_0^\infty (1 - e^{-H^2 \omega^2 t}) \frac{\sin \omega x}{\omega} d\omega$$

$$= \frac{2v_0}{\pi} \int_0^\infty \frac{\sin \omega x}{\omega} d\omega - \frac{2v_0}{\pi} \int_0^\infty e^{-H^2 \omega^2 t} \frac{\sin \omega x}{\omega} d\omega. \tag{10.70}$$

The first term on the right side of Eq. (10.70) is $\pi/2$. The integral of the second term on the right side of Eq. (10.70) is

$$\int_0^\infty e^{-H^2 \omega^2 t} \frac{\sin \omega x}{\omega} d\omega = \frac{\pi}{2} erf\left(\frac{x}{2H\sqrt{t}} \right). \tag{10.71}$$

Thus, Eq. (10.70) is derived as

$$v(x,t) = v_0 \left[1 - erf\left(\frac{x}{2H\sqrt{t}} \right) \right] = v_0 \; erf \, c\left(\frac{x}{2H\sqrt{t}} \right). \tag{10.72}$$

Example 10.3. Solve the wave equation in Chap. 12,

$$\frac{\partial^2 \eta}{\partial x^2} = \frac{1}{C^2} \frac{\partial^2 \eta}{\partial t^2}. \tag{1}$$

The boundary conditions are

$$\eta(x,0) = f(x), \quad \frac{\partial \eta}{\partial t}(x,0) = 0, \tag{2}$$

in which η = deviation of water surface profile, x = coordinate system in the horizontal, t = time, and C = celerity.

Solution. Apply the Fourier transform to Eq. (1). Then, we have

$$\bar{\eta}(\omega,t) = F\{\eta(x,t)\} = \int_{-\infty}^{\infty} \eta(x,t)e^{-i\omega x}\,dx, \tag{3}$$

in which

$$\eta|_{x=\infty} = \eta|_{x=-\infty} = \left.\frac{\partial \eta}{\partial x}\right|_{x=\infty} = \left.\frac{\partial \eta}{\partial x}\right|_{x=-\infty} = 0 \tag{4}$$

are assumed. The Fourier transform of Eq. (1) is

$$-\omega^2\bar{\eta} = \frac{1}{C^2}\frac{d^2\bar{\eta}}{dt^2}. \tag{5}$$

The general solution of Eq. (5) is

$$\bar{\eta} = A(\omega)\sin\omega Ct + B(\omega)\cos\omega Ct. \tag{6}$$

The Fourier transforms of Eq. (2) are

$$\bar{\eta}(\omega,0) = F\{\eta(x,0)\} = \int_{-\infty}^{\infty} f(x)e^{-i\omega x}\,dx = \bar{f}(\omega), \tag{7}$$

$$F\left\{\frac{\partial \eta}{\partial t}(x,0)\right\} = \left.\frac{d\bar{\eta}}{dt}\right|_{t=0} = 0. \tag{8}$$

Equations (6), (7) and (8) indicate

$$\bar{\eta} = \bar{f}(\omega)\cos\omega Ct. \tag{9}$$

The inverse Fourier transform of Eq. (9)

$$\eta(x,t) = \frac{1}{2\pi}\int_{-\infty}^{\infty} \bar{f}(\omega)\cos\omega Ct\, e^{i\omega x}\,dx. \tag{10}$$

In Eq. (10), we have

$$\cos \omega Ct = \frac{e^{i\omega Ct} + e^{-i\omega Ct}}{2}. \tag{11}$$

Substituting Eq. (11) into Eq. (10), we get

$$\eta(x,t) = \frac{1}{4\pi} \int_{-\infty}^{\infty} [e^{i\omega(x+Ct)} + e^{i\omega(x-Ct)}] \bar{f}(\omega) d\omega. \tag{12}$$

Since the inverse Fourier transform of Eq. (7) is

$$f(x) = \frac{1}{2\pi} \int_{-\infty}^{\infty} \bar{f}(\omega) e^{i\omega x} d\omega, \tag{13}$$

we have

$$f(x \pm Ct) = \frac{1}{2\pi} \int_{-\infty}^{\infty} e^{i\omega(x \pm Ct)} \bar{f}(\omega) d\omega. \tag{14}$$

Substituting Eq. (13) into Eq. (11), we gain

$$\eta(x,t) = \frac{1}{2}[f(x + Ct) + f(x - Ct)]. \tag{15}$$

Equation (15) shows that the derivation of the water surface profile $f(x)$ propagates in the $\pm x$ directions with the celerity C.

Problem 10.3. Check whether Eq. (10.60) can be solved using the Fourier sine transform and Fourier cosine transform with the boundary conditions Eqs. (16) and (17).

$$\frac{\partial v}{\partial x} = -p_0 \text{ (constant)}, \quad \text{for } x = 0, \quad t > 0, \tag{16}$$

$$v = 0 \quad \text{at } x = 0 \quad \text{and} \quad t = 0. \tag{17}$$

Chapter 11

Laplace Transform

There are many integral transforms such as the Fourier transform, Fourier sine transform, Fourier cosine transform, Mellin transform, Hilbert transform, Hankel transform, Bessel transform, etc. These integral transforms have been developing in order to integrate linear ordinary differential equations. The resultant integral transform of a linear ordinary differential equation becomes an algebraic equation. After the integral transform, a linear partial differential equation becomes a linear ordinary differential equation when the number of independent variables is two. The most famous integral transform is a Laplace transform which was introduced by Laplace in 1879, and used by Heaviside from 1887 to 1898 in electrical engineering. Since the governing equation in hydraulic engineering is usually nonlinear, the Laplace transform is not often used. Small and capital letters indicate the Laplace transform and the original variable, respectively, in this chapter.

11.1. Definition of Laplace Transform

The Laplace transform as an operational method that Heaviside suggested, is defined by

$$f(s) = \int_0^\infty e^{-st} F(t) dt = L\{F(t)\}, \qquad (11.1)$$

in which s is a complex number of which the real part is positive. This fact indicates that the applicability of the Laplace transform becomes easier than the Fourier transform. The function $f(s)$ is defined all on the s-plane by using an analytic continuation (Ahlfors, 1966). When an original function is $F(t)$, $f(p)$ or $f(s)$ are defined by the Laplace transform of the function $F(t)$. If the Laplace transform of a constant is defined a constant, the following Heaviside's definition (Pipes and Harrill, 1971) is used:

$$f(p) = \int_0^\infty p e^{-pt} F(t) dt. \tag{11.2}$$

This is called the p-multiplied Laplace transform.

We calculate the Laplace transform by Eq. (11.1) and use a variable s in this book. Thus, when $F(t) = 1$, the Laplace transform (Churchill, 1958) is

$$f(s) = L\{F(t)\} = \int_0^\infty e^{-st} dt = \frac{1}{s}, \quad (\Re\{s\} > 0). \tag{11.3}$$

We must be careful that the Laplace transform of $F(t) = 1$ is $1/s$. When $F(t) = e^{kt}$ and k is a constant, the Laplace transform of $F(t)$ is

$$f(s) = L\{F(t)\} = \int_0^\infty e^{-st} e^{kt} dt = \frac{1}{s-k}, \tag{11.4}$$

in which $t > 0$ and $\Re\{s\} > \Re\{k\}$. In the same way, we have

$$L\{t\} = \frac{1}{s^2}, \tag{11.5}$$

$$L\{\sin kt\} = \frac{k}{s^2 + k^2}. \tag{11.6}$$

The Laplace transform of the derivative $F'(t)$ is given by

$$L\{F'(t)\} = \int_0^\infty e^{-st} F'(t) dt$$

$$= \left[e^{-st} F(t)\right]_0^\infty + s \int_0^\infty e^{-st} F(t) dt = -F(0) + s f(s), \tag{11.7}$$

in which "$'$" $= d/dt$. In the same way, we get

$$L\{F''(t)\} = s^2 f(s) - s F(0) - F'(0). \tag{11.8}$$

Defining

$$L\left\{\int_0^t F(\tau) d\tau\right\} = L\{G(t)\}, \tag{11.9}$$

we have

$$L\{G'(t)\} = s L\{G(t)\} - G(0). \tag{11.10}$$

When we use Eq. (11.9) and

$$G(0) = 0, \quad G'(t) = F(t), \tag{11.11}$$

Equation (11.10) leads to

$$L\{F(t)\} = sL\{G(t)\}. \tag{11.12}$$

Thus, the Laplace transform of $G(t)$ is given by

$$L\{G(t)\} = L\left\{\int_0^t F(\tau)d\tau\right\} = \frac{f(s)}{s}. \tag{11.13}$$

Example 11.1. Obtain the Laplace transform of $F(t) = \sin t \cos t$.

Solution.

$$L\{F(t)\} = \int_0^\infty e^{-st} \sin t \cos t \, dt$$

$$= \int_0^\infty e^{-st} \frac{\sin 2t}{2} dt = \frac{1}{2} \cdot \frac{2}{s^2 + 2^2} = \frac{1}{s^2 + 4}. \tag{1}$$

Problem 11.1. Obtain the Laplace transform of $F(t) = e^{-t} \sin t$.

11.2. Inverse Laplace Transform

When we obtain an original function $F(t)$ from the Laplace transform $f(s)$, we call an inverse Laplace transform. It is defined by

$$F(t) = L^{-1}\{f(s)\}. \tag{11.14}$$

On the achievement of the inverse Laplace transform, there are two important rules. They are

Principle (i): The replacement of s in $f(s)$ to $s - a$ corresponds to the multiplication of e^{at} to the original function $F(t)$. It is

$$f(s - a) = \int_0^\infty e^{-(s-a)t} F(t)dt = \int_a^\infty e^{-st} e^{at} F(t)dt. \tag{11.15}$$

Principle (ii): The rational function can be fractional and transform each fractional one by one. For example, consider the following inverse Laplace transform:

$$L^{-1}\left\{\frac{s+1}{s^2 + 2s}\right\}. \tag{11.16}$$

Equation (11.16) is rewritten in the fractionals as

$$\frac{s+1}{s^2+2s} = \frac{1/2}{s} + \frac{1/2}{s+2}. \tag{11.17}$$

The inverse Laplace transforms on the right side of Eq. (11.17) are

$$L^{-1}\left\{\frac{1}{s}\right\} = 1, \tag{11.18}$$

$$L^{-1}\left\{\frac{1}{s+2}\right\} = e^{-2t}. \tag{11.19}$$

Thus, Eq. (11.16) is

$$L^{-1}\left\{\frac{s+1}{s^2+2s}\right\} = \frac{1}{2} + \frac{1}{2}e^{-2t}. \tag{11.20}$$

The inverse Laplace transform of a simple function can be done by using Table 11.1.

Consider the computational method of the inverse Laplace transform. Using Eq. (10.48), we use

$$F(t) = \frac{1}{2\pi}\int_{-\infty}^{\infty} e^{ivt}dv \int_{0}^{\infty} F(u)e^{-ivu}du, \tag{11.21}$$

in which

$$F(t) = 0, \quad \text{for } t < 0. \tag{11.22}$$

Let us assume $\phi(t)$ as follows:

$$\phi(t) = e^{-ct}F(t), \tag{11.23}$$

in which c is a positive real number. Substituting $\phi(t)$ instead of $F(t)$ into Eq. (11.21), we have

$$e^{-ct}F(t) = \frac{1}{2\pi}\int_{-\infty}^{\infty} e^{ivt}dv \int_{0}^{\infty} F(u)e^{-uc}e^{-ivu}du. \tag{11.24}$$

Substituting

$$s = c + iv \tag{11.25}$$

Table 11.1. Inverse Laplace transforms.

$F(t)$	$f(s)$	$\alpha(s > \alpha)$
1	$\dfrac{1}{s}$	0
e^{at}	$\dfrac{1}{s-a}$	a
$t^n\,(n = 1, 2, \ldots)$	$\dfrac{n!}{s^{n+1}}$	0
$t^n e^{at}\,(n = 1, 2, \ldots)$	$\dfrac{n!}{(s-a)^{n+1}}$	a
$\sin kt$	$\dfrac{k}{s^2+k^2}$	0
$\cos kt$	$\dfrac{s}{s^2+k^2}$	0
$\sinh kt$	$\dfrac{k}{s^2-k^2}$	$|k|$
$\cosh kt$	$\dfrac{s}{s^2-k^2}$	$|k|$
$e^{-at}\sin kt$	$\dfrac{k}{(s+a)^2+k^2}$	$-a$
$e^{-at}\cos kt$	$\dfrac{s+a}{(s+a)^2+k^2}$	$-a$
$\sqrt{t}$	$\dfrac{\sqrt{\pi}}{2\sqrt{s^3}}$	0
$\dfrac{1}{\sqrt{t}}$	$\sqrt{\dfrac{\pi}{s}}$	0
$t^k\,(k > -1)$	$\dfrac{\Gamma(k+1)}{s^{k+1}}$	0
$t^k e^{at}\,(k > -1)$	$\dfrac{\Gamma(k+1)}{(s-a)^{k+1}}$	a
$U(t-k)$: unit step function	$\dfrac{e^{-ks}}{s}$	0
$e^{at} - e^{bt}$	$\dfrac{a-b}{(s-a)(s-b)}$	a
$\dfrac{1}{a}\sin at - \dfrac{1}{b}\sin bt$	$\dfrac{b^2-a^2}{(s^2+a^2)(s^2+b^2)}$	0
$\cos at - \cos bt$	$\dfrac{(b^2-a^2)s}{(s^2+a^2)(s^2+b^2)}$	0

into Eq. (11.24), we obtain

$$e^{-ct}F(t) = \frac{1}{2\pi i} \int_{c-i\infty}^{c+i\infty} e^{(s-c)t} ds \int_0^\infty F(u)e^{-su} du. \qquad (11.26)$$

Multiplying e^{ct} both sides of Eq. (11.26), we get

$$F(t) = \frac{1}{2\pi i} \int_{c-i\infty}^{c+i\infty} e^{st} ds \int_0^\infty F(u)e^{-su} du, \qquad (11.27)$$

in which $c > c_0 > 0$ and $\int_0^\infty |F(t)| dt$ exists. Thus, assuming

$$f(s) = \int_0^\infty F(u)e^{-su} du, \qquad (11.28)$$

we have

$$F(t) = \frac{1}{2\pi i} \int_{c-i\infty}^{c+i\infty} f(s)e^{st} ds. \qquad (11.29)$$

Equations (11.28) and (11.29) are called a Fourier–Mellin equation. If we use Eq. (11.28), $f(s)$ can be derived from $F(t)$. When the Laplace transform $f(s)$ satisfies

$$\lim_{|s|\to\infty} |f(s)| = 0 \qquad (11.30)$$

and on a semi-circle C_0 and $R = |s - c|$ in Fig. 11.1,

$$\lim_{R\to\infty} \left| \int_{C_0} e^{st} f(s) ds \right| = 0, \quad t > 0, \quad \Re s \le 0. \qquad (11.31)$$

Let us consider an integral along a closed curve $C + C_0$ in Fig. 11.1. That is, Eq. (11.29) is

$$F(t) = \frac{1}{2\pi i} \int_{c-i\infty}^{c+i\infty} f(s)e^{st} ds = \lim_{R\to\infty} \frac{1}{2\pi i} \oint_{C+C_0} f(s)e^{st} ds. \qquad (11.32)$$

According to the Cauchy theorem and the residues computation (Chap. 6), we obtain

$$\oint_{C+C_0} f(s)e^{st} ds = 2\pi i \sum_{\text{inside } C+C_0} Res[e^{st} f(s)]. \qquad (11.33)$$

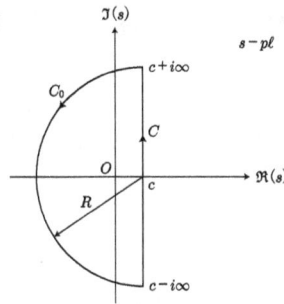

Figure 11.1.   Integral for an inverse Laplace transform.

Thus, the inverse Laplace transform of $f(s)$ is given by

$$L^{-1}\{f(s)\} = F(t) = \sum_{\text{inside } C+C_0} Res\left[e^{st}f(s)\right]. \tag{11.34}$$

in which the closed curve $C + C_0$ contains all singular points. As an example, consider the Laplace transform function

$$f(s) = \frac{k}{s^2 + k^2}. \tag{11.35}$$

The inverse Laplace transform of Eq. (11.35) is given by Eq. (11.34) as

$$F(t) = \sum_{\text{inside } C+C_0} Res\left[\frac{ke^{st}}{s^2 + k^2}\right]. \tag{11.36}$$

Since the poles of Eq. (11.35) are $s = \pm ik$, the residues are computed by

$$Res|_{s=ik}\left[\frac{ke^{st}}{s^2 + k^2}\right] = \frac{e^{ikt}}{2i} \tag{11.37}$$

and

$$Res|_{s=-ik}\left[\frac{ke^{st}}{s^2 + k^2}\right] = -\frac{e^{-ikt}}{2i}. \tag{11.38}$$

Thus, Eq. (11.34) indicates

$$F(t) = \frac{e^{ikt} - e^{-ikt}}{2i} = \sin kt. \tag{11.39}$$

The next example is $f(s) = s^{-m}(m > 0)$. The Gamma function, Eq. (9.1), is defined as

$$\Gamma(n) = \int_0^\infty x^{n-1}e^{-x}dx, \quad \text{for } n > 0, \tag{11.40}$$

in which Eq. (9.4) shows

$$\Gamma(1) = 1. \tag{11.41}$$

From Eq. (9.3) we get

$$\Gamma(n+1) = n\Gamma(n). \tag{11.42}$$

Substituting $x = st$ into Eq. (11.40), since

$$\Gamma(n) = s^n \int_0^\infty e^{-st}t^{n-1}dt, \tag{11.43}$$

we have

$$L\{t^{n-1}\} = \frac{\Gamma(n)}{s^n}, \quad \text{for } n > 0. \tag{11.44}$$

Substituting $n - 1 = m$ into Eq. (11.44), we obtain

$$L^{-1}\left\{\frac{1}{s^{m+1}}\right\} = \frac{t^m}{\Gamma(m+1)}, \quad \text{for } m > -1. \tag{11.45}$$

If m is a positive integer, we have

$$\Gamma(m+1) = m! \tag{11.46}$$

From Eq. (9.7) we have

$$\Gamma(1/2) = \sqrt{\pi}, \quad \Gamma(-1/2) = \frac{\Gamma(1/2)}{-1/2} = -2\sqrt{\pi}. \tag{11.47}$$

From Eq. (11.45) we get

$$L^{-1}\left\{\frac{1}{s^{1/2}}\right\} = \frac{t^{-1/2}}{\Gamma(-1/2)} = \frac{1}{\sqrt{\pi t}}. \tag{11.48}$$

In the same way, we obtain

$$L^{-1}\{s^{-3/2}\} = \frac{t^{1/2}}{\Gamma(1/2)} = \frac{2\sqrt{t}}{\sqrt{\pi}}, \tag{11.49}$$

$$L^{-1}\{s^{1/2}\} = \frac{t^{-3/2}}{\Gamma(-1/2)} = -\frac{t^{-3/2}}{2\sqrt{\pi}}. \tag{11.50}$$

As the application of the Laplace transform, Faltung's theorem or convolution function exist. Since

$$L\{F_1(t)\} = f_1(s), \quad L\{F_2(t)\} = f_2(s) \tag{11.51}$$

are assumed, we get

$$L\left\{\int_0^t F_1(x)F_2(t-x)dx\right\} = L\left\{\int_0^t F_2(x)F_1(t-x)dx\right\} = f_1(s)f_2(s). \tag{11.52}$$

From Eq. (11.29), the inverse Laplace transform of $f_1(s)f_2(s)$ is given by

$$F_3(t) = \frac{1}{2\pi i}\int_{c-i\infty}^{c+i\infty} f_1(s)f_2(s)e^{st}ds. \tag{11.53}$$

Substituting

$$f_2(s) = \int_0^\infty e^{-sx}F_2(x)dx \tag{11.54}$$

into Eq. (11.53), we obtain

$$F_3(t) = \frac{1}{2\pi i}\int_0^\infty F_2(x)dx \int_{c-i\infty}^{c+i\infty} f_1(s)e^{s(t-x)}ds. \tag{11.55}$$

Since the inverse Laplace transform of $f_1(s)e^{-sx}$ is

$$F_1(t-x) = \frac{1}{2\pi i}\int_{c-i\infty}^{c+i\infty} f_1(s)e^{s(t-x)}ds, \tag{11.56}$$

Thus, Eq. (11.55) is

$$F_3(t) = \int_0^\infty F_2(x)F_1(t-x)dx. \tag{11.57}$$

Since we usually assume

$$F_1(t) = 0, \quad \text{for } t < 0, \tag{11.58}$$

Eq. (11.57) becomes

$$F_3(t) = \int_0^t F_2(x)F_1(t - x)dx \tag{11.59}$$

or

$$F_3(t) = \int_0^t F_2(t - x)F_1(x)dx. \tag{11.60}$$

Example 11.2. When the Laplace transform is

$$f(s) = \frac{k}{s^2 - k^2}, \tag{1}$$

obtain the inverse Laplace transform of $f(s)$.

Solution. Since the poles of $e^{st}f(s)$ are $s = \pm k$, the residues at the poles are

$$Res\big|_{s=k}\left[\frac{ke^{st}}{s^2 - k^2}\right] = \frac{e^{kt}}{2}, \tag{2}$$

$$Res\big|_{s=-k}\left[\frac{ke^{st}}{s^2 - k^2}\right] = -\frac{e^{-kt}}{2}. \tag{3}$$

Thus, we have

$$F(t) = \frac{e^{kt} - e^{-kt}}{2} = \sinh kt. \tag{4}$$

Problem 11.2. When the Laplace transform is

$$f(s) = \frac{s}{s^2 + k^2}, \tag{5}$$

obtain the inverse Laplace transform of $f(s)$.

11.3. Applications to Ordinary Differential Equations

The most important application of the Laplace transform in engineering is to solve ordinary differential equations. As an example, let us solve the following ordinary differential equation:

$$\frac{d^2Y}{dt^2} - \frac{dY}{dt} - 6Y = 2 \tag{11.61}$$

with initial conditions $Y(0) = 1$ and $Y'(0) = 0$. The Laplace tranform of Eq. (11.61) is given by

$$s^2 y - s - sy + 1 - 6y = \frac{2}{s}, \tag{11.62}$$

in which $y(s) = L\{Y(t)\}$. Solving Eq. (11.62) about $y(s)$, we have

$$y = \frac{s^2 - s + 2}{s(s-3)(s+2)} = \frac{A}{s} + \frac{B}{s-3} + \frac{C}{s+2}. \tag{11.63}$$

The constants A, B, and C are $-1/3$, $8/15$, and $4/5$, respectively. $y(s)$ is

$$y = -\frac{1}{3}\frac{1}{s} + \frac{8}{15}\frac{1}{s-3} + \frac{4}{5}\frac{1}{s+2}. \tag{11.64}$$

The inverse Laplace transform of Eq. (11.64) is

$$Y = L^{-1}\{y\} = -\frac{1}{3} + \frac{8}{15}e^{3t} + \frac{4}{5}e^{-2t}. \tag{11.65}$$

Example 11.3. Solve the ordinary differential equation

$$Y''(t) + 4Y(t) = \sin t, \tag{1}$$

in which "$'$" $= d/dt$. With the initial conditions $Y(0) = Y'(0) = 0$.

Solution. The Laplace transform of Eq. (1) is given by

$$s^2 y + 4y = \frac{1}{s^2 + 1}. \tag{2}$$

Thus, $y(s)$ is

$$y = \frac{1}{(s^2 + 4)(s^2 + 1)} = \left(-\frac{1}{s^2 + 4} + \frac{1}{s^2 + 1}\right)\frac{1}{3}. \tag{3}$$

The inverse Laplace transform of $y(s)$ is derived by Eqs. (11.35) and (11.39) as

$$Y = L^{-1}\{y\} = -\frac{1}{6}\sin 2t + \frac{1}{3}\sin t. \tag{4}$$

Problem 11.3. Solve the following differential equation:

$$Y''(t) + Y'(t) + 3Y(t) = \cos t \tag{5}$$

with $Y(0) = 1$ and $Y'(0) = 1$.

11.4. Applications to Partial Differential Equations

Apply the Laplace transform to partial differential equations. The equation to express water infiltration into the ground is given by

$$\frac{\partial V}{\partial t} = H^2 \frac{\partial^2 V}{\partial x^2}, \tag{11.66}$$

in which V = water content, H^2 = transmissivity, x = coordinate system in downward direction, and t = time. The transmissivity is defined by the product of the hydraulic conductivity and the thickness of an aquifer. The boundary conditions are given by

$$V = V_0 \,(\text{const}), \quad \text{for } x = 0 \quad \text{and} \quad t > 0, \tag{11.67}$$

$$V = 0, \quad \text{for } x > 0 \quad \text{and} \quad t = 0. \tag{11.68}$$

The Laplace transform of V about t is given by

$$v(x, s) = L\{V(x, t)\}. \tag{11.69}$$

Equation (11.68) indicates

$$L\left\{ \frac{\partial V}{\partial t} \right\} = sv. \tag{11.70}$$

The Laplace transform of $\partial^2 V / \partial x^2$ is

$$L\left\{ \frac{\partial^2 V}{\partial x^2} \right\} = \frac{d^2 v}{dx^2}. \tag{11.71}$$

Thus, Eq. (11.66) leads to the ordinary differential equation

$$sv = H^2 \frac{d^2 v}{dx^2}. \tag{11.72}$$

The solution of Eq. (11.72) is

$$v = Ae^{-\sqrt{sx}/H} + Be^{\sqrt{sx}/H}, \tag{11.73}$$

in which A and B are integral constants. In order that V is finite for $x \to \infty$, we have

$$B = 0. \tag{11.74}$$

Thus, v is considered to be

$$v = Ae^{-\sqrt{sx}/H}. \tag{11.75}$$

Since $V = V_0$ at $x = 0$, we have

$$A = V_0/s. \tag{11.76}$$

Thus, v is given by

$$v = \frac{V_0 e^{-\sqrt{sx}/H}}{s} = L\{V\}. \tag{11.77}$$

The inverse Laplace transform of v is

$$V = L^{-1}\left\{\frac{V_0 e^{-x\sqrt{s}/H}}{s}\right\}. \tag{11.78}$$

When m is a real number, the Maclaurin series e^m is

$$e^m = 1 + \frac{m}{1!} + \frac{m^2}{2!} + \frac{m^3}{3!} + \cdots . \tag{11.79}$$

Applying the Maclaurin series to $e^{-a\sqrt{s}}/s$, we obtain

$$\frac{e^{-a\sqrt{s}}}{s} = \frac{1}{s}\left[1 - as^{1/2} + \frac{a^2 s}{2!} - \frac{a^3 s^{3/2}}{3!} + \frac{a^4 s^2}{4!} - \frac{a^5 s^{5/2}}{5!} + \cdots\right], \tag{11.80}$$

in which $a = x/H$. The inverse Laplace transform on the left side of Eq. (11.80) is obtained by the summation of the inverse Laplace transform of each term on the right side of Eq. (11.80) and by using the following relationship:

$$L^{-1}\{s^{-n-1}\} = \frac{t^n}{\Gamma(n+1)}, \quad \text{for } t > 0. \tag{11.81}$$

Then, the inverse Laplace transform of Eq. (11.80) is

$$L^{-1}\left\{\frac{e^{-a\sqrt{s}}}{s}\right\} = 1 - \frac{2}{\sqrt{\pi}}\left[\frac{a}{2\sqrt{t}} - \frac{1}{3}\left(\frac{a}{2\sqrt{t}}\right)^3\right.$$

$$\left. + \frac{1}{2! \times 5}\left(\frac{a}{2\sqrt{t}}\right)^5 - \cdots\right]. \tag{11.82}$$

The Maclaurin series of the error function is

$$erf(w) = \frac{2}{\sqrt{\pi}}\int_0^w e^{-\theta^2}\,d\theta$$

$$= \frac{2}{\sqrt{\pi}}\left[w - \frac{w^3}{3 \times 1!} + \frac{w^5}{5 \times 2!} - \frac{w^7}{7 \times 3!} + \cdots\right]. \tag{11.83}$$

When

$$w = \frac{a}{2\sqrt{t}} \tag{11.84}$$

is substituted into Eq. (11.83), the comparison of Eqs. (11.82) and (11.83) gives

$$L^{-1}\left\{\frac{e^{-a\sqrt{s}}}{s}\right\} = 1 - erf\left(\frac{a}{2\sqrt{t}}\right). \tag{11.85}$$

Thus, the inverse Laplace transform of v is

$$V = V_0\left[1 - erf\left(\frac{x}{2H\sqrt{t}}\right)\right]. \tag{11.86}$$

This is equal to Eq. (10.72). As the application of the inverse Laplace transform to hydraulic engineering, Kikkawa *et al.* (1978) solved the transient phenomenon when water is abruptly and selectively withdrawn in a lake.

Example 11.4. Solve the following partial differential equation:

$$\frac{\partial U}{\partial t} = H^2\frac{\partial^2 U}{\partial x^2} \tag{1}$$

with the initial condition

$$U(x, 0) = 0, \quad \text{for } x > 0, \tag{2}$$

and the boundary condition

$$-K\frac{\partial U}{\partial x}(0, t) = p_0 \text{ (constant)}, \quad \text{for } t > 0. \tag{3}$$

Solution. The Laplace transform of U about t is

$$u(x, s) = L\{U(x, t)\}. \tag{4}$$

The Laplace transform of Eq. (1) is

$$su(x, s) = H^2 u_{xx}(x, s). \tag{5}$$

The Laplace transform of Eq. (3) is

$$-Ku_x(0, s) = \frac{p_0}{s}. \tag{6}$$

Since the solution of Eq. (5) is finite and Eq. (6) is considered, we obtain

$$u(x, s) = \frac{p_0}{Ks}\frac{H}{\sqrt{s}}e^{-x\sqrt{s}/H}. \tag{7}$$

When the Laplace transform

$$L\left\{\frac{1}{\sqrt{\pi t}}\exp\left(-\frac{k^2}{4t}\right)\right\} = \frac{1}{\sqrt{s}}e^{-k\sqrt{s}} \tag{8}$$

and Eq. (11.13) are used, the inverse Laplace transform of Eq. (7) is given by

$$U(x, t) = \frac{p_0}{K}\sqrt{\frac{k}{\pi}}\int_0^t e^{-x^2/(4k\tau)}\frac{d\tau}{\sqrt{\tau}} = \frac{p_0 x}{K\sqrt{\pi}}\int_{x/(2H\sqrt{t})}^\infty \frac{1}{\lambda^2}e^{-\lambda^2}d\lambda. \tag{9}$$

The partial integral of Eq. (9) shows

$$U(x, t) = \frac{p_0}{K}\left[2H\sqrt{\frac{t}{\pi}}e^{-x^2/(4H^2 t)} - x \ \mathrm{erf} \ c\left(\frac{x}{2H\sqrt{t}}\right)\right], \tag{10}$$

in which

$$erf\ c\ X = \frac{2}{\sqrt{\pi}} \int_X^\infty e^{-p^2} dp = 1 - erf\ X. \tag{11}$$

Problem 11.4. Show that the solution of Eq. (1) with the boundary conditions (12) and (13)

$$U(0,t) = F_0, \quad \text{for } 0 < t \le t_0, \tag{12}$$

$$U(0,t) = 0, \quad \text{for } t > t_0, \tag{13}$$

is

$$U(x,t) = F_0\ erf\ c\left(\frac{x}{2H\sqrt{t}}\right), \quad \text{for } 0 < t \le t_0, \tag{14}$$

$$U(x,t) = F_0\left[erf\left(\frac{x}{2H\sqrt{t-t_0}}\right) - erf\left(\frac{x}{2H\sqrt{t}}\right)\right], \quad \text{for } t > t_0, \tag{15}$$

instead of Eqs. (2) and (3).

11.5. Green's Functions

In application of the Laplace transform method in solving boundary value problems, there is another method which is known as the method of Green's functions. The Green's function is a solution for a system excited by a point-source of forcing function. When the Green's function is known, the solution for any arbitrary forcing function may be expressed in the form of a convolution-type integral. The usage of the Green's function is related to the Laplace transform. As our first example, we consider the following differential equation:

$$\frac{d^2Y}{dx^2} + k^2Y = -F(x) \tag{11.87}$$

with the boundary conditions

$$Y(0) = Y(a) = 0. \tag{11.88}$$

Equation (11.87) represents the vibrations of a string subject to a distributed forcing function. To solve Eq. (11.87) for the Green's function by the Laplace transform, we replace $F(x)$ with an impulse function that occurs at an arbitrary point ξ; that is, Eq. (11.87) becomes

$$\frac{d^2Y}{dx^2} + k^2Y = -\delta(x - \xi),\qquad(11.89)$$

in which $\delta(x - \xi) = $ a delta function. The delta function $\delta(x)$ is defined as

$$\delta(x) = \begin{cases} 0, & \text{for } x \neq 0, \\ \infty, & \text{for } x = 0, \end{cases}\qquad(11.90)$$

$$\int_{-\infty}^{\infty} \delta(x)dx = 1.\qquad(11.91)$$

Thus, for an arbitrary function $F(x)$ we get

$$\int_{-\infty}^{\infty} F(x)\delta(x)dx = F(0), \quad \int_{-\infty}^{\infty} F(x)\delta(x - \xi)dx = F(\xi).\quad(11.92)$$

In solving Eq. (11.89) by the Laplace transform, it should be noted that both $Y(0)$ and $Y'(0)$ are required. To apply this we have $Y'(0)$ as an unknown parameter which is evaluated by the boundary condition $Y(a) = 0$ to the inverted solution. To proceed, let

$$y = L\{Y(x)\},\qquad(11.93)$$

as hence

$$L\{Y''(x)\} = s^2y - Y'(0),\qquad(11.94)$$

in which "$'$" $= d/dx$. Since $Y(0) = 0$. The Laplace transform of Eq. (11.89) is

$$s^2y - Y'(0) + k^2y = -e^{-\xi s}\qquad(11.95)$$

and the transformed solution is

$$y = \frac{Y'(0)}{s^2 + k^2} - \frac{e^{-\xi s}}{s^2 + k^2}.\qquad(11.96)$$

The inverse Laplace transform is found to be

$$Y(x;\xi) = \frac{Y'(0)\sin kx}{k} - \frac{\sin k(x - \xi)U(x - \xi)}{k},\qquad(11.97)$$

in which $U(x - \xi)$ is the unit step function. It is defined by

$$U(x - \xi) = \begin{cases} 0, & \text{for } x < \xi, \\ 1/2, & \text{for } x = \xi, \\ 1, & \text{for } x > \xi. \end{cases} \tag{11.98}$$

The delta function is also defined by

$$\frac{dU(x - \xi)}{dx} = \delta(x - \xi). \tag{11.99}$$

We may eliminate the unknown $Y'(0)$ in the mannner described above. Applying $Y(a) = 0$ yields

$$0 = \frac{Y'(0) \sin ka}{k} - \frac{\sin k(a - \xi)}{k} \tag{11.100}$$

and therefore

$$Y'(0) = \frac{\sin k(a - \xi)}{\sin ka}. \tag{11.101}$$

Hence the described solution is

$$Y(x; \xi) = \frac{\sin kx \sin k(a - \xi)}{k \sin ka} - \frac{\sin k(x - \xi) U(x - \xi)}{k}. \tag{11.102}$$

The solution of Eq. (11.89), Eq. (11.102) is called the Green's function of Eq. (11.87) and physically represents the displacement of a point x when a unit load has been acted at a point ξ. It is more common to write it in two separate expressions as follows:

$$Y(x; \xi) = \frac{\sin k\xi \sin k(a - x)}{k \sin ka}, \quad \text{for } 0 \leq \xi \leq x, \tag{11.103}$$

$$Y(x; \xi) = \frac{\sin kx \sin k(a - \xi)}{k \sin ka}, \quad \text{for } x \leq \xi \leq a. \tag{11.104}$$

Now, if we are given an arbitrary forcing function as in Eq. (11.87), the answer is written as

$$Y(x) = \int_0^a F(\xi) Y(x; \xi) d\xi, \tag{11.105}$$

in which it should be noted that the integral must be divided into two parts, one ranging from 0 to x and the other from x to a. $Y(x;\xi)$ represents the displacement of the string at the point x due to a unit load applied at the point ξ. The term $F(\xi)Y(x;\xi)d\xi$ is simply the incremental displacement at x due to the actual load intensity at ξ. The integral over ξ from 0 to a sums the contributions of the entire distributed load to yield the total displacement at the point x. This is the same as a method of unit hydrograph when $F(\xi)$ and $Y(x;\xi)$ are rainfall and a unit hydrograph, respectively. Next, let us consider the application of the Green's function to the deflection of a cantilever beam as shown in Fig. 11.2. The deflection of the beam from the unloaded position is measured as positive downward and is denoted by $Y(x)$. The basic beam relations are

$$Y(x) = \text{deflection}$$

$$\theta(x) = \frac{dY}{dx} = \text{slope}$$

$$M(x) = EI\frac{d^2Y}{dx^2} = \text{bending moment}$$

$$S(x) = EI\frac{d^3Y}{dx^3} = \text{shear force}$$

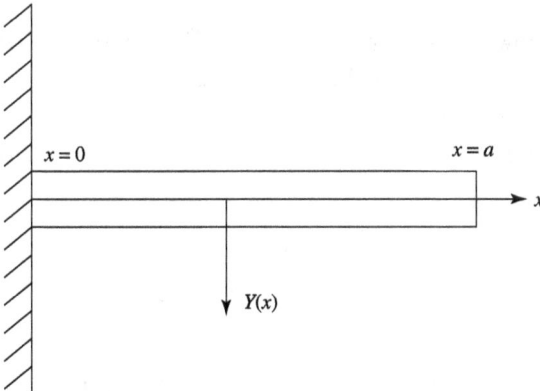

Figure 11.2. Deflection of a cantilever beam.

in which E = Young's modulus force and I = moment of inertia. When the applied load is denoted as $w(x)$ and is positive in the download direction, the beam equation for static deflection is

$$\frac{dS(x)}{dx} = -w(x) \qquad (11.106)$$

or in terms of the deflection, this equation becomes

$$EI\frac{d^4Y}{dx^4} = w(x). \qquad (11.107)$$

In order to determine the Green's function for the beam, we consider an impulse load applied at an arbitrary point ξ; that is, $w(x) = \delta(x - \xi)$. Hence the equation for the Green's function becomes

$$EI\frac{d^4Y}{dx^4} = \delta(x - \xi). \qquad (11.108)$$

The Laplace transform of Eq. (11.108) is

$$EI\left[s^4y - s^3Y(0) - s^2Y'(0) - sY''(0) - Y'''(0)\right] = e^{-s\xi}. \qquad (11.109)$$

The boundary conditions are that the deflection and slope must vanish at $x = 0$, while the moment and shear vanish at $x = a$; that

$$Y(0) = 0, \quad Y'(0) = 0, \quad Y''(a) = 0, \quad Y'''(a) = 0. \qquad (11.110)$$

We do not know values for $Y''(0)$ and $Y'''(0)$, which necessitate leaving them as parameters to be eliminated from the inverted solution by application of the conditions at $x = a$. Applying the two known conditions to Eq. (11.109) and solving for the transform solution gives

$$y = \frac{e^{-s\xi}}{s^4EI} + \frac{Y'''(0)}{s^4EI} + \frac{Y''(0)}{s^3EI}. \qquad (11.111)$$

Inverting this equation yields

$$Y(x;\xi) = \frac{(x - \xi)^3U(x - \xi)}{6EI} + \frac{Y'''(0)x^3}{6EI} + \frac{Y''(0)x^2}{2EI}, \qquad (11.112)$$

in which $U(x)$ indicates the unit step function.

We eliminate $Y''(0)$ and $Y'''(0)$ by applying the boundary conditions at $x = a$ which give the desired Green's function:

$$Y(x,\xi) = \frac{(x-\xi)^3 U(x-\xi)}{6EI} - \frac{x^3}{6EI} + \frac{\xi x^2}{2EI}. \qquad (11.113)$$

The familiar form of Eq. (11.113) is written as

$$Y(x,\xi) = -\frac{x^3}{6EI} + \frac{\xi x^2}{2EI}, \qquad \text{for } x \le \xi \le a, \qquad (11.114)$$

$$Y(x,\xi) = \frac{(x-\xi)^3}{6EI} - \frac{x^3}{6EI} + \frac{\xi x^2}{2EI}, \qquad \text{for } 0 \le \xi \le x. \qquad (11.115)$$

Using the Green's function, we write the solution to the deflection of a cantilever beam subject to an arbitrary loading distribution $w(x)$:

$$Y(x) = \int_0^a w(\xi) Y(x,\xi) d\xi. \qquad (11.116)$$

To illustrate this solution of Eq. (11.116) let us consider the determination of the static deflection of the cantilever beam subjected to a constant distributed load:

$$w(x) = w_0 U(x). \qquad (11.117)$$

Substituting Eq. (11.117) along with Eq. (11.113) into Eq. (11.116), we have

$$Y(x) = \frac{w_0}{6EI} \left[-x^3 \int_0^a d\xi + 3x^2 \int_0^a \xi d\xi + \int_0^x (x-\xi)^3 d\xi \right]. \qquad (11.118)$$

Performing the indicated integration and combining terms yields the solution

$$Y(x) = \frac{w_0 x^2}{24EI}(6a^2 - 4ax + x^2). \qquad (11.119)$$

The final example is the application of the Green's function to the linear partial differential equation of rainfall and runoff process. It is

$$\frac{\partial H}{\partial t} + v_0 \frac{\partial H}{\partial x} = r(x,t), \qquad (11.120)$$

in which H = water depth, v_0 = constant velocity, and $r(x,t)$ = rainfall. This is the continuity equation of the open channel flow and called the kinematic wave model. The boundary and initial conditions are

$$H = 0, \quad \text{at } x = 0, \tag{11.121}$$

$$H = H_0, \quad \text{at } t = 0, \tag{11.122}$$

in which H_0 = a constant. To obtain the Green's function of Eq. (11.120), we have

$$\frac{\partial H}{\partial t} + v_0 \frac{\partial H}{\partial x} = \delta(t - \tau)\delta(x - \xi). \tag{11.123}$$

The Laplace transform of H about x is

$$h(s) = L\{H\}, \tag{11.124}$$

$$L\left\{\frac{\partial H}{\partial x}\right\} = sh - H(x = 0) = sh. \tag{11.125}$$

The Laplace transform of Eq. (11.123) about x is

$$\frac{dh}{dt} + v_0 sh = \delta(t - \tau)e^{-\xi s}. \tag{11.126}$$

The solution of Eq. (11.126) is

$$h = e^{-v_0 st}\left[\int_0^t e^{v_0 st}e^{-s\xi}\delta(t - \tau)dt + c_1\right], \tag{11.127}$$

in which c_1 = an integral constant. Thus, $c_1 = H_0/s$ is obtained from the condition of $h = H_0/s$ at $t = 0$. Thus,

$$H(x,t;\xi,\tau) = L^{-1}\{h\} = L^{-1}\left\{e^{s(v_0\tau - v_0 t - \xi)} + H_0\frac{e^{-v_0 st}}{s}\right\}$$

$$= \delta[x - \xi - v_0(t - \tau)] + H_0 U(x - v_0 t). \tag{11.128}$$

This shows that the rainfall at $x = \xi$ and $t = \tau$ progresses downstream with the velocity v_0. The solution is

$$H = \int_0^\ell \int_0^T r(\xi,\tau)H(x,t;\xi,\tau)d\xi d\tau, \tag{11.129}$$

in which ℓ = the length of the watershed and T = time base of hydrograph. Equation (11.129) indicates that the integrated rainfall propagates with the velocity v_0.

Example 11.5. Solving the following equation:

$$\frac{d^2Y}{dx^2} + k^2Y = -\sin \omega x \tag{1}$$

with the boundary conditions

$$Y = 0, \quad \text{at } x = 0, \tag{2}$$

$$Y = 0, \quad \text{at } x = a. \tag{3}$$

Solution. The Green's function of Eq. (1) is given by Eq. (11.105). The solution is

$$Y(x) = \int_0^a F(\xi)Y(x,\xi)d\xi$$

$$= \int_0^x \sin \omega\xi \, \frac{\sin k\xi \sin k(a-x)}{k \sin ka} \, d\xi$$

$$+ \int_x^a \sin \omega\xi \, \frac{\sin kx \sin k(a-\xi)}{k \sin ka} \, d\xi$$

$$= \frac{\sin k(a-x)}{2k \sin ka} \left[\frac{\sin (k+\omega)x}{k+\omega} + \frac{\sin (\omega-k)x}{\omega-k} \right]$$

$$+ \frac{\sin kx}{2k \sin ka} \left\{ \frac{\sin \omega a - \sin [(\omega-k)x + ka]}{\omega-k} \right.$$

$$\left. + \frac{\sin \omega a - \sin [(\omega+k)x - ka]}{\omega+k} \right\}. \tag{4}$$

Problem 11.5. Solve the following equation using the Green's function:

$$\frac{d^2Y}{dx^2} + 3\frac{dY}{dx} + 2Y = \sin \omega x. \tag{5}$$

Chapter 12

Wave Equations

One of the most fundamental and common phenomena that occur in nature is the phenomenon of wave motion. When a stone is dropped into a pond, the surface of the water is disturbed and waves of displacement travel radially outward. When a bell is struck, sound waves propagate from the source of sound. We study wave equations such as the long wave equation and the Saint Venant equation. The method of solution is the separation of variables, series expansion, integral transform, and the method of characteristics. These are used for the computations of tidal waves, tsunamis, and flood propagations.

12.1. Classification of Partial Differential Equations of the Second-Order

When independent variables are x and y and a dependent variable is ϕ, the partial differential equations of the second-order are written as

$$A\frac{\partial^2 \phi}{\partial x^2} + B\frac{\partial^2 \phi}{\partial x \partial y} + C\frac{\partial^2 \phi}{\partial y^2} = f\left(x, y, \phi, \frac{\partial \phi}{\partial x}, \frac{\partial \phi}{\partial y}\right), \qquad (12.1)$$

in which A, B, and C are smooth functions of x, y, ϕ, $\partial \phi/\partial x$, or $\partial \phi/\partial y$. Then, Eq. (12.1) is classified to the following three types:

$B^2 - AC > 0 \cdots \cdots$ hyperbolic type (wave equation)

$B^2 - AC < 0 \cdots \cdots$ elliptic type (potential equation)

$B^2 - AC = 0 \cdots \cdots$ parabolic type (diffusion or heat

conduction equation).

As an example, the wave equation

$$\frac{\partial^2 \phi}{\partial x^2} = \frac{\partial^2 \phi}{\partial y^2} \tag{12.2}$$

corresponds to $A = 1$, $B = 0$, and $C = -1$ in Eq. (12.1). Since $B^2 - AC > 0$, Eq. (12.2) belongs to the hyperbolic type. The potential equation

$$\frac{\partial^2 \phi}{\partial x^2} + \frac{\partial^2 \phi}{\partial y^2} = 0 \tag{12.3}$$

has $A = 1$, $B = 0$, and $C = 1$ in Eq. (12.1). Since $B^2 - AC < 0$, Eq. (12.3) belongs to the elliptic type. The equation of diffusion or heat conduction

$$\frac{\partial \phi}{\partial x} = \frac{\partial^2 \phi}{\partial y^2} \tag{12.4}$$

has $A = B = 0$ and $C = 1$ in Eq. (12.1). Since $B^2 - AC = 0$, Eq. (12.4) belongs to the parabolic type. In the same way as ordinary differential equations, when we solve partial differential equations, boundary conditions are necessary. As the boundary conditions, when a value ϕ is given on the boundary, we call a Dirichlet's problem for partial differential equations. When derivatives $\partial\phi/\partial x$ or $\partial\phi/\partial y$ are given on the boundary, we call a Neumann's problem for partial differential equations.

Example 12.1. When h and u are water depth and velocity, respectively, the momentum equation and continuity equation to indicate long waves of infinitesimally small amplitude are given by

$$\frac{\partial u}{\partial t} + g\frac{\partial h}{\partial x} = 0, \tag{1}$$

$$\frac{\partial h}{\partial t} + \frac{\partial hu}{\partial x} = 0. \tag{2}$$

Show that Eqs. (1) and (2) belong to the hyperbolic type.

Solution. Differentiate Eq. (1) about t, we have

$$\frac{\partial^2 u}{\partial t^2} + g\frac{\partial^2 h}{\partial x \partial t} = 0. \tag{3}$$

Differentiating Eq. (2) about x, we get

$$\frac{\partial^2 h}{\partial t \partial x} + u \frac{\partial^2 h}{\partial x^2} + 2 \frac{\partial h}{\partial x} \frac{\partial u}{\partial x} + h \frac{\partial u}{\partial x^2} = 0. \tag{4}$$

Eliminating $\frac{\partial^2 h}{\partial x \partial t}$ from Eqs. (3) and (4), we obtain

$$\frac{1}{g} \frac{\partial^2 u}{\partial t^2} - u \frac{\partial^2 h}{\partial x^2} - 2 \frac{\partial h}{\partial x} \frac{\partial u}{\partial x} - h \frac{\partial^2 u}{\partial x^2} = 0. \tag{5}$$

From Eq. (1), we derive

$$\frac{\partial h}{\partial x} = -\frac{1}{g} \frac{\partial u}{\partial t}, \quad \frac{\partial^2 h}{\partial x^2} = -\frac{1}{g} \frac{\partial^2 u}{\partial t \partial x}. \tag{6}$$

Substituting Eq. (6) into Eq. (5), we have

$$\frac{1}{g} \frac{\partial^2 u}{\partial t^2} - h \frac{\partial^2 u}{\partial x^2} + \frac{u}{g} \frac{\partial^2 u}{\partial t \partial x} + \frac{2}{g} \frac{\partial u}{\partial t} \frac{\partial u}{\partial x} = 0. \tag{7}$$

Since $A = 1/g$, $B = u/g$, and $C = -h$ in Eq. (7), we get

$$B^2 - AC = \frac{u^2 + gh}{g^2} > 0. \tag{8}$$

Thus, the system of Eqs. (1) and (2) is the hyperbolic type.

Problem 12.1. Determine the type of the following partial differential equation:

$$\frac{\partial u}{\partial t} + u \frac{\partial u}{\partial x} = \frac{\partial^2 u}{\partial x^2} + \frac{\partial^2 u}{\partial y^2}. \tag{9}$$

If u is not a function of y, determine the type of

$$\frac{\partial u}{\partial t} + u \frac{\partial u}{\partial x} = \frac{\partial^2 u}{\partial x^2}. \tag{10}$$

Determine the type of the following partial differential equation:

$$\frac{\partial^2 u}{\partial x \partial y} = 0. \tag{11}$$

12.2. Fundamental Partial Differential Equations

There are various wave phenomena such as water waves, pressure waves (sound), etc. These are expressed by the wave equation. The three-dimensional linear waves are described by

$$\frac{\partial^2 \zeta}{\partial t^2} = C^2 \left(\frac{\partial^2 \zeta}{\partial x^2} + \frac{\partial^2 \zeta}{\partial y^2} + \frac{\partial^2 \zeta}{\partial z^2} \right) = C^2 \nabla^2 \zeta, \qquad (12.5)$$

in which C = celerity, x, y, and z = Cartesian coordinate system, and $\nabla^2 = \frac{\partial^2}{\partial x^2} + \frac{\partial^2}{\partial y^2} + \frac{\partial^2}{\partial z^2}$. When the coordinate system is three-dimensional, ζ indicates pressure. When the coordinate system is two-dimensional, ζ in Eq. (12.5) may be the water surface displacement and satisfies

$$\frac{\partial^2 \zeta}{\partial t^2} = C^2 \left(\frac{\partial^2 \zeta}{\partial x^2} + \frac{\partial^2 \zeta}{\partial y^2} \right), \qquad (12.6)$$

in which $C = \sqrt{gh_0}$, the celerity and h_0 = a constant water depth. Equation (12.6) describes the wave propagation in two dimensions with the celerity C. In this case the vertical acceleration is negligible in comparison with the horizontal acceleration. This indicates that the pressure is hydrostatic. This is the long wave approximation. Equation (12.6) indicates wave propagation when the water depth is much larger than the water surface deviation. Equations (12.5) or (12.6) are linear and called the hyperbolic partial differential equation. The methods of solutions are the separation of variables, series expansion, integral transforms, etc. We use the method of the separation of variables. For the steady state, Eq. (12.6) becomes the potential equation as explained in Chap. 13. The continuity equation for long waves of infinitesimally small amplitude is

$$\frac{\partial \zeta}{\partial t} + h \frac{\partial u}{\partial x} = 0, \qquad (12.7)$$

in which h = uniform water depth and u = velocity. The momentum equation for long waves of infinitesimally small amplitude is

$$\frac{1}{g} \frac{\partial u}{\partial t} + \frac{\partial \zeta}{\partial x} = 0. \qquad (12.8)$$

Eliminating u in Eqs. (12.7) and (12.8), we obtain

$$\frac{\partial^2 \zeta}{\partial t^2} = gh\frac{\sqrt{gh}^2 \zeta}{\partial x^2}. \tag{12.9}$$

The general solution of Eq. (12.9) is given by $\zeta = f_1(x - \sqrt{gh}t) + f_2(x + \sqrt{gh}t)$ when f_1 and f_2 are arbitrary functions. Eliminating ζ in Eqs. (12.7) and (12.8), we have

$$\frac{\partial^2 u}{\partial t^2} = gh\frac{\partial^2 u}{\partial x^2}. \tag{12.10}$$

As the application of linear long waves, let us consider the wave propagation problem when a submerged offshore breakwater is located in a uniform water depth in Fig. 12.1. Assume that the water depth without the submerged offshore breakwater is h, the water depth on the top and the width of the submerged offshore breakwater are h' and 2ℓ, respectively. When the length and period of incident long waves are L and T, respectively, an angular wave number and an angular frequency are $k = 2\pi/L$ and $\omega = 2\pi/T$, respectively. The incident long waves propagate with a constant wave period. The wave length L becomes L' and the angular wave number k becomes k' over the top of the submerged offshore breakwater. The celerity of the incident long waves is $C = \omega/k$ and the celerity of the transformed long waves on the top of the submerged offshore breakwater is $C' = \omega/k'$. The celerities are, respectively, $C = \sqrt{gh}$ and $C' = \sqrt{gh'}$. Thus, the water surface displacement in the regions I, II, and III is obtained by Eq. (12.9). The water depth is h in the regions I and III and h' in the region II are substituted in Eq. (12.9). The solutions of Eq. (12.9) are

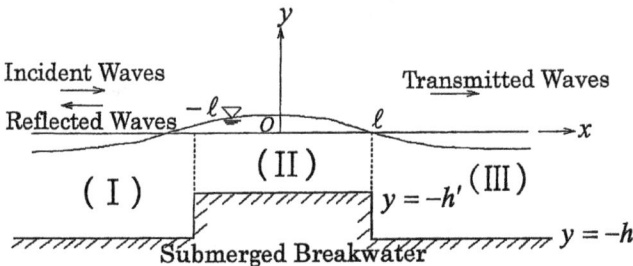

Figure 12.1. Wave propagation over a submerged offshore breakwater.

$$\zeta_I = Ae^{i(kx+\omega t)} + Be^{-i(kx-\omega t)}, \tag{12.11}$$

$$\zeta_{II} = De^{i(k'x+\omega t)} + Ee^{-i(k'x-\omega t)}, \tag{12.12}$$

$$\zeta_{III} = Fe^{-i(kx-\omega t)}, \tag{12.13}$$

in which A, B, D, E, and F are integral constants. The terms of $e^{i(kx+\omega t)}$ and $e^{-i(kx-\omega t)}$ show the propagating waves in $-x$ and x directions, respectively. One is a reflected wave and the other are either incident or transmitted waves. Substituting $h_0 = h$ into Eq. (12.7), we have

$$u = -\frac{1}{h}\int \frac{\partial \zeta}{\partial t} dx. \tag{12.14}$$

The velocities in the regions *I*, *II*, and *III* are from Eqs. (12.11), (12.12), (12.13) and (12.14),

$$u_I = -\frac{\omega}{hk}[Ae^{i(kx+\omega t)} - Be^{-i(kx-\omega t)}], \tag{12.15}$$

$$u_{II} = -\frac{\omega}{h'k'}[De^{i(k'x+\omega t)} - Ee^{-i(k'x-\omega t)}], \tag{12.16}$$

$$u_{III} = \frac{\omega}{hk}Fe^{-i(kx-\omega t)}. \tag{12.17}$$

The integrals of the velocities over a wave period indicate that integral constants are zero. Note that the water depths in Eqs. (12.15) and (12.17), and Eq. (12.16) are h and h', respectively. A, B, and F are the complex amplitudes of the reflected, incident, and transmitted waves, respectively. The integral constants D and E are the complex amplitudes of the reflected and transmitted waves over the submerged offshare breakwater, respectively. Four unknowns A, D, E, and F are determined from the continuity conditions of the water surface level and velocity at $x = -\ell$ and ℓ as follows:

$$Ae^{i(kx+\omega t)} + Be^{-i(kx-\omega t)} = De^{i(k'x+\omega t)} + Ee^{-i(k'x-\omega t)}, \tag{12.18}$$

$$De^{i(k'x+\omega t)} + Ee^{-i(k'x-\omega t)} = Fe^{-i(kx-\omega t)}, \tag{12.19}$$

$$Ae^{i(kx+\omega t)} - Be^{-i(kx-\omega t)} = \lambda[De^{i(k'x+\omega t)} - Ee^{-i(k'x-\omega t)}], \tag{12.20}$$

$$-[De^{i(k'x+\omega t)} - Ee^{-i(k'x-\omega t)}] = Fe^{-i(kx-\omega t)}/\lambda, \tag{12.21}$$

in which $\lambda = kh/(k'h') = \sqrt{h/h'}$ because $L = TC$ and $L' = TC'$. The ratio of the complex amplitude F/B is

$$\frac{F}{B} = \frac{4e^{2ik\ell}}{\left(1 - \frac{1}{\lambda}\right)^2 e^{-2ik'\ell} + \left(1 + \frac{1}{\lambda}\right)^2 e^{2ik'\ell}}$$

$$= \frac{4e^{2ik\ell}}{2\left(1 + \frac{1}{\lambda^2}\right)\cos 2k'\ell + \frac{4i}{\lambda}\sin 2k'\ell}. \tag{12.22}$$

The absolute value of F/B is computed

$$\left|\frac{F}{B}\right| = \frac{4}{\sqrt{4\left(1 + \frac{1}{\lambda^2}\right)^2 \cos^2 2k'\ell + \frac{16}{\lambda^2}\sin^2 2k'\ell}}. \tag{12.23}$$

Equation (12.23) is the transmission coefficient of the long waves due to the submerged offshore breakwater.

Example 12.2. Obtain the reflection coefficient $|A/B|$ due to the submerged offshore breakwater.

Solution. From Eqs. (12.18), (12.19), (12.20), and (12.21), we have

$$\frac{A}{B}e^{-ik\ell} = \frac{2\left(1 - \frac{1}{\lambda}\right)e^{ik\ell - 2ik'\ell} + 2\left(1 + \frac{1}{\lambda}\right)e^{ik\ell + 2ik'\ell}}{\left(1 - \frac{1}{\lambda}\right)^2 e^{-2ik'\ell} + \left(1 + \frac{1}{\lambda}\right)^2 e^{2ik'\ell}} - e^{ik\ell}. \tag{1}$$

Thus, we get

$$\frac{A}{B} = \frac{2\left(1 - \frac{1}{\lambda}\right)e^{2ik\ell - 2ik'\ell} + 2\left(1 + \frac{1}{\lambda}\right)e^{2ik\ell + 2ik'\ell}}{\left(1 - \frac{1}{\lambda}\right)^2 e^{-2ik'\ell} + \left(1 + \frac{1}{\lambda}\right)^2 e^{2ik'\ell}} - e^{2ik\ell}$$

$$= \frac{2e^{2ik\ell}\left(1 - \frac{1}{\lambda^2}\right)\cos 2k'\ell}{\left(1 - \frac{1}{\lambda}\right)^2 e^{-2ik'\ell} + \left(1 + \frac{1}{\lambda}\right)^2 e^{2ik'\ell}}. \tag{2}$$

The magnitude of A/B is the reflection coefficient as

$$\left|\frac{A}{B}\right| = \frac{2\left|\left(1 - \frac{1}{\lambda^2}\right)\cos 2k'\ell\right|}{\sqrt{4\left(1 + \frac{1}{\lambda}\right)^2 \cos^2 2k'\ell + \frac{16}{\lambda^2}\sin^2 2k'\ell}}. \tag{3}$$

Equation (3) is the reflection coefficient of the long waves due to the submerged offshore breakwater.

Problem 12.2. Show that $\zeta = \sin k(x - Ct)$ satisfies Eq. (12.9) for $C = \sqrt{gh}$ and obtain the velocity u. Note that h is the uniform water depth.

12.3. One-Dimensional Wave Equation

When the one-dimensional long wave of infinitesimally small amplitude propagates in an open channel of which length is ℓ in the x direction, the governing equation is given by

$$\frac{\partial^2 \zeta}{\partial t^2} = C^2 \frac{\partial^2 \zeta}{\partial x^2}. \qquad (12.24)$$

To solve Eq. (12.24), we assume

$$\zeta = e^{i\omega t} v(x). \qquad (12.25)$$

Substituting Eq. (12.25) into Eq. (12.24), we get

$$\frac{d^2 v}{dx^2} + \left(\frac{\omega}{C}\right)^2 v = 0. \qquad (12.26)$$

The solution of Eq. (12.26) is given by

$$v = A \sin \frac{\omega}{C}x + B \cos \frac{\omega}{C}x. \qquad (12.27)$$

When there are two vertical walls at $x = 0$ and ℓ as the boundaries, the wave forms an anti-node at $x = 0$ and ℓ. At the anti-node, the water surface slope is zero. Thus, we get

$$\frac{\partial \zeta}{\partial x} = 0. \qquad (12.28)$$

This denotes from Eq. (12.25) as

$$\frac{dv}{dx} = 0, \quad \text{at } x = 0, \ \ell. \qquad (12.29)$$

From Eq. (12.27) this denotes

$$A = 0 \qquad (12.30)$$

and

$$0 = -B\frac{\omega}{C} \sin \frac{\omega}{C}\ell. \qquad (12.31)$$

If $B = 0$, the solution is trivial. Eq. (12.31) indicates

$$\sin \frac{\omega}{C}\ell = 0. \tag{12.32}$$

From Eq. (12.32), we obtain

$$\frac{\omega}{C}\ell = k\pi, \quad \text{for } k = 0, 1, 2, \ldots. \tag{12.33}$$

Since ω is the function of k, ω is defined ω_k. Then,

$$\omega_k = \frac{k\pi C}{\ell}, \quad \text{for } k = 0, 1, 2, \ldots. \tag{12.34}$$

If v is defined v_k for each k, v_k is expressed by

$$v_k = B_k \cos \frac{k\pi x}{\ell}, \tag{12.35}$$

in which B_k is an integral constant. Thus, the solution is

$$\zeta_k = e^{i\omega_k t} B_k \cos \frac{k\pi x}{\ell}. \tag{12.36}$$

Since the real and imaginary parts of Eq. (12.36) satisfy Eq. (12.24), Eq. (12.25) leads to

$$\zeta_k = \left(C_k \cos \frac{k\pi Ct}{\ell} + D_k \sin \frac{k\pi Ct}{\ell} \right) \cos \frac{k\pi x}{\ell}, \tag{12.37}$$

in which C_k and D_k are integral constants. The two integral constants are derived from the initial condition. Since Eq. (12.24) is linear, the general solution is given by the sum of ζ_k about k. That is,

$$\zeta = \sum_{k=0}^{\infty} \zeta_k = \sum_{k=0}^{\infty} \left(C_k \cos \frac{k\pi Ct}{\ell} + D_k \sin \frac{k\pi Ct}{\ell} \right) \cos \frac{k\pi x}{\ell}. \tag{12.38}$$

As the initial condition, we assume

$$\zeta = \zeta_0(x), \quad \text{for } t = 0, \tag{12.39}$$

or

$$\frac{\partial \zeta}{\partial t} = v_0(x), \quad \text{for } t = 0. \tag{12.40}$$

The displacement of the water surface profile is generally caused by winds or low pressure systems. Equation (12.39) is written by

$$\zeta_0(x) = \sum_{k=0}^{\infty} C_k \cos \frac{k\pi x}{\ell}. \tag{12.41}$$

Equation (12.41) shows that the initial displacement of the water surface is expressed by the Fourier cosine series. Multiplying $\cos \frac{r\pi x}{\ell}$ to both sides of Eq. (12.41) and integrating it from 0 to ℓ, we obtain

$$\int_0^{\ell} \zeta_0(x) \cos \frac{r\pi x}{\ell} dx = \sum_{k=1}^{\infty} C_k \int_0^{\ell} \cos \frac{k\pi x}{\ell} \cos \frac{r\pi x}{\ell} dx,$$

$$\text{for } r = 0, 1, 2, 3, \ldots. \tag{12.42}$$

From the orthogonality property of the cosine functions, we have

$$\int_0^{\ell} \cos \frac{k\pi x}{\ell} \cos \frac{r\pi x}{\ell} dx = 0, \quad \text{if } k \neq r, \tag{12.43}$$

$$\int_0^{\ell} \cos \frac{k\pi x}{\ell} \cos \frac{r\pi x}{\ell} dx = \frac{\ell}{2}, \quad \text{if } k = r, \tag{12.44}$$

$$\int_0^{\ell} \zeta_0(x) = C_0 \ell, \quad \text{for } r = 0. \tag{12.45}$$

Thus, Eq. (12.42) shows

$$\int_0^{\ell} \zeta_0(x) \cos \frac{r\pi x}{\ell} dx = \frac{\ell}{2} C_r, \quad \text{for } r = 1, 2, 3, \ldots. \tag{12.46}$$

Equations (12.45) and (12.46) determine C_k for $k = 0, 1, 2, \ldots$. In the same way, applying Eq. (12.40) into Eq. (12.38), we have

$$v_0(x) = \sum_{k=1}^{\infty} D_k \frac{k\pi C}{\ell} \cos \frac{k\pi x}{\ell}. \tag{12.47}$$

Multiplying $\cos \frac{k\pi x}{\ell}$ to both sides of Eq. (12.47) and integrating it from $x = 0$ to ℓ, we gain

$$D_r = \frac{2}{r\pi C} \int_0^{\ell} v_0(x) \cos \frac{r\pi x}{\ell} dx, \quad \text{for } r = 1, 2, 3, \ldots. \tag{12.48}$$

The substitution of Eqs. (12.45), (12.46), and (12.48) into Eq. (12.38) gives the solution of Eq. (12.24).

Example 12.3. When the initial condition of the water surface profile, $\zeta_0(x) = 0.1x - 0.05\ell$ in Eq. (12.39) is used, obtain C_k.

Solution. Substituting $\zeta_0(x)$ into Eq. (12.46), we have

$$
\begin{aligned}
C_k &= \frac{2}{\ell} \int_0^\ell \zeta_0(x) \cos \frac{k\pi x}{\ell} dx \\
&= \frac{2}{\ell} \int_0^\ell (0.1x - 0.05\ell) \cos \frac{k\pi x}{\ell} dx \\
&= \frac{0.2}{k\pi} \left[\frac{\ell}{k\pi} \cos \frac{k\pi x}{\ell} \right]_0^\ell = \frac{0.2\ell}{k^2\pi^2} [(-1)^k - 1],
\end{aligned}
$$

$$\text{for } k = 1, 2, 3, \dots. \tag{1}$$

Substituting $\zeta_0(x)$ into Eq. (12.45), we get

$$C_0 = \frac{1}{\ell} \int_0^\ell (0.1x - 0.05\ell) dx = 0. \tag{2}$$

Problem 12.3. Consider the small coplanar oscillation of a uniform flexible chain hanging from a support under the action of gravity as shown in Fig. 12.2. We consider small deviations y from the equilibrium position. The variable x is measured from the free end of the chain. The governing equation is given by

$$x \frac{\partial^2 y}{\partial x^2} + \frac{\partial y}{\partial x} = \frac{1}{g} \frac{\partial^2 y}{\partial t^2}. \tag{3}$$

The boundary condition is

$$y = 0 \quad \text{at } x = \ell. \tag{4}$$

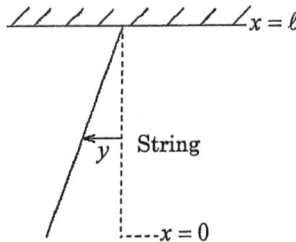

Figure 12.2.   Vibration of a string.

12.4. Water Surface Oscillations of Rectangular Lake

Consider the water surface oscillation of a rectangular lake as shown in Fig. 12.3. The governing equation is

$$\frac{\partial^2 \zeta}{\partial t^2} = C^2 \left(\frac{\partial^2 \zeta}{\partial x^2} + \frac{\partial^2 \zeta}{\partial y^2} \right), \tag{12.49}$$

in which ζ = displacement of water surface. The length and the width of the rectangular lake are a and b, respectively. The boundary conditions are the formation of complete reflection at $x = 0$, a, $y = 0$, and b. They are

$$\frac{\partial \zeta}{\partial x} = 0 \quad \text{at } x = 0, \ a, \tag{12.50}$$

$$\frac{\partial \zeta}{\partial y} = 0 \quad \text{at } y = 0, \ b. \tag{12.51}$$

Since the ground surface slope at a shoreline of lakes is not vertical, the wave reflection is not complete. The ground surface slope at the shoreline is usually very small and the incident waves break on the shore. As the initial condition, the initial deviation of the water surface profile is given by

$$\zeta = \zeta_0(x, y). \tag{12.52}$$

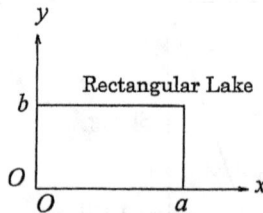

Figure 12.3.   Oscillation of a rectangular lake.

Assume that the solution of Eq. (12.49) forms

$$\zeta = v(x, y)e^{i\omega t}. \tag{12.53}$$

Substituting Eq. (12.53) into Eq. (12.45), we get

$$\frac{\partial^2 v}{\partial x^2} + \frac{\partial^2 v}{\partial y^2} + k^2 v = 0, \tag{12.54}$$

in which

$$k = \frac{\omega}{C}. \tag{12.55}$$

Assume the following separation of variables:

$$v(x, y) = F_1(x) \cdot F_2(y). \tag{12.56}$$

The substitution of Eq. (12.56) into Eq. (12.54) leads to

$$F_2 \frac{\partial^2 F_1}{\partial x^2} + F_1 \frac{\partial^2 F_2}{\partial y^2} + k^2 F_1 F_2 = 0. \tag{12.57}$$

Equation (12.57) can be written as

$$\frac{1}{F_1} \frac{d^2 F_1}{dx^2} = -\frac{1}{F_2} \frac{d^2 F_2}{dy^2} - k^2. \tag{12.58}$$

The left and right sides of Eq. (12.58) are the function of x and y, respectively. When x or y vary, in order that Eq. (12.58) is meaningful, Eq. (12.58) must be constant. Define Eq. (12.58) $-m^2$. The reason of $-m^2$ is the convenience of the following procedure. Then, Eq. (12.58) is defined as

$$\frac{1}{F_1} \frac{d^2 F_1}{dx^2} = -\frac{1}{F_2} \frac{d^2 F_2}{dy^2} - k^2 = -m^2. \tag{12.59}$$

Equation (12.59) produces the following two ordinary differential equations:

$$\frac{d^2 F_1}{dx^2} + m^2 F_1 = 0, \tag{12.60}$$

$$\frac{d^2 F_2}{dy^2} + q^2 F_2 = 0, \tag{12.61}$$

in which

$$q^2 = k^2 - m^2. \tag{12.62}$$

The solution of Eq. (12.60) with Eq. (12.50) gives

$$F_1 = A_1 \cos mx, \tag{12.63}$$

in which $A_1 =$ an integral constant. Equation (12.50) determines m as follows:

$$ma = n\pi, \quad \text{for } n = 0, 1, 2, \ldots. \tag{12.64}$$

The solution of Eq. (12.61) with Eq. (12.51) shows

$$F_2 = A_2 \cos qy, \tag{12.65}$$

in which $A_2 =$ an integral constant. Equation (12.51) restricts q as follows:

$$qb = r\pi, \quad \text{for } r = 0, 1, 2, \ldots. \tag{12.66}$$

Equation (12.62) derives the angular frequency as

$$k^2 = m^2 + q^2 = \pi^2 \left(\frac{n^2}{a^2} + \frac{r^2}{b^2} \right). \tag{12.67}$$

The angular frequency k is the function of n and r. The angular wave number ω is

$$\omega = \frac{2\pi}{T} = \frac{L}{T} \cdot \frac{2\pi}{L} = C \cdot k. \tag{12.68}$$

Since ω is the function of n and r from Eqs. (12.67) and (12.68), we write $\omega = \omega_{nr}$ as follows:

$$\omega_{nr} = C\pi \sqrt{\frac{n^2}{a^2} + \frac{r^2}{b^2}}, \quad \text{for } n = 0, 1, 2, \ldots, \quad r = 0, 1, 2, \ldots. \tag{12.69}$$

Substituting Eqs. (12.63) and (12.65) into Eq. (12.56) and defining $v = v_{nr}$, we obtain

$$v_{nr} = A_1 A_2 \cos \frac{n\pi x}{a} \cos \frac{r\pi y}{b}, \tag{12.70}$$

Defining $A_{nr} = A_1 A_2$, we obtain

$$v_{nr} = A_{nr} \cos \frac{n\pi x}{a} \cos \frac{r\pi y}{b}, \tag{12.71}$$

in which A_1 and A_2 are n-th and r-th amplitudes, respectively. When the initial displacement of the water surface is given by ζ_0 and $v_0(x)$ is zero at $t = 0$, we substitute Eq. (12.52) into Eq. (12.53) and sum Eq. (12.71) for all n and r. Since the function to satisfy the initial condition is the cosine function, we have

$$\zeta = \sum_{n=0}^{\infty} \sum_{r=0}^{\infty} A_{nr} \cos \omega_{nr} t \cos \frac{n\pi x}{a} \cos \frac{r\pi y}{b}. \tag{12.72}$$

Since the initial condition is $\zeta = \zeta_0(x, y)$ at $t = 0$, we get

$$\zeta_0(x, y) = \sum_{n=0}^{\infty} \sum_{r=0}^{\infty} A_{nr} \cos \frac{n\pi x}{a} \cos \frac{r\pi y}{b}. \tag{12.73}$$

The Fourier cosine series of ζ_0 about x in Eq. (12.73) shows

$$\zeta_0(x, y) = \sum_{n=0}^{\infty} A_n(y) \cos \frac{n\pi x}{a}. \tag{12.74}$$

Equation (12.74) derives

$$A_n(y) = \frac{2}{a} \int_0^a \zeta_0(x, y) \cos \frac{n\pi x}{a} dx. \tag{12.75}$$

From Eq. (12.73), we get

$$A_n(y) = \sum_{r=0}^{\infty} A_{nr} \cos \frac{r\pi y}{b}, \tag{12.76}$$

in which

$$A_{nr} = \frac{2}{b} \int_0^b A_n(y) \cos \frac{r\pi y}{b} dy. \tag{12.77}$$

Substituting Eq. (12.75) into Eq. (12.77), we have

$$A_{nr} = \frac{4}{ab} \int_0^b dy \int_0^a \zeta_0(x, y) \cos \frac{n\pi x}{a} \cos \frac{r\pi y}{b} dx,$$

$$\text{for } n = 0, 1, 2, \ldots, \quad r = 0, 1, 2, \ldots. \tag{12.78}$$

Equation (12.78) determines the coefficients A_{nr}. The substitution of Eq. (12.78) into Eq. (12.72) shows the solution.

Example 12.4. When $\zeta_0(x, y) = 0.1xy$, obtain A_{nr} in Eq. (12.78).

Solution. From Eq. (12.78), we get

$$
A_{nr} = \frac{4}{ab} \int_0^b y\, dy \int_0^a 0.1x \cos \frac{n\pi x}{a} \cos \frac{r\pi y}{b}\, dx
$$

$$
= \frac{0.4}{ab} \left(\frac{b}{r\pi} \right)^2 \left[\cos \frac{r\pi y}{b} + \frac{r\pi y}{b} \sin \frac{r\pi y}{b} \right]_0^b
$$

$$
\times \left(\frac{a}{n\pi} \right)^2 \left[\cos \frac{n\pi x}{a} + \frac{n\pi x}{a} \sin \frac{n\pi x}{a} \right]_0^a
$$

$$
= \frac{0.4ab}{r^2 n^2 \pi^4} [(-1)^r - 1][(-1)^n - 1], \quad \text{for } r \neq 0,\ n \neq 0, \quad (1)
$$

$$
A_{00} = \frac{4}{ab} \int_0^b 0.1y\, dy \int_0^a x\, dx = \frac{0.1ab}{4}, \quad \text{for } r = n = 0, \quad (2)
$$

$$
A_{0r} = \frac{2}{ab} \int_0^a 0.1x\, dx \int_0^b y \cos \frac{r\pi y}{b}\, dy
$$

$$
= \frac{0.1ab}{r^2 \pi^2} [(-1)^r - 1], \quad \text{for } r \neq 0,\ n = 0, \quad (3)
$$

$$
A_{n0} = \frac{2}{ab} \int_0^a 0.1x \cos \frac{n\pi x}{a}\, dx \int_0^b y\, dy
$$

$$
= \frac{0.1ab}{n^2 \pi^2} [(-1)^n - 1], \quad \text{for } r = 0,\ n \neq 0. \quad (4)
$$

Problem 12.4. The equation of oscillation of a membrane as shown in Fig. 12.4 is given by

$$
\frac{\partial^2 u}{\partial x^2} + \frac{\partial^2 u}{\partial y^2} = \frac{1}{C^2} \frac{\partial^2 u}{\partial t^2}, \quad (5)
$$

in which $C = \sqrt{T/m}$, celerity, T = tension, and m = specific weight of gravity of the membrane. The membrane is fixed at $x = 0,\ a$, $y = 0$, and b. When the initial conditions are $u(t = 0, x, y) = u_0(x, y)$ and $du(t = 0, x, y)/dt = 0$, solve Eq. (5).

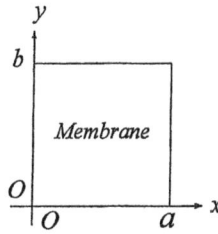

Figure 12.4.   Vibration of membrane.

12.5.   Water Surface Oscillations of Circular Lake

The governing equation of the water surface oscillation in a circular lake is given by

$$\nabla^2 \zeta = \frac{1}{C^2} \frac{\partial^2 \zeta}{\partial t^2}, \tag{12.79}$$

in which ζ = displacement of water surface, t = time and ∇^2 = Laplacian. When the boundary shape is circular, the coordinate system is suitable to be cylindrical. Then, the left side of Eq. (12.79) is transformed to

$$\nabla^2 \zeta = \frac{1}{r} \frac{\partial}{\partial r} \left(r \frac{\partial \zeta}{\partial r} \right) + \frac{1}{r^2} \frac{\partial^2 \zeta}{\partial \theta^2}. \tag{12.80}$$

Assume that the radius of the circular lake is a and waves reflect completely at the perimeter of the circular lake. When the wave is symmetrical about the center of the circular lake and the origin $r = 0$ is the center of the circular lake, the governing equation is independent of θ. Then, we get

$$\frac{1}{r} \frac{\partial}{\partial r} \left(r \frac{\partial \zeta}{\partial r} \right) = \frac{1}{C^2} \frac{\partial^2 \zeta}{\partial t^2}. \tag{12.81}$$

As the initial water surface profile, we have

$$\zeta = \zeta_0(r). \tag{12.82}$$

The water surface profile is assumed to be

$$\zeta = v(r)e^{i\omega t}. \tag{12.83}$$

Substituting Eq. (12.83) into Eq. (12.81), we have

$$\frac{1}{r}\frac{d}{dr}\left(r\frac{dv}{dr}\right) + \left(\frac{\omega}{C}\right)^2 v = 0. \tag{12.84}$$

This equation is rewritten as

$$\frac{d^2v}{dr^2} + \frac{1}{r}\frac{dv}{dr} + k^2 v = 0, \tag{12.85}$$

in which

$$k = \frac{\omega}{C}. \tag{12.86}$$

Equation (12.85) is the Bessel differential equation and its general solution is given by

$$v(r) = AJ_0(kr) + BY_0(kr), \tag{12.87}$$

in which A and B are arbitrary constants. $J_0(kr)$ and $Y_0(kr)$ are the first and second kind Bessel functions of order zero. Since the water surface displacement is finite at $r = 0$, we have

$$B = 0. \tag{12.88}$$

Since the wave reflects at the perimeter of the circular lake $r = a$, we have

$$\left.\frac{dv}{dr}\right|_{r=a} = AkJ_0'(kr) = 0. \tag{12.89}$$

From the property of the Bessel functions Eq. (9.61), we obtain

$$J_0'(ka) = -J_1(ka) = 0, \tag{12.90}$$

in which "$'$" $= d/dr$ and $J_1(ka)$ is the first kind Bessel function of order one. The roots of Eq. (12.90) are shown as

$$ka = 1.2197, \quad 2.2330, \quad 3.2383, \ldots. \tag{12.91}$$

Thus, since $\omega = kC$ and from Eq. (12.86), we get

$$\omega_1 = \frac{1.2197C}{a}, \tag{12.92}$$

$$\omega_2 = \frac{2.2330C}{a}, \tag{12.93}$$

$$\omega_3 = \frac{3.2383C}{a}, \tag{12.94}$$

$$\vdots$$

When the water surface displacement $\zeta_0(r)$ at $t = 0$ is at rest, the water surface displacement is

$$\zeta = \sum_{n=1}^{\infty} A_n J_0\left(\frac{\omega_n}{C}r\right) \cos \omega_n t. \tag{12.95}$$

Thus, the water surface displacement at $t = 0$ is

$$\zeta_0(r) = \sum_{n=1}^{\infty} A_n J_0\left(\frac{\omega_n}{C}r\right). \tag{12.96}$$

The coefficient A_n is determined as

$$A_n = \frac{2}{a^2 J_1^2(\omega_n a/C)} \int_0^a r J_0\left(\frac{\omega_n r}{C}\right) \zeta_0(r) dr. \tag{12.97}$$

Substituting Eq. (12.97) into Eq. (12.95), we have the solution ζ.

Example 12.5. Obtain the solution of the water surface displacement when the solution of Eqs. (12.79) and (12.80) is dependent on θ.

Solution. When s is a positive integer, we assume the following solution:

$$\zeta = f(r)e^{i(\omega t + s\theta)}. \tag{1}$$

Substituting Eq. (1) into Eqs. (12.79) and (12.80), we have

$$f'' + \frac{1}{r}f' + \left(k^2 - \frac{s^2}{r^2}\right)f = 0, \tag{2}$$

in which $k = \omega/C$. The solution of Eq. (2) is the Bessel function of order s. Since the solution of this phenomenon is finite at $r = 0$, the solution is written

$$\zeta_s = A_s J_s(kr)e^{(\omega t + s\theta)}. \tag{3}$$

The eigen value k is determined by the reflection condition at $r = a$ as follows:

$$J_s'(k_n a) = 0, \quad \text{for } n = 1, 2, 3, \ldots, \tag{4}$$

in which $J_s(k_n a)$ is the first kind Bessel function of order s. From Eq. (4) we obtain k_n or ω_n. The solution for $s = 0$ indicates the solution of Eq. (12.85).

$$\zeta = \sum_n \sum_s (A_s \cos s\theta + B_s \sin s\theta) J_s\left(\frac{\omega_{sn} r}{C}\right) e^{i\omega t}. \tag{5}$$

This is the water surface displacement.

Problem 12.5. When the initial water surface in the circular lake is horizontal and $\partial \zeta / \partial t = u_0(r)$, obtain A_n.

12.6. Saint Venant Equations

When the open channel flow is unsteady, the phenomenon is expressed by Saint Venant equations. This consists of the continuity equation

$$\frac{\partial h}{\partial t} + \frac{\partial hv}{\partial x} = 0 \tag{12.98}$$

and the momentum equation

$$\frac{1}{g}\frac{\partial v}{\partial t} + \frac{v}{g}\frac{\partial v}{\partial x} + \frac{\partial h}{\partial x} = S_0 - S_f, \tag{12.99}$$

in which h = water depth, v = velocity, g = the gravitational acceleration, S_0 = bottom slope, S_f = energy slope, t = time, and x = coordinate system in the flow direction. By using Manning's formula in the wide open channel, the energy slope is given by

$$S_f = \frac{n^2 v^2}{h^{4/3}}, \tag{12.100}$$

in which n = Manning's roughness coefficient. When Eqs. (12.98) and (12.99) are linearized as the long waves of infinitesimally small amplitude, Eq. (12.98) is approximated by

$$\frac{\partial \zeta}{\partial t} + h_0 \frac{\partial v}{\partial x} = 0. \tag{12.101}$$

Equation (12.99) is also approximated by

$$\frac{1}{g}\frac{\partial v}{\partial t} + \frac{\partial \zeta}{\partial x} = 0, \tag{12.102}$$

in which ζ = water surface displacement from the still water level and h_0 = the uniform water depth. Combining Eqs. (12.101) and (12.102), we have

$$\frac{\partial^2 \zeta}{\partial t^2} = gh_0\frac{\partial^2 \zeta}{\partial x^2}. \tag{12.103}$$

Equation (12.103) corresponds to Eq. (12.24). Since Eq. (12.103) is the wave equation, Eqs. (12.98) and (12.99) are also the wave equation. According to the mathematical theory of partial differential equations, there exists two characteristics on the $x - t$ plane in the wave equation. Let us derive the equations of characteristic of Eqs. (12.98) and (12.99). Since independent variables are t and x, the complete differentiations of dependent variables h and v are written by

$$\frac{\partial h}{\partial t}dt + \frac{\partial h}{\partial x}dx = dh, \tag{12.104}$$

$$\frac{\partial v}{\partial t}dt + \frac{\partial v}{\partial x}dx = dv. \tag{12.105}$$

From Eqs. (12.98), (12.99), (12.104), and (12.105) we get

$$\begin{bmatrix} 1 & v & 0 & h \\ 0 & 1 & 1/g & v/g \\ dt & dx & 0 & 0 \\ 0 & 0 & dt & dx \end{bmatrix}\begin{bmatrix} \partial h/\partial t \\ \partial h/\partial x \\ \partial v/\partial t \\ \partial v/\partial x \end{bmatrix} = \begin{bmatrix} 0 \\ S_0 - S_f \\ dh \\ dv \end{bmatrix}. \tag{12.106}$$

The derivatives $\partial h/\partial t$, $\partial h/\partial x$, $\partial v/\partial t$, and $\partial v/\partial x$ are calculated by multiplying an inverse matrix of the matrix on the left side of Eq. (12.106). The condition that the determinant of the matrix on the left side of Eq. (12.106) becomes 0 is obtained by

$$\frac{dx}{dt} = v \pm \sqrt{gh}. \tag{12.107}$$

The condition constructs a quadratic equation of dx/dt and shows two equations of characteristic on the $x - t$ plane. When $\partial h/\partial t$, $\partial h/\partial x$, $\partial v/\partial t$, or $\partial v/\partial x$ are computed, the determinant of the matrix on the left side of Eq. (12.106) forms the denominator of the solutions. Since on the characteristic the denominator becomes zero, the numerator must be zero. Otherwise, the solutions of $\partial h/\partial t$, $\partial h/\partial x$, $\partial v/\partial t$, or $\partial v/\partial x$ are infinite on the characteristic. This is a contradiction. Thus, the numerators of the solutions $\partial h/\partial t$, $\partial h/\partial x$, $\partial v/\partial t$, or $\partial v/\partial x$ are zero on the characteristic. The condition that the numerator becomes zero on the characteristic is given by

$$\frac{dv}{dt} \pm \sqrt{\frac{g}{h}} \cdot \frac{dh}{dt} = g(S_0 - S_f). \tag{12.108}$$

The $\pm$ in Eq. (12.107) corresponds to $\pm$ in Eq. (12.108). Equation (12.108) is formed on the characteristic Eq. (12.107). The characteristics from points R and L reach a point P as shown in Fig. 12.5. On the characteristics LP and RP, the plus and minus signs term in Eqs. (107) and (108) are formed, respectively. Unknowns at the point P are h, v, x, and t and the number of equations are four. The four nonlinear equations about h, v, x, and t are solvable by an iteration method (Chap. 15). The method of the solution is called the method of characteristics, because the characteristics are used. This method is used for the computation of floods and water hammer analysis in hydraulics, and shock wave in aerodynamics. Differentiating Eq. (12.98) about x and Eq. (12.99) about t and eliminating

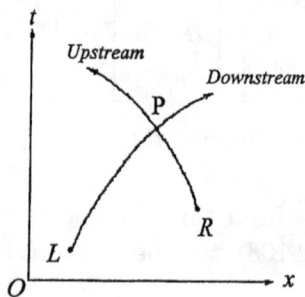

Figure 12.5.   Method of characteristics.

$\partial^2 h/\partial x \partial t$, we have

$$\frac{1}{g}\frac{\partial^2 v}{\partial t^2} + \frac{1}{g}\left(\frac{\partial v}{\partial t}\frac{\partial v}{\partial x} + v\frac{\partial^2 v}{\partial x \partial t}\right) = v\frac{\partial^2 h}{\partial x^2} + 2\frac{\partial h}{\partial x}\frac{\partial v}{\partial x} + h\frac{\partial^2 v}{\partial x^2}. \quad (12.109)$$

From Eq. (12.99), we obtain

$$\frac{\partial h}{\partial x} = S_0 - S_f - \frac{1}{g}\frac{\partial v}{\partial t} - \frac{v}{g}\frac{\partial v}{\partial x}. \quad (12.110)$$

Differentiating Eq. (12.110) about x, we obtain

$$\frac{\partial^2 h}{\partial x^2} = -\frac{\partial S_f}{\partial x} - \frac{1}{g}\frac{\partial^2 v}{\partial t \partial x} - \frac{1}{g}\left(\frac{\partial v}{\partial x}\right)^2 - \frac{v}{g}\frac{\partial^2 v}{\partial x^2}. \quad (12.111)$$

Substituting Eq. (12.111) into Eq. (12.108), we gain

$$\frac{1}{g}\frac{\partial^2 v}{\partial t^2} + \frac{1}{g}\left(\frac{\partial v}{\partial t}\frac{\partial v}{\partial x} + v\frac{\partial^2 v}{\partial x \partial t}\right)$$

$$= -v\left[\frac{\partial S_f}{\partial x} + \frac{1}{g}\frac{\partial^2 v}{\partial t \partial x} + \frac{1}{g}\left(\frac{\partial v}{\partial x}\right)^2 + \frac{v}{g}\frac{\partial^2 v}{\partial x^2}\right]$$

$$+ 2\frac{\partial v}{\partial x}\left(S_0 - S_f - \frac{1}{g}\frac{\partial v}{\partial t} - \frac{v}{g}\frac{\partial v}{\partial x}\right) + h\frac{\partial^2 v}{\partial x^2}. \quad (12.112)$$

Rewriting Eq. (12.112), we have

$$\frac{1}{g}\frac{\partial^2 v}{\partial t^2} + \frac{2v}{g}\frac{\partial^2 v}{\partial x \partial t} + \left(\frac{v^2}{g} - h\right)\frac{\partial^2 v}{\partial x^2} + \frac{1}{g}\frac{\partial v}{\partial t}\frac{\partial v}{\partial x} + v\frac{\partial S_f}{\partial x}$$

$$+ \frac{v}{g}\left(\frac{\partial v}{\partial x}\right)^2 - 2S_0\frac{\partial v}{\partial x} + 2S_f\frac{\partial v}{\partial x} + \frac{2}{g}\frac{\partial v}{\partial x}\frac{\partial v}{\partial t} + \frac{2v}{g}\left(\frac{\partial v}{\partial x}\right)^2 = 0. \quad (12.113)$$

Using the classification of the partial differential equations of the second-order in Sec. 12.1, we get

$$\frac{1}{4}(B^2 - AC) = \left(\frac{v}{g}\right)^2 - \frac{1}{g}\left(\frac{v^2}{g} - h\right) = \frac{h}{g} > 0, \quad (12.114)$$

in which $A = 1/g$, $B = 2v/g$, and $C = v^2/g - h$. This is the wave equation. Mizumura (1984) applied the method of characteristics to compute pressure distribution among the concrete blocks of the offshore breakwater due to wave action.

Example 12.6. Solving $\frac{\partial v}{\partial x}$ in Eq. (12.106) and substituting Eq. (12.107) into $\frac{\partial v}{\partial x}$, derive Eq. (12.108).

Solution. Define the following determinant in Eq. (12.106):

$$
\begin{vmatrix}
1 & v & 0 & h \\
0 & 1 & 1/g & v/g \\
dt & dx & 0 & 0 \\
0 & 0 & dt & dx
\end{vmatrix} = D. \tag{1}
$$

We obtain $\frac{\partial v}{\partial x}$ as

$$
\frac{\partial v}{\partial x} = \frac{1}{D}
\begin{vmatrix}
1 & v & 0 & 0 \\
0 & 1 & 1/g & S_0 - S_f \\
dt & dx & 0 & dh \\
0 & 0 & dt & dv
\end{vmatrix}, \tag{2}
$$

$$
\text{numerator of } \frac{\partial v}{\partial x} = dxdt(S_0 - S_f) - \frac{dvdx}{g} - dhdt
$$

$$
+ dt\left[\frac{vdv}{g} - v(S_0 - S_f)dt\right]. \tag{3}
$$

Assuming that the numerator of $\frac{\partial v}{\partial x}$ is zero and dividing it by dt^2, we have

$$
\left[(S_0 - S_f) - \frac{1}{g}\frac{dv}{dt}\right]\frac{dx}{dt} - \frac{dh}{dt} + \frac{v}{g}\frac{dv}{dt} - v(S_0 - S_f) = 0. \tag{4}
$$

Substituting Eq. (12.107) into Eq. (4), we get

$$
\pm\sqrt{gh}(S_0 - S_f) - \frac{1}{g}\frac{dv}{dt}(v \pm \sqrt{gh}) - \frac{dh}{dt} + \frac{v}{g}\frac{dv}{dt} = 0. \tag{5}
$$

Multiplying $\sqrt{g/h}$ to Eq. (5), we obtain

$$
\pm g(S_0 - S_f) \mp \frac{dv}{dt} = \sqrt{\frac{g}{h}}\frac{dh}{dt}. \tag{6}
$$

Rewriting Eq.(6), we get Eq. (12.108) as

$$\frac{dv}{dt} \pm \sqrt{\frac{g}{h}\frac{dh}{dt}} = g(S_0 - S_f). \tag{7}$$

Problem 12.6. When we assume that

$$S_f = S_0 - \frac{\partial h}{\partial x} \quad \text{and} \quad S_0 \gg \left|\frac{\partial h}{\partial x}\right|, \tag{8}$$

in Eq. (12.99) and using

$$S_f^{1/2} \cong S_0^{1/2}\left(1 - \frac{1}{2S_0}\frac{\partial h}{\partial x}\right) \tag{9}$$

and Manning's formula

$$v = \frac{\sqrt{S_f}}{n}h^{2/3}, \tag{10}$$

we get

$$\frac{\partial h}{\partial t} + \frac{\partial}{\partial x}\left(\frac{\sqrt{S_f}}{n}h^{5/3}\right) = 0. \tag{11}$$

Show that Eq. (11) is the diffusion type.

12.7. Long Wave Transform on Slope

There is a steep slope as shown in Fig. 12.6. The surface of the steep slope is very rough such that long waves dissipate and do not break on it. As the incident waves we assume the linear long waves

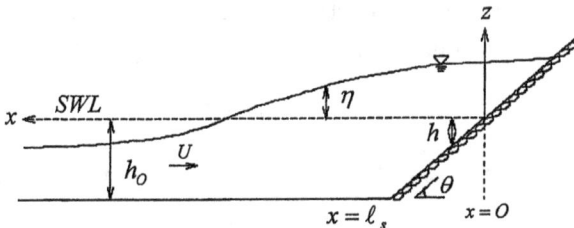

Figure 12.6. Wave transform on the slope.

of infinitesimally small amplitude. The continuity and momentum equations for $x \geq \ell_s$ are (Madsen and White, 1976)

$$\frac{\partial \eta}{\partial t} + \frac{\partial h_0 U}{\partial x} = 0, \tag{12.115}$$

$$\frac{\partial U}{\partial t} + g \frac{\partial \eta}{\partial x} = 0, \tag{12.116}$$

in which t = time, x = the coordinate system from the shoreline, η = deviation of water surface profile, U = horizontal velocity, and g = the gravitational acceleration. For $x < \ell_s$ the water depth h decreases linearly. Thus, we have

$$h = x \tan \theta, \tag{12.117}$$

in which θ = an angle of the slope. For the region $0 \leq x \leq \ell_s$ the continuity and momentum equations are, respectively,

$$\frac{\partial \eta}{\partial t} + \frac{\partial h U}{\partial x} = 0, \tag{12.118}$$

$$\frac{\partial U}{\partial t} + g \frac{\partial \eta}{\partial x} + \frac{\tau_b}{\rho h} = 0, \tag{12.119}$$

in which τ_b = shear stress on the bottom and ρ = water density. The shear stress τ_b is expressed by

$$\tau_b = \frac{1}{2} \rho f_w |U| U, \tag{12.120}$$

in which f_w = friction coefficient due to waves. Equation (12.120) is linearized as follows:

$$\frac{1}{2} \frac{f_w |U| U}{h} \cong f_s \omega U, \tag{12.121}$$

in which f_s = linearized friction coefficient and ω = an angular frequency. Substituting Eq. (12.121) into Eq. (12.119), we get

$$\frac{\partial U}{\partial t} + g \frac{\partial \eta}{\partial x} + f_s \omega U = 0. \tag{12.122}$$

Assume U and η in Eqs. (12.115) and (12.116), in the following forms:

$$U = \Re[u(x) e^{i\omega t}], \tag{12.123}$$

$$\eta = \Re[\zeta(x) e^{i\omega t}], \tag{12.124}$$

in which $\Re$ = real part, $\omega = 2\pi/T$, and T = wave period. After substituting Eqs. (12.123) and (12.124) into Eqs. (12.115), (12.116), (12.118), and (12.122) and eliminating $u(x)$ from Eqs. (12.115) and (12.116), we obtain

$$\frac{d^2\zeta}{dx^2} + \frac{\omega^2}{gh_0}\zeta = 0, \quad \text{for } x > \ell_s, \tag{12.125}$$

$u(x)$ is given by

$$u = -\frac{g}{i\omega}\frac{d\zeta}{dx}, \quad \text{for } x > \ell_s. \tag{12.126}$$

Substituting Eqs. (12.123) and (12.124) into Eqs. (12.118) and (12.122), we have

$$\frac{d}{dx}\left(h\frac{d\zeta}{dx}\right) + \frac{\omega^2(1-if_s)}{g}\zeta = 0, \quad \text{for } 0 < x \le \ell_s, \tag{12.127}$$

$$u = -\frac{g}{i\omega(1-if_s)}\frac{d\zeta}{dx}, \quad \text{for } 0 < x \le \ell_s. \tag{12.128}$$

The solution of Eq. (12.125) is written as

$$\zeta = a_i e^{ik_0 x} + a_r e^{-ik_0 x}, \tag{12.129}$$

in which $k_0 = 2\pi/L$, L = the wave length, a_i = a complex amplitude of incident waves, and a_r = a complex amplitude of reflected waves. Equation (12.127) is rewritten by Eq. (12.117) as follows:

$$\frac{d^2\zeta}{dx^2} + \frac{1}{x}\frac{d\zeta}{dx} + \frac{\omega^2(1-if_s)}{xg\tan\theta}\zeta = 0. \tag{12.130}$$

The general Bessel differential equation of a complex variable z (Eq. (9.64)) is

$$\frac{d^2\zeta}{dz^2} + \frac{1-2\alpha}{z}\frac{d\zeta}{dz} + \left[(\beta\gamma z^{\gamma-1})^2 + \frac{\alpha^2 - \nu^2\gamma^2}{z^2}\right]\zeta = 0. \tag{12.131}$$

The solution of Eq. (12.131) is given by

$$\zeta = z^\alpha Z_\nu(\beta z^\gamma), \tag{12.132}$$

in which

$$\alpha = 0, \quad \beta = 2\omega\sqrt{\frac{1 - if_s}{g\tan\theta}}, \quad \gamma = \frac{1}{2}, \quad \nu = 0. \qquad (12.133)$$

The usage of Eqs. (12.133) shows that Eq. (12.130) is the same as Eq. (12.131). Thus, the general solution of Eq. (12.130) is given by

$$\zeta = AJ_0\left[2\omega\sqrt{\frac{1 - if_s}{g\tan\theta}}x^{1/2}\right], \qquad (12.134)$$

in which J_0 is the first kind Bessel function of order zero. The first kind Bessel function was selected based on the finiteness at $x = 0$. From the continuity conditions of ζ and u, the complex amplitude of the reflected wave a_r and the complex amplitude on the slope A are obtained. Using the Bessel functions, Mizumura and Yamamoto (1991) analytically computed the reflection and transmission coefficients due to the offshore breakwater of concrete blocks instead of the rough and steep slope.

Example 12.7. Obtain a_r and A in Eqs. (12.129) and (12.134), respectively.

Solution. From the continuity conditions of ζ and u at $x = \ell_s$

$$a_i e^{ik_0\ell_s} + a_r e^{-ik_0\ell_s} = AJ_0\left[2\omega\sqrt{\frac{1 - if_s}{g\tan\theta}}\ell_s^{1/2}\right], \qquad (1)$$

$$ik_0(a_i e^{ik_0\ell_s} - a_r e^{-ik_0\ell_s}) = -iA\sqrt{\frac{g}{(1 - if_s)\ell_s\tan\theta}}J_1$$

$$\times J_1\left[2\omega\sqrt{\frac{1 - if_s}{g\tan\theta}}\ell_s^{1/2}\right]. \qquad (2)$$

Using $h_0 = \ell_s\tan\theta$ and $\omega = \sqrt{gh_0}\cdot k_0$, we get

$$\omega\sqrt{\frac{1 - if_s}{g\tan\theta}}\ell_s^{1/2} = k_0\ell_s\sqrt{1 - if_s}. \qquad (3)$$

Therefore, we have

$$\frac{a_r}{a_i} = \frac{J_0(2k_0\ell_s\sqrt{1-if_s}) - \frac{i}{\sqrt{1-if_s}}J_1(2k_0\ell_s\sqrt{1-if_s})}{J_0(2k_0\ell_s\sqrt{1-if_s}) + \frac{i}{\sqrt{1-if_s}}J_1(2k_0\ell_s\sqrt{1-if_s})}e^{2ik_0\ell_s}, \quad (4)$$

$$\frac{A}{2a_i} = \frac{e^{ik_0\ell_s}}{J_0(2k_0\ell_s\sqrt{1-if_s}) + \frac{i}{\sqrt{1-if_s}}J_1(2k_0\ell_s\sqrt{1-if_s})}. \quad (5)$$

Problem 12.7. Derive the velocity u from Eqs. (12.128) and (12.134).

12.8. Side Outflow from Steep Open Channel Flow

Mizumura (2003b, 2005b) studied the side outflow on a flood plain from a side outlet of the main channel. The side outlet as a model simulates a failure of a river bank in a prototype. When the Froude number of the main channel is larger than 1, let us consider the flood from the breach OO' in Fig. 12.7. When the flow reaches the point O, the negative shock wave is produced from point O to refract the flow direction. The negative and positive shock waves in an open channel flow correspond to the decrease and increase of the water

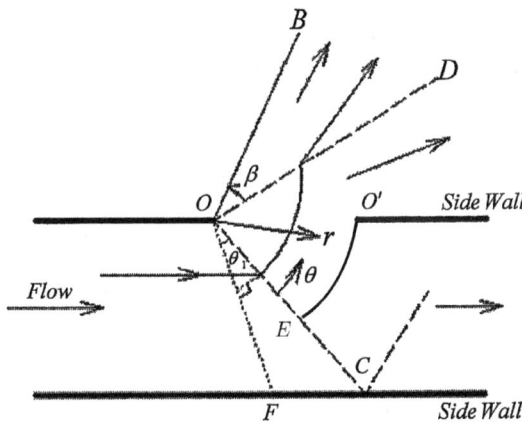

Figure 12.7. Side outflow from steep open channel flow.

level, respectively. The main flow intends to flow to the OB direction in Fig. 12.7. The negative shock waves are produced from point O between the lines OC and OD. The analysis is conducted in the domain of the angle COD. The Saint Venant equations in the cylindrical coordinate system (r, θ) are written as

$$v_r \frac{\partial v_r}{\partial r} + \frac{v_\theta}{r} \frac{\partial v_r}{\partial \theta} - \frac{v_\theta^2}{r} = -g \frac{\partial h}{\partial r} + g(S_{or} - S_{fr}), \quad (12.135)$$

$$v_r \frac{\partial v_\theta}{\partial r} + \frac{v_\theta}{r} \frac{\partial v_\theta}{\partial \theta} + \frac{v_r v_\theta}{r} = -\frac{g}{r} \frac{\partial h}{\partial \theta} + g(S_{o\theta} - S_{f\theta}), \quad (12.136)$$

in which v_r = velocity in the r direction, v_θ = velocity in the θ direction, h = water depth, g = the gravitational acceleration, S_{or} = bottom slope in the r direction, S_{fr} = energy slope in the r direction, $S_{o\theta}$ = bottom slope in the θ direction, and $S_{f\theta}$ = energy slope in the θ direction. The velocities v_r and v_θ are depth-averaged velocities. The depth-averaged continuity equation is

$$\frac{\partial(rv_r h)}{\partial r} + \frac{(v_\theta h)}{\partial \theta} = 0. \quad (12.137)$$

The Saint Venant equations, Eqs. (12.135), (12.136), and (12.137) are the wave equations. Since the negative shock waves propagate from the point O in the r direction, the velocity and water depth in the r direction are conserved. This indicates that the velocity and water depth are the functions of θ only. Since the transient region is very small in comparison with the length of the main channel flow, the bottom slope S_{or} and energy slope S_{fr} are neglected in this computation. Equation (12.135) is rewritten as

$$\frac{dv_r}{d\theta} - v_\theta = 0. \quad (12.138)$$

In the same way, since $S_{o\theta} = S_{f\theta} = 0$, Eq. (12.136) is

$$v_\theta \frac{dv_\theta}{d\theta} + v_r v_\theta = -g \frac{dh}{d\theta}. \quad (12.139)$$

Equation (12.137) is also written as

$$v_r h + \frac{d(v_\theta h)}{d\theta} = 0. \quad (12.140)$$

Substituting Eq. (12.138) into Eq. (12.139), we have

$$\frac{1}{2}\frac{d}{d\theta}(v_\theta^2 + v_r^2) = -g\frac{dh}{d\theta}. \qquad (12.141)$$

Multiplying v_θ/g to Eq. (12.141) and using Eq. (12.140), we get

$$\frac{v_\theta}{g}\left(v_\theta\frac{dv_\theta}{d\theta} + v_r\frac{dv_r}{d\theta}\right) = h\left(v_r + \frac{dv_\theta}{d\theta}\right). \qquad (12.142)$$

Using Eq. (12.138) and substituting Eq. (12.141) into Eq. (12.142), we obtain

$$\left(v_r + \frac{dv_\theta}{d\theta}\right)\left(\frac{v_\theta^2}{g} - h\right) = 0. \qquad (12.143)$$

Equation (12.143) has two solutions. One is

$$v_\theta = \sqrt{gh} \qquad (12.144)$$

and the other is

$$v_r + \frac{dv_\theta}{d\theta} = 0. \qquad (12.145)$$

Combining Eqs. (12.139) and (12.145), we derive

$$\frac{dh}{d\theta} = 0. \qquad (12.146)$$

Equation (12.146) indicates that the water depth h is constant in the θ direction. This is a contradiction. Integrating Eq. (12.141), we get

$$\frac{v_\theta^2 + v_r^2}{2g} + h = E, \qquad (12.147)$$

in which E is the specific energy. Combining Eqs. (12.138), (12.144), and (12.147), we obtain

$$\left(\frac{dv_r}{d\theta}\right)^2 = \frac{2gE}{3} - \frac{v_r^2}{3}, \qquad (12.148)$$

Integrating Eq. (12.148) in the condition of $v_r = 0$ at $\theta = -\theta_1$, we have

$$v_r = \sqrt{2gE}\sin\left(\frac{\theta + \theta_1}{\sqrt{3}}\right), \qquad (12.149)$$

in which $\theta_1 = $ an integral constant.

From Eq. (12.138) we obtain

$$v_\theta = \sqrt{\frac{2gE}{3}} \cos\left(\frac{\theta + \theta_1}{\sqrt{3}}\right). \tag{12.150}$$

The condition $v_\theta = 0$ shows the line OB in Fig. 12.7. Equations (12.147), (12.149), and (12.150) indicate

$$\frac{h}{E} = \frac{2}{3} \cos^2\left(\frac{\theta + \theta_1}{\sqrt{3}}\right). \tag{12.151}$$

Since the Froude number is $F_r = \sqrt{v_r^2 + v_\theta^2}/\sqrt{gh}$, Eq. (12.147) becomes

$$\frac{h}{E} = \frac{2}{2 + F_r^2}. \tag{12.152}$$

Equations (12.151) and (12.152) indicate

$$\theta + \theta_1 = \sqrt{3} \cos^{-1}\left(\sqrt{\frac{3}{2 + F_r^2}}\right). \tag{12.153}$$

Since $\theta = 0$ on the line OC and $F_r = F_{ro}$ is formed in Eq. (12.153), we have

$$\theta_1 = \sqrt{3} \cos^{-1}\left(\sqrt{\frac{3}{2 + F_{ro}^2}}\right), \tag{12.154}$$

in which F_{ro} is the Froude number in the main channel flow. Next, consider the discharge ratio Q_2/Q in which Q_2 is the side outflow from the breach OO' of the main channel flow and Q is the discharge of the main channel flow. The main flow starts to curve at the line OC by the negative shock waves from the point O and it flows out of the levee through the breach OO' to the flood plain. Thus, the streamline at the point E in Fig. 12.7 divides the main flow and the side outflow. Assuming that the flow is uniform in the width direction, from the geometrical consideration we obtain the following discharge ratio:

$$\frac{Q_2}{Q} = \frac{OE}{OC}. \tag{12.155}$$

The equation of the streamline in the cylindrical coordinate system is given by

$$\frac{1}{r}\frac{dr}{d\theta} = \frac{v_r}{v_\theta}. \tag{12.156}$$

The integration of Eq. (12.156) with the usage of Eqs. (12.149) and (12.150) and the condition of $\theta = \theta_1 + \beta_0$ at $r = OO'$ induces

$$OO' = r_o \left(\cos \frac{\theta_1 + \beta_0}{\sqrt{3}} \right)^{-3}, \tag{12.157}$$

in which $r = r_0$ at $\theta + \theta_1$. The angle β_0 is determined by the Froude number of the main channel flow as

$$\beta_0 = \sin^{-1} \frac{1}{F_{r0}}. \tag{12.158}$$

Since the angle θ starts from the line OC, θ corresponds to zero for the line OE in Fig. 12.7. Thus, we have

$$OE = r_o \left(\cos \frac{\theta_1}{\sqrt{3}} \right)^{-3}. \tag{12.159}$$

From the geometrical consideration in Fig. 12.7, we have

$$OC = \frac{B}{\sin \beta_0}. \tag{12.160}$$

Therefore, the discharge ratio is

$$\frac{Q_2}{Q} = \frac{\ell}{BF_{r0}} \left(\cos \frac{\theta_1 + \beta_0}{\sqrt{3}} \right)^3 \bigg/ \left(\cos \frac{\theta_1}{\sqrt{3}} \right)^3. \tag{12.161}$$

The function

$$\cos \frac{\theta_1 + \beta_0}{\sqrt{3}} \bigg/ \cos \frac{\theta_1}{\sqrt{3}} \tag{12.162}$$

is almost constant 0.65 for $1 < F_{r0} < 3$. Since $\ell/B = 0.847$ in the experiments, we get the discharge ratio as

$$\frac{Q_2}{Q} = \frac{\ell}{B} \frac{0.275}{F_{r0}} = \frac{0.23}{F_{r0}}. \tag{12.163}$$

Figure 12.8 compares Eq. (12.163) with the experimental data. Although the fluid is assumed to be perfect, the experimental result is in good agreement with the theory.

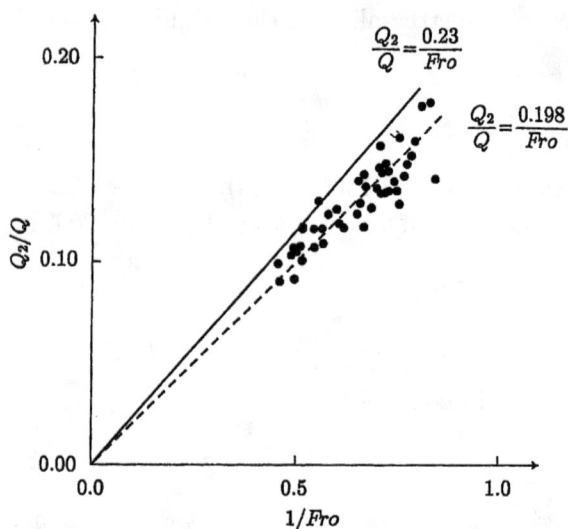

Figure 12.8.   Comparison of discharge ratio.

Example 12.8. Derive Eq. (12.143).

Solution. Equation (12.140) is given by

$$v_r h + v_\theta \frac{dh}{d\theta} + h \frac{dv_\theta}{d\theta} = 0. \tag{1}$$

Multiplying v_θ/g to Eq. (12.141), we get

$$\frac{v_\theta}{g} \left(v_\theta \frac{dv_\theta}{d\theta} + v_r \frac{dv_r}{d\theta} \right) = -v_\theta \frac{dh}{d\theta}. \tag{2}$$

Equation (1) leads to

$$v_\theta \frac{dh}{d\theta} = -v_r h - h \frac{dv_\theta}{d\theta}. \tag{3}$$

Substituting Eq. (3) into Eq. (2), we have

$$\frac{v_\theta}{g} \left(v_\theta \frac{dv_\theta}{d\theta} + v_r \frac{dv_r}{d\theta} \right) = \left(v_r + \frac{dv_\theta}{d\theta} \right) h. \tag{4}$$

Deriving $dv_r/d\theta = v_\theta$ from Eq. (12.138) and substituting it into Eq. (4), we get

$$\frac{v_\theta^2}{g}\left(\frac{dv_\theta}{d\theta} + v_r\right) = \left(v_r + \frac{dv_\theta}{d\theta}\right)h. \tag{5}$$

Thus, we obtain

$$\left(v_r + \frac{dv_\theta}{d\theta}\right)\left(\frac{v_\theta^2}{g} - h\right) = 0. \tag{6}$$

Mizumura (2005b) investigated the effect of the bottom slope and width of the breach of the steep open channel flow to the discharge ratio using the Taylors series. Mizumura (1993) and Mizumura and Yamasaka (1998) analyzed the meandering water rivulet on the smooth slope using the Saint Venant equations in the cylindrical coordinate system. The analysis is based on the linear and nonlinear methods. He found that the surface tension plays an important role in the meandering water rivulet. Mizumura (2005a) also investigated the characteristics of the master recession curves using the Saint Venant equations.

Problem 12.8. Derive Eq. (12.152).

Chapter 13

Potential Equations

The potential equation is called the Laplace equation, or the partial differential equation of the elliptic type, and is frequently used in engineering fields. This does not explicitly contain a time-varying term. In hydraulic engineering, groundwater flow, water waves, and overland flows are analyzed by the potential equation. When the time-varying term in the two-dimensional wave and diffusion equations is zero, the wave and diffusion equations reduce to the potential equation. Similar equations to the potential equation are a Poisson equation or the Helmholtz equation. The two-dimensional Navier–Stokes equations are the diffusion equation and become the potential equation if the flow is steady. The one-dimensional Navier–Stokes equation is the diffusion equation.

13.1. Fundamental Equations

The three-dimensional potential equation is given by

$$\nabla^2 \phi = \frac{\partial^2 \phi}{\partial x^2} + \frac{\partial^2 \phi}{\partial y^2} + \frac{\partial^2 \phi}{\partial z^2} = 0, \tag{13.1}$$

in which $\nabla^2 = \Delta$ and is called Laplacian. The velocity potential ϕ satisfies Eq. (13.1) in the three-dimensional flow. For the two-dimensional flow, the stream function ψ satisfies Eq. (13.2).

$$\frac{\partial^2 \psi}{\partial x^2} + \frac{\partial^2 \psi}{\partial y^2} = 0. \tag{13.2}$$

The stream function ψ is used in the two-dimensional flow. This is the non-rotational condition in fluid mechanics. When the right side of Eq. (13.2) is not zero, Eq. (13.2) is called the Poisson equation. Equation (13.1) is expressed in the cylindrical coordinate system as

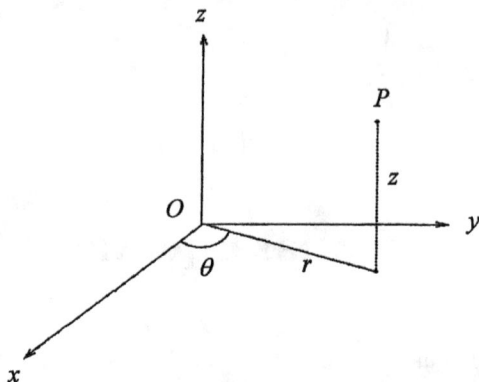

Figure 13.1.   Cylindrical coordinate system.

shown in Fig. 13.1:

$$\nabla^2 \phi = \frac{1}{r} \frac{\partial}{\partial r} \left(r \frac{\partial \phi}{\partial r} \right) + \frac{1}{r^2} \frac{\partial^2 \phi}{\partial \theta^2} + \frac{\partial^2 \phi}{\partial z^2} = 0, \qquad (13.3)$$

in which the coordinate system r, θ, and z are explained in Fig. 13.1. For the spherical coordinate system, we have

$$\nabla^2 \phi = \frac{1}{r^2} \frac{\partial}{\partial r} \left(r^2 \frac{\partial \phi}{\partial r} \right) + \frac{1}{r^2 \sin \theta} \frac{\partial}{\partial \theta} \left(\sin \theta \frac{\partial \phi}{\partial \theta} \right) + \frac{1}{r^2 \sin^2 \theta} \frac{\partial^2 \phi}{\partial \omega^2} = 0,$$

$$(13.4)$$

in which the coordinate system r, θ, and ω are explained in Fig. 13.2. The function which satisfies Eqs. (13.1), (13.2), (13.3), or (13.4) is

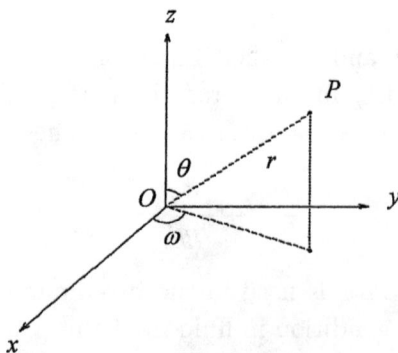

Figure 13.2.   Spherical coordinate system.

called a harmonic function. To understand the physical property of the potential equation, Eq. (13.2) is expressed by

$$\frac{\partial^2 \psi}{\partial x^2} = -\frac{\partial^2 \psi}{\partial y^2}. \tag{13.5}$$

This indicates that the signs of the curvatures in the x and y directions are opposite. This is the geometrical property of the two-dimensional potential equation.

Example 13.1. When the complex potential w is expressed by the sum of the velocity potential ϕ and the stream function ψ as $w = \phi + i\psi$, show that w satisfies $\nabla^2 w = 0$.

Solution. Since a complex variable $z = x + iy$ is used, we get

$$\frac{\partial}{\partial x} = \frac{\partial z}{\partial x} \cdot \frac{d}{dz}, \quad \frac{\partial^2}{\partial x^2} = \left(\frac{\partial z}{\partial x}\right)^2 \cdot \frac{d^2}{dz^2}, \tag{1}$$

$$\frac{\partial}{\partial y} = \frac{\partial z}{\partial y} \cdot \frac{d}{dz}, \quad \frac{\partial^2}{\partial y^2} = \left(\frac{\partial z}{\partial y}\right)^2 \cdot \frac{d^2}{dz^2}. \tag{2}$$

Since $\nabla^2 \phi = \nabla^2 \psi = 0$, $\nabla^2 w = \nabla^2 \phi + i\nabla^2 \psi = 0$ is not formed. Because w is the function of z and ∇^2 is the operations of x and y, we have

$$\frac{\partial^2 w}{\partial x^2} + \frac{\partial^2 w}{\partial y^2} = \frac{d^2 w}{dz^2} - \frac{d^2 w}{dz^2} = 0. \tag{3}$$

Noting the following relations:

$$\frac{\partial z}{\partial x} = 1, \quad \frac{\partial z}{\partial y} = i. \tag{4}$$

Problem 13.1. Show that the solution of the following partial differential equation:

$$\frac{1}{r}\frac{\partial}{\partial r}\left(r\frac{\partial \phi}{\partial r}\right) + \frac{1}{r^2}\frac{\partial^2 \phi}{\partial \theta^2} = 0 \tag{5}$$

is

$$\phi_0 = (A_0\theta + B_0)(C_0 \ln r + D_0), \quad \text{for } n = 0, \tag{6}$$

or

$$\phi_n = (A_n \cos n\theta + B_n \sin n\theta)(C_n r^n + D_n r^{-n}), \quad \text{for } n \neq 0. \quad (7)$$

13.2. Surface Wave of Infinitesimally Small Amplitude

The phenomenon of the long wave is expressed by the wave equation. But if the vertical component of the wave motion is not negligible in comparison with the horizontal component of the wave motion, the wave motion is expressed by the potential equation in the coordinate system in Fig. 13.3. It is

$$\frac{\partial^2 \phi}{\partial x^2} + \frac{\partial^2 \phi}{\partial y^2} = 0, \quad (13.6)$$

in which $x =$ the coordinate system in the horizontal direction and $y =$ the coordinate system in the vertically upward direction. To solve Eq. (13.6), the boundary condition at the bed is

$$\frac{\partial \phi}{\partial y} = 0, \quad \text{at } y = -h. \quad (13.7)$$

The linearized dynamic boundary condition at the water surface is given by

$$\frac{\partial \phi}{\partial t} + g\eta = 0, \quad \text{at } y = 0. \quad (13.8)$$

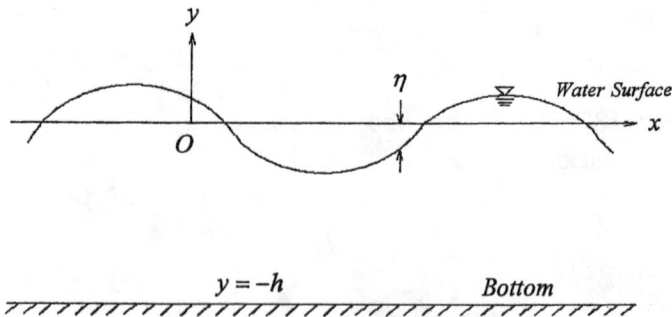

Figure 13.3.   Water waves.

The linearized kinematic boundary condition at the water surface is

$$\frac{\partial \eta}{\partial t} = \frac{\partial \phi}{\partial y}, \quad \text{at } y = 0. \tag{13.9}$$

Equations (13.8) and (13.9) are linearized beforehand. Combining Eqs. (13.8) and (13.9), we obtain

$$\frac{\partial^2 \phi}{\partial t^2} + g\frac{\partial \phi}{\partial y} = 0, \quad \text{at } y = 0. \tag{13.10}$$

This is the combined boundary condition at the water surface. To solve the Laplace equation Eq. (13.6), we assume that the velocity potential moves with the celerity C when the water depth h is constant. Thus, we can assume

$$\phi = f(y)\cos k(x - Ct), \tag{13.11}$$

in which k = angular wave number, $2\pi/$wave length. This corresponds to the method of separation of variables. Substituting Eq. (13.11) into Eq. (13.6), we have

$$\frac{d^2 f}{dy^2} - k^2 f = 0. \tag{13.12}$$

The solution of Eq. (13.12) is given by

$$f(y) = Ae^{ky} + Be^{-ky}, \tag{13.13}$$

in which A and B are integral constants. From the boundary condition on the bed, Eq. (13.7), we obtain

$$Ae^{-kh} = Be^{kh}. \tag{13.14}$$

Eliminating A in Eq. (13.13) by Eq. (13.14), we have

$$f(y) = B(e^{k(y+2h)} + e^{-ky}) = Be^{kh}\left(e^{k(y+h)} + e^{-k(y+h)}\right)$$
$$= 2Be^{kh}\cosh k(y + h). \tag{13.15}$$

Substituting Eq. (13.15) into Eq. (13.11), we obtain

$$\phi = 2Be^{kh}\cosh k(y + h)\cos k(x - Ct). \tag{13.16}$$

Substituting Eq. (13.16) into Eq. (13.10), we have

$$C = \sqrt{\frac{g}{k} \tanh kh}. \tag{13.17}$$

This is the equation to describe the celerity of the surface water wave. When H is the wave height, the wave profile is given by

$$\eta = \frac{H}{2} \sin k(x - Ct). \tag{13.18}$$

Combining Eqs. (13.8) and (13.16), we get

$$2Be^{kh} = \frac{Hg}{2kh \cosh kh}. \tag{13.19}$$

Therefore, the velocity potential Eq. (13.11) is

$$\phi = \frac{Hg}{2kh \cosh kh} \cosh k(y + h) \cos k(x - Ct). \tag{13.20}$$

Example 13.2. Derive Eq. (13.11), applying the method of separation of variables to Eq. (13.6).

Solution. When ϕ is the velocity potential as follows:

$$\phi = \Phi(x, y)e^{ikCt}. \tag{1}$$

Assume the separation of variables as

$$\Phi = X(x)Y(y). \tag{2}$$

Substituting Eq. (2) into Eq. (13.6), we obtain

$$\frac{\ddot{X}}{X} = -\frac{\ddot{Y}}{Y} = k^2, \tag{3}$$

in which $\dot{X}(x) = dX/dx$ and $\dot{Y}(y) = dY/dy$. This derives the following two ordinary differential equations:

$$\ddot{X} + k^2 X = 0, \quad \ddot{Y} - k^2 Y = 0. \tag{4}$$

Thus, the solutions of the two ordinary differential equations are respectively

$$X = C_1 e^{ikx} + C_2 e^{-ikx}, \quad Y = C_3 e^{ky} + C_4 e^{-ky}. \tag{5}$$

Thus, the velocity potential is written by

$$\begin{aligned}
\phi &= \left(C_1 e^{ikx} + C_2 e^{-ikx} \right) \left(C_3 e^{ky} + C_4 e^{-ky} \right) e^{ikCt} \\
&= f(y) \left[C_1 e^{ik(x+Ct)} + C_2 e^{-ik(x-Ct)} \right],
\end{aligned} \tag{6}$$

in which $f(y) = C_3 e^{ky} + C_4 e^{-ky}$. On the right side of Eq. (6),

$$\text{term of } C_1 \cdots \text{ Propagating waves in the } -x \text{ direction} \tag{7}$$

$$\text{term of } C_2 \cdots \text{ Propagating waves in the } +x \text{ direction} \tag{8}$$

When the progressing wave as a *cos* function is considered, the velocity potential is given by

$$\phi = f(y) \cos k(x - Ct). \tag{9}$$

Problem 13.2. The piezometric head ϕ in the confined groundwater flow is expressed by the potential equation. When the boundary condition is given by

$$\phi = 0, \quad \text{at } x = 0, \tag{10}$$

$$\phi = 0, \quad \text{at } x = \ell, \tag{11}$$

$$\phi = G(x), \quad \text{at } y = 0, \tag{12}$$

$$\phi = F(x), \quad \text{at } y = h. \tag{13}$$

Solve

$$\nabla^2 \phi = 0. \tag{14}$$

If we assume

$$\phi = U(x, y) + W(x, y), \tag{15}$$

then, show that the solutions U and W are given by

$$U(x, y) = \sum_{n=1}^{\infty} a_n \frac{\sinh(n\pi y/\ell)}{\sinh(n\pi h/\ell)} \sin \frac{n\pi x}{\ell}, \tag{16}$$

in which

$$a_n = \frac{2}{\ell} \int_0^\ell F(x) \sin \frac{n\pi x}{\ell} \, dx \tag{17}$$

and

$$W(x,y) = \sum_{n=1}^\infty b_n \frac{\sinh(n\pi(h-y)/\ell)}{\sinh(n\pi h/\ell)} \sin \frac{n\pi x}{\ell}, \tag{18}$$

in which

$$b_n = \frac{2}{\ell} \int_0^\ell G(x) \sin \frac{n\pi x}{\ell} \, dx. \tag{19}$$

13.3. Groundwater Flow

The piezometric head h consists of the pressure head p/γ and the potential head z. It is written as

$$h = \frac{p}{\gamma} + z, \tag{13.21}$$

in which γ is the specific weight of water and z is the height from the base line. Using the piezometric head and the Darcy law, we have the continuity equation

$$\frac{\partial^2 h}{\partial x^2} + \frac{\partial^2 h}{\partial y^2} = 0. \tag{13.22}$$

This is the potential equation. The boundary conditions

$$h = 0, \quad \text{at } x = 0, \tag{13.23}$$

$$h = 1, \quad \text{at } y = 0, \tag{13.24}$$

$$h = 0, \quad \text{at } y \to \infty \tag{13.25}$$

are assumed. The piezometric head $h(x,y)$ is also assumed to be

$$h(x,y) = X(x) \cdot Y(y). \tag{13.26}$$

Substituting Eq. (13.26) into Eq. (13.22), we have

$$\frac{\ddot{X}}{X} = -\frac{\ddot{Y}}{Y} = -k^2, \tag{13.27}$$

in which $\ddot{X} = d^2X/dx^2$ and $\ddot{Y} = d^2Y/dy^2$. When the parameter k is a constant, we obtain

$$\ddot{X} + k^2 X = 0, \tag{13.28}$$

$$\ddot{Y} - k^2 Y = 0. \tag{13.29}$$

Considering Eq. (13.23), the solution of Eq. (13.28) is given by

$$X = C_1 \sin \frac{n\pi x}{a}. \tag{13.30}$$

Considering Eq. (13.25), we have the solution of Eq. (13.29) as follows:

$$Y = C_2 e^{-\frac{n\pi y}{a}}. \tag{13.31}$$

The product of Eqs. (13.30) and (13.31) indicates $C_1 C_2 = A_n$. Then, the solution h is written as

$$h = \sum_{n=1}^{\infty} A_n e^{-\frac{n\pi y}{a}} \sin \frac{n\pi x}{a}. \tag{13.32}$$

The condition of Eq. (13.24) shows

$$1 = \sum_{n=1}^{\infty} A_n \sin \frac{n\pi x}{a}. \tag{13.33}$$

Multiplying $\sin \frac{r\pi x}{a}$ to Eq. (13.33) and integrating from $x = 0$ to a, we get

$$\int_0^a \sin \frac{r\pi x}{a} dx = A_r \frac{a}{2}, \qquad \text{for } r = 1, 2, 3, \ldots . \tag{13.34}$$

Thus, we obtain

$$A_r = \frac{2}{ar\pi} [1 - (-1)^r]. \tag{13.35}$$

The substitution of Eq. (13.35) into Eq. (13.32) gives the solution.

Example 13.3. When the boundary condition Eq. (1) instead of Eq. (13.24) is

$$h = x, \qquad \text{at } y = 0, \tag{1}$$

obtain A_r in Eq. (13.34).

Solution. Equation (13.33) is changed to

$$x = \sum_{n=1}^{\infty} A_n \sin \frac{n\pi x}{a}. \tag{2}$$

Multiplying $\sin \frac{r\pi x}{a}$ to both sides of Eq. (2) and integrating from 0 to a, we get

$$\int_0^a x \sin \frac{r\pi x}{a} \, dx = A_r \frac{a}{2}. \tag{3}$$

Thus, we obtain

$$A_r = \frac{2(-1)^{r+1} a}{r\pi}. \tag{4}$$

The substitution of Eq. (4) into Eq. (13.32) gives the solution.

Problem 13.3. Solve Eq. (13.22) with the boundary conditions explained as

$$h = h_0 \text{ (const)}, \quad \text{at } x = 0, \tag{5}$$

$$h = 0, \quad \text{at } x = b, \tag{6}$$

$$\frac{\partial h}{\partial y} = 0, \quad \text{at } y = 0, \tag{7}$$

$$\frac{\partial h}{\partial y} = 0, \quad \text{at } y = a. \tag{8}$$

13.4. Motion of a Sphere in Fluid

When a sphere of radius a moves with the velocity U in the perfect fluid at rest, the spherical coordinate system based on the center of the sphere is (r, θ, w) (Lamb, 1945). Then, the flow is independent of the coordinate w. The velocity potential ϕ is expressed by Eq. (13.4) as

$$\frac{\partial}{\partial r}\left(r^2 \frac{\partial \phi}{\partial r}\right) + \frac{1}{\sin \theta} \frac{\partial}{\partial \theta}\left(\sin \theta \frac{\partial \phi}{\partial \theta}\right) = 0. \tag{13.36}$$

To solve Eq. (13.36), we assume

$$\phi = R(r) \cdot \Theta(\theta). \tag{13.37}$$

Substituting Eq. (13.37) into Eq. (13.36), we obtain

$$\frac{1}{R}\frac{d}{dr}\left(r^2\frac{dR}{dr}\right) = -\frac{1}{\Theta}\frac{d}{\sin\theta\, d\theta}\left(\sin\theta\frac{d\Theta}{d\theta}\right). \tag{13.38}$$

Since the term on the left side of Eq. (13.38) is the function of r and the term on the right side is the function of θ, the term on each side is equal to a constant. We define the constant $n(n+1)$, in which n is a positive integer. Then, Eq. (13.38) becomes the following two equations:

$$\frac{d}{dr}\left(r^2\frac{dR}{dr}\right) - n(n+1)R = 0 \tag{13.39}$$

and

$$\frac{d}{\sin\theta\, d\theta}\left(\sin\theta\frac{d\Theta}{d\theta}\right) + n(n+1)\Theta = 0. \tag{13.40}$$

The solution of Eq. (13.39) is given by

$$R = Ar^n + Br^{-n-1}, \tag{13.41}$$

in which A and B are integral constants. Defining $\mu = \cos\theta$ in Eq. (13.40), we transform Eq. (13.40) as

$$\frac{d}{d\mu}\left[(1-\mu^2)\frac{d\Theta}{d\mu}\right] + n(n+1)\Theta = 0. \tag{13.42}$$

This is the Legendre differential equation in Chap. 9. The general solution of Eq. (13.42) is

$$\Theta(\theta) = CP_n(\mu) + DQ_n(\mu), \tag{13.43}$$

in which C and D are integral constants. In Eq. (13.43) $P_n(\mu)$ and $Q_n(\mu)$ are the first and second kind Legendre functions of order n. Since Eq. (13.36) is linear, we can apply the principle of superposition. Thus, the general solution of Eq. (13.36) is

$$\phi = \sum_{n=1}^{\infty}\left(A_n r^n + \frac{B_n}{r^{n+1}}\right)[C_n P_n(\mu) + D_n Q_n(\mu)]. \tag{13.44}$$

The boundary conditions are

$$\frac{\partial \phi}{\partial r} = U \cos \theta, \quad \text{at } r = a \tag{13.45}$$

and

$$\frac{\partial \phi}{\partial r} = \frac{1}{r} \frac{\partial \phi}{\partial \theta} = 0, \quad \text{at } r = \infty. \tag{13.46}$$

From Eq. (13.46), we have

$$A_n = 0. \tag{13.47}$$

Considering

$$P_0(\mu) = 1, \quad P_1(\mu) = \mu, \quad P_2(\mu) = \frac{1}{2}(3\mu^2 - 1), \dots, \tag{13.48}$$

$$Q_0(\mu) = \frac{1}{2} \log \frac{1+\mu}{1-\mu}, \quad Q_1(\mu) = \frac{1}{2}\mu \log \frac{1+\mu}{1-\mu} - 1,$$

$$Q_2(\mu) = \frac{1}{4}(3\mu^2 - 1) \log \frac{1+\mu}{1-\mu} - \frac{3}{2}\mu, \dots \tag{13.49}$$

and Eq. (13.45), we get $n = 1$. Then,

$$\phi = \frac{B_1 C_1}{r^2} P_1(\cos \theta) = \frac{B_1 C_1}{r^2} \cos \theta. \tag{13.50}$$

Equation (13.45) indicates

$$-\frac{2B_1 C_1}{a^3} \cos \theta = U \cos \theta. \tag{13.51}$$

Thus, we obtain

$$B_1 C_1 = -\frac{U a^3}{2}. \tag{13.52}$$

Finally, we get the velocity potential

$$\phi = -\frac{U a^3}{2r^2} \cos \theta. \tag{13.53}$$

This is the velocity potential when the sphere of radius a moves with the velocity U.

Example 13.4. A fluid of velocity U flows past a sphere of radius a at rest. Calculate the velocity on the sphere v_θ.

Solution. The velocity potential of Eq. (13.53) is obtained by the condition of the moving sphere in the still fluid. Therefore, the motion that the sphere is at rest and the fluid moves corresponds to the motion that the origin is at the center of the sphere and the fluid moves. We add the velocity potential of the uniform flow $\phi = -Ux$ to Eq. (13.53). Thus, we have

$$\phi = -Ur\cos\theta - \frac{Ua^3\cos\theta}{2r^2}, \tag{1}$$

in which

$$v_r|_{r=a} = \left.\frac{\partial\phi}{\partial r}\right|_{r=a} = 0, \tag{2}$$

$$v_\theta|_{r=a} = \left.\frac{1}{r}\frac{\partial\phi}{\partial\theta}\right|_{r=a} = \frac{3}{2}U\sin\theta. \tag{3}$$

Problem 13.4. When the velocity potential $\phi(R,\theta) = F(\theta)$ is given on the perimeter of a circle of radius R, show that the velocity potential in the circle is given by

$$\phi(r,\theta) = \frac{1}{\pi}\int_{-\pi}^{\pi}\left[\frac{1}{2} + \sum_{n=1}^{\infty}\left(\frac{r}{R}\right)^n\cos n(\theta-u)\right]F(u)\,du. \tag{4}$$

13.5. Two-Dimensional Stratified Flow into a Sink

The two-dimensional flow of a stratified fluid between two horizontal planes into a line sink shows the effect of stratification. Yih (1958) analyzed the withdrawal from a line sink at the bottom of a stratified lake where the density is linearly distributed in the vertical direction as shown in Fig. 13.4. When the density is constant along the streamline, the combined equation of continuity equation and momentum equations is written by the following partial differential equation of

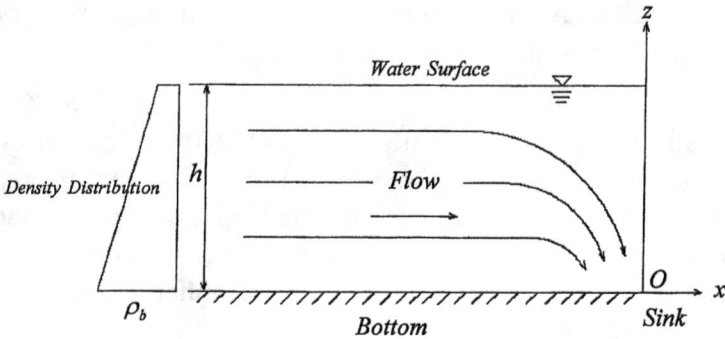

Figure 13.4.   Two-dimensional stratified flow into sink.

second-order of Helmholtz type:

$$\frac{\partial^2 \psi}{\partial \xi^2} + \frac{\partial^2 \psi}{\partial \eta^2} - \frac{\eta}{F^2} = -\frac{\psi}{F^2}, \tag{13.54}$$

in which $\xi = x/h, \eta = z/h, \psi = \psi'/(Ah), A = \sqrt{\rho/\rho_b} \cdot U, \psi' =$ stream function, $F^2 = \frac{q^2}{gh^4\beta}, q = \sqrt{\rho/\rho_b} \cdot Uh, \beta = \frac{\rho_b - \rho}{\rho_b h}$, and $U =$ velocity for $x \to -\infty$. The boundary conditions of Eq. (13.54) are

$\psi = 0, \quad$ at $\eta = 0$ (flow bottom), $\hspace{3cm}$ (13.55)

$\psi = 1, \quad$ at $\eta = 1 \quad$ and $\quad \xi = 0 \; (\eta \neq 0)$ (water surface), $\hspace{0.5cm}$ (13.56)

$\psi = \eta, \quad$ at $\xi = -\infty.$ $\hspace{5cm}$ (13.57)

If $\psi = \psi_0 + \eta$, Eq. (13.54) is written by

$$\frac{\partial^2 \psi_0}{\partial \xi^2} + \frac{\partial^2 \psi_0}{\partial \eta^2} + \frac{\psi_0}{F^2} = 0. \tag{13.58}$$

This is the Helmholtz equation. We assume the following solution of Eq. (13.58), considering the boundary conditions Eqs. (13.55) and (13.56):

$$\psi_0 = \sum_{n=1}^{\infty} A_n(\xi) \sin n\pi\eta. \tag{13.59}$$

Substituting Eq. (13.59) into Eq. (13.58), we get

$$\ddot{A}_n + \left(\frac{1}{F^2} - n^2\pi^2\right) A_n = 0, \tag{13.60}$$

in which "$'$" $= d/d\xi$. Considering Eq. (13.57), the solution of Eq. (13.59) for $F > 1/\pi$ is

$$A_n = C_n \exp\left[\left(n^2\pi^2 - \frac{1}{F^2}\right)^{1/2}\xi\right], \quad \text{for } n = 1, 2, \ldots. \quad (13.61)$$

Since $\psi = 1$ at $\xi = 0$, we get $\psi_0 = 1 - \eta$. Therefore,

$$1 - \eta = \sum_{n=1}^{\infty} C_n \sin n\pi\eta. \quad (13.62)$$

This is the Fourier sine series. The coefficient becomes $C_n = \frac{2}{n\pi}$. If we solve Eq. (13.54) under this condition for $F > 1/\pi$, we obtain

$$\psi = \eta + \frac{2}{\pi}\sum_{n=1}^{\infty}\frac{1}{n}\exp\left[\left(n^2\pi^2 - \frac{1}{F^2}\right)^{1/2}\xi\right]\sin n\pi\eta. \quad (13.63)$$

Figure 13.5 represents the flow patterns for $F = 0.5, 0.32$, and $1/\pi = 0.3183$. For $F = 0.32\,(>1/\pi)$, the closed zone is generated over the line sink and the closed zone expands to the upstream for $1/\pi$. This

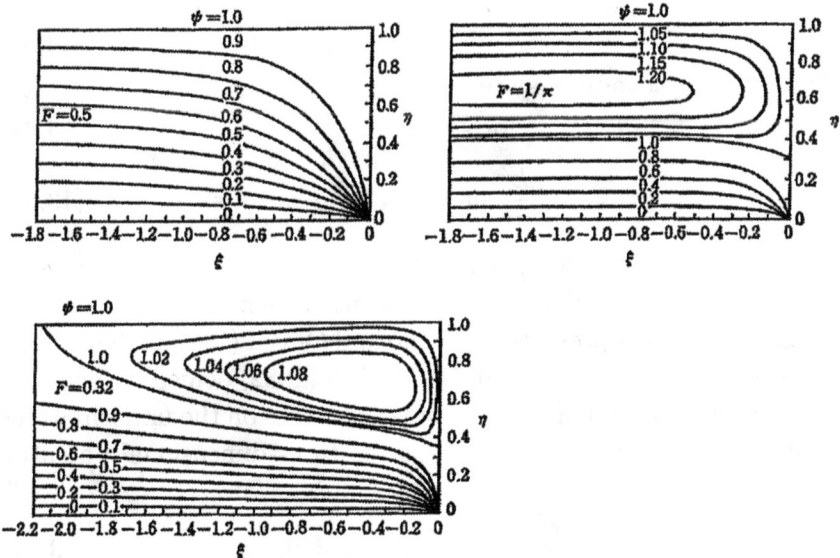

Figure 13.5. Solutions for Eq. (13.54).

violates the boundary condition of Eq. (13.57). For $F < 1/\pi$, Yih also obtained

$$
\psi = \eta + \frac{2}{\pi} \left\{ \sum_{n=1}^{N} \frac{\sin n\pi\eta}{n} \left[\cos\left[(F^{-2} - n^2\pi^2)^{1/2} \xi \right] \right. \right.
$$

$$
\left. + f(n) \sin\left[(F^{-2} - n^2\pi^2)^{1/2} \xi \right] \right]
$$

$$
\left. + \sum_{n=N+1}^{\infty} \frac{\sin n\pi\eta}{n} \exp\left[(n^2\pi^2 - F^{-2})^{1/2} \xi \right] \right\}, \qquad (13.64)
$$

in which $f(n)$ is any function of n. The function is selected to satisfy the upstream conditions. This shows the solution for

$$
\frac{1}{(N+1)\pi} < F < \frac{1}{N\pi}. \qquad (13.65)
$$

For example,

$$
\frac{1}{2\pi} < F < \frac{1}{\pi}, \qquad (13.66)
$$

we have

$$
\psi = \eta - \frac{2}{\pi} \sum_{n=1}^{\infty} \left(n - \frac{1}{n} \right) \sin n\pi\eta \cdot \exp\left[(n^2\pi^2 - F^{-2})^{1/2} \xi \right].
$$

$$
(13.67)
$$

The boundary conditions are

$$
\psi = 1, \quad \text{for } \epsilon \leq \eta \leq 1, \qquad (13.68)
$$

$$
\psi = 1 - \frac{k^2}{\pi^3} \sin k\eta, \quad \text{for } 0 < \eta \leq \epsilon, \qquad (13.69)
$$

$$
\psi = 0, \quad \text{at } \eta = 0. \qquad (13.70)
$$

An analysis on an axisymmetric flow into a point sink in a linearly stratified fluid was presented by Hino and Onishi (1967). They used a perturbation method in Chap. 19, but Yih (1982) denied it. The next example is a diffraction problem of water waves on the uniform water depth h. The governing equation is the three-dimensional potential equation.

$$
\frac{\partial^2 \phi}{\partial x^2} + \frac{\partial^2 \phi}{\partial y^2} + \frac{\partial^2 \phi}{\partial z^2} = 0, \qquad (13.71)
$$

in which $\phi =$ the velocity potential, x and $y =$ the horizontal coordinate system at the free water surface, and $z =$ the vertically upward coordinate system. The boundary condition at the bed is

$$\frac{\partial \phi}{\partial z} = 0, \quad \text{at } z = -h. \tag{13.72}$$

Therefore, the velocity potential ϕ is assumed to be

$$\phi = \Phi(x, y) \cosh k(z + h). \tag{13.73}$$

Equation (13.73) satisfies Eq. (13.72). Therefore, the function Φ satisfies the following Helmholtz equation:

$$\frac{\partial^2 \Phi}{\partial x^2} + \frac{\partial^2 \Phi}{\partial y^2} + k^2 \Phi = 0. \tag{13.74}$$

Equation (13.74) is the governing equation for the water wave diffraction. The method of the solution is not covered in this book. When Eq. (13.74) is rewritten in the cylindrical coordinate, the resultant differential equation leads to the Bessel differential equation by the method of the separation of variables.

Example 13.5. Show that Eq. (13.63) is not right for $F < 1/\pi$.

Solution. In order that the solution of Eq. (13.60) is an exponential type, the Froude number satisfies $\frac{1}{F^2} - n^2\pi^2 < 0$. Since the minimum integer of n is 1, the solution is always the exponential type for $\frac{1}{F^2} < \pi^2$.

Problem 13.5. Solve the following Helmholtz equation using the method of the separation of variables:

$$\frac{\partial^2 \phi}{\partial x^2} + \frac{\partial^2 \phi}{\partial y^2} + \phi = 0, \tag{1}$$

with the boundary conditions

$$\phi = 0, \quad \text{at } x = 0, \ \ell, \tag{2}$$

$$\phi = 5, \quad \text{at } y = 0, \tag{3}$$

$$\phi = 0, \quad \text{at } y = \infty. \tag{4}$$

13.6. Analysis of Sand Waves

A bed in rivers, lakes, or seas consists of sand or soil, and movable fluid motion. Thus, fluid motion affects materials on the movable

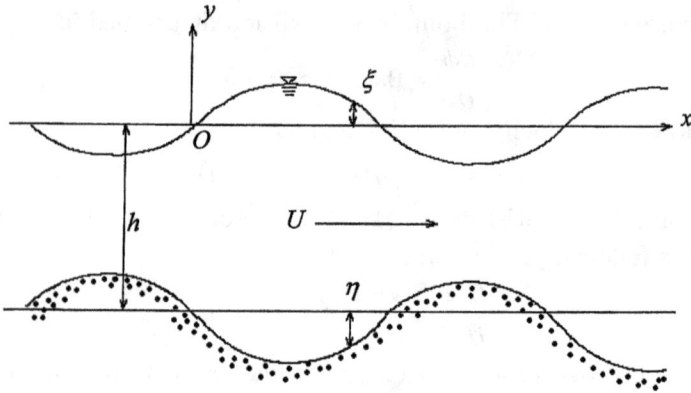

Figure 13.6.   Flow over wavy bed.

bed and the movable bed form influences the fluid motion. There are sand waves, dunes, antidunes, and sand ripples on the movable bed. The dunes or ripples move in the flow direction, on the contrary, the antidunes move opposite to the flow direction. They are also found on deserts under wind action. They are very important for desertification. Kennedy (1963) conducted the pioneering work to study the stability of sand surface in the open channel flow. He used the potential flow theory for the open channel flow over the wavy bed and showed that the theory predicted the formation of sand waves. Referring to Fig. 13.6, we use the potential equation as the governing equation.

$$\frac{\partial^2 \phi}{\partial x^2} + \frac{\partial^2 \phi}{\partial y^2} = 0.$$
(13.75)

The water surface profile is assumed to be

$$y = \xi(x,t) = A(t) \sin k(x - U_b t),$$
(13.76)

in which $A(t)$ = amplitude of the water wave, k = an angular wave frequency, and U_b = movement velocity of sand waves. Generally U_b is much smaller than U. The bed form is also assumed to be

$$y = -h + \eta(x,t) = -h + a(t) \sin k(x - U_b t),$$
(13.77)

in which $a(t)$ = amplitude of sand waves on the bed and h = water depth. The kinematic boundary condition on the water surface is

$$\frac{\partial \xi}{\partial t} + \left(U + \frac{\partial \phi}{\partial x} \right) \frac{\partial \xi}{\partial x} = \frac{\partial \phi}{\partial y}, \quad \text{at } y = \xi(x,t).$$
(13.78)

The dynamic boundary condition on the water surface is

$$\frac{\partial \phi}{\partial t} + g\xi + \frac{1}{2}\left[\left(\frac{\partial \phi}{\partial x}\right)^2 + \left(\frac{\partial \phi}{\partial y}\right)^2\right] = \text{const}, \quad \text{at } y = \xi(x,t),$$

(13.79)

in which $U =$ the uniform velocity in the open channel flow without sand waves. The linearized kinematic boundary condition is

$$\frac{\partial \eta}{\partial t} + U\frac{\partial \eta}{\partial x} = \frac{\partial \phi}{\partial y}, \quad \text{at } y = -h + \eta(x,t).$$

(13.80)

The solution of Eq. (13.75) is assumed to be

$$\phi = Ux - f(y)\cos k(x - U_b t).$$

(13.81)

Thus, $f(y)$ satisfies

$$\frac{d^2 f(y)}{dy^2} - k^2 f(y) = 0.$$

(13.82)

The solution of Eq. (13.82) is

$$f(y) = C_1 \sinh ky + C_2 \cosh ky.$$

(13.83)

Equation (13.78) is linearized as the small amplitude waves. Substituting Eqs. (13.76) and (13.81) into Eq. (13.78) and assuming that

$$U \gg \frac{\partial \phi}{\partial x}, \quad \frac{da(t)}{dt} \cong 0, \quad U_b \ll U, \quad \frac{dA(t)}{dt} \cong 0,$$

(13.84)

we have Eq. (13.78) as

$$AUk = -\frac{df}{dy}.$$

(13.85)

Equation (13.79) is also linearized as the small amplitude waves and assuming that

$$\left|\frac{\partial \phi}{\partial t} + g\xi\right| \gg \left(\frac{\partial \phi}{\partial x}\right)^2, \quad \left|\frac{\partial \phi}{\partial t} + g\xi\right| \gg \left(\frac{\partial \phi}{\partial y}\right)^2,$$

(13.86)

we get

$$gA + fkU = 0.$$

(13.87)

Eliminating A from Eqs. (13.85) and (13.87), we have

$$\frac{df}{dy} = k^2 hF^2 f, \tag{13.88}$$

in which $F = U/\sqrt{gh}$, the Froude number. Equation (13.80) is also written as

$$aUk = -\frac{df}{dy}. \tag{13.89}$$

Substituting Eq. (13.83) into Eq. (13.88), we obtain

$$C_1 = khF^2 C_2. \tag{13.90}$$

Substituting Eq. (13.83) into Eq. (13.89) and using Eq. (13.90), we have

$$C_1 = \frac{aUkhF^2}{\sinh kh - khF^2 \cosh kh} \tag{13.91}$$

and

$$C_2 = \frac{aU}{\sinh kh - khF^2 \cosh kh}. \tag{13.92}$$

From Eqs. (13.83) and (13.81), the velocity potential ϕ is given by

$$\phi = Ux - Ua\frac{\cosh ky + F^2 kh \sinh ky}{\sinh kh - khF^2 \cosh kh} \cos k(x - U_b t). \tag{13.93}$$

Applying Eq. (13.93) into the continuity equation of sand, we can investigate stability of the sand surface. This theory determines the generation or disappearance of sand waves, dunes, or antidunes. From Eq. (13.80) and (13.93), we obtain the following equation:

$$\frac{a(t)}{A(t)} = \left(1 - \frac{\tanh kh}{khF^2}\right) \cosh kh. \tag{13.94}$$

When $F > \sqrt{\frac{\tanh kh}{kh}}$, $a(t)/A(t) < 0$ and the antidune is formed. When $F < \sqrt{\frac{\tanh kh}{kh}}$, $a(t)/A(t) > 0$ and the dune is formed. The antidune and dune are formed when the flow is supercritical and subcritical, respectively. When $a(t)/A(t) < 0$, the phase difference of the water surface profile and the bed form is almost π. When $a(t)/A(t) > 0$, the phase difference between the water surface profile and the bed form is almost 0. Mizumura (1989) studied the stability condition of the sand surface when the bed is assumed to be porous using the potential flow in the open channel flow and the Darcy flow in the porous media. Mizumura (1995b, 1998) also investigated the water surface profile over the wavy bed when the Froude number is larger than 1 and smaller than 1. For subcritical flow the phase lag between the bed form and the water surface profile is almost π. The peak and trough of the water surface profile are flatter and steeper than the sine curve, respectively. For supercritical flow the phase lag between the bed form and the water surface profile is almost 0. The peak and trough of the water surface profile are steeper and flatter than the sine curve, respectively. This corresponds to the physical property of a Stokes wave in Chap. 16.

Example 13.6. The condition that the denominator of Eqs. (13.91) and (13.92) equal zero indicates $F^2 = \frac{\tanh kh}{kh}$. What does it mean physically?

Solution. If $F^2 = \frac{\tanh kh}{kh}$, $a(t)/A(t) = 0$ from Eq. (13.94). This indicates that $a(t)$ is zero and the bed form becomes flat.

Problem 13.6. Kennedy solved the stability problem for the zero-th order. Increase the order of the solution.

Chapter 14

Diffusion Equations

This partial differential equation which is called the diffusion or heat conduction equation belongs to the parabolic type. In hydraulic engineering,
we use the Navier–Stokes equations to analyze flow. One-dimensional
Navier–Stokes equation is the diffusion equation. It is called a Burgers model. Two-dimensional Navier–Stokes equations are the potential
equation. The Navier–Stokes equations are usually simplified with the
boundary-layer theory in Chap. 17. As a result, the two-dimensional
steady Navier–Stokes equations become the diffusion equation. This type
is applicable for the analysis of oscillatory motion of a body in fluid,
distribution of suspended sediment in rivers, groundwater flow motion,
consolidation model, or shoreline change model.

14.1. Governing Equations

The diffusion equation or heat conduction equation are often used in
hydraulic engineering. This is written as

$$\frac{\partial C}{\partial t} = K\nabla^2 C, \tag{14.1}$$

in which $\nabla^2 = \partial^2/\partial x^2 + \partial^2/\partial y^2 + \partial^2/\partial z^2$, $C =$ concentration or
temperature, and $K =$ the diffusion coefficient or heat conduction
coefficient. Equation (14.1) represents fluid motion over the abruptly
moving plate (Rayleigh problem), fluid motion over an oscillatory
moving plate, or computation of vorticity in fluid motion. In the
consolidation theory, Eq. (14.1) is derived from the Darcy law when
C is pore pressure. The governing equation of confined groundwater
flow is also Eq. (14.1) when the Darcy law is formed. In the confined
groundwater flow, C indicates the piezometric head and K is the ratio
of storativity to tranmissivity. Equation (14.1) is also applicable to

the computation of concentration of suspended load in rivers. Equation (14.1) in three-dimension is rewritten in the two-dimension as

$$\frac{\partial C}{\partial t} + \frac{\partial (Cu)}{\partial x} + \frac{\partial (Cv)}{\partial y} = \frac{\partial}{\partial x}\left(K_x\frac{\partial C}{\partial x}\right) + \frac{\partial}{\partial y}\left(K_y\frac{\partial C}{\partial y}\right) + v_s\frac{\partial C}{\partial y},$$

$$(14.2)$$

in which x = the coordinate system in the flow direction, y = the coordinate system in the vertical direction, u and v = the velocities of x and y directions, respectively, K_x and K_y = the diffusion coefficients in the x and y directions, respectively, v_s = the settling velocity of sand particles. Unconfined groundwater flow on the hillslope is expressed by a Boussinesq equation (Brutsaert, 2005; Mizumura, 2008a). The Boussinesq equation which plays an important role in hillslope hydrology is derived from Fig. 14.1 and written as

$$\lambda\frac{\partial h}{\partial t} = k\frac{\partial}{\partial x}\left[h\left(\frac{\partial h}{\partial x}\cos\theta - \sin\theta\right)\right].$$

$$(14.3)$$

Equation (14.3) is also a nonlinear diffusion equation. In Eq. (14.3), λ = porosity, x = the coordinate system in the flow or downslope direction, h = the water depth normal to the bottom, k = the hydraulic conductivity, and θ = an angle of the bottom slope to the horizontal. When we investigate coastal changes due to the effect of wave motion and longshore currents, we use a simple model, one-line theory. This is also a linear diffusion equation. The diffusion model in flood routing is also a linear partial differential equation. It is originally a nonlinear wave equation, but the linearization makes it the linear diffusion equation.

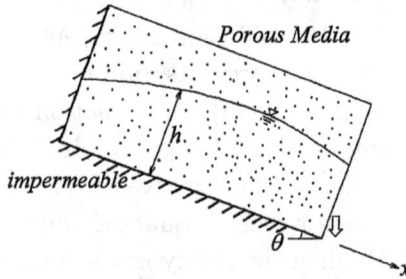

Figure 14.1.   Boussinesq problem.

Example 14.1. Show that when $|\partial h/\partial x|$ in Eq. (14.3) is infinitesimally small and the water depth h is almost constant, Eq. (14.3) becomes linear.

Solution. Equation (14.3) is rewritten as

$$\lambda\frac{\partial h}{\partial t} = kh\frac{\partial^2 h}{\partial x^2}\cos\theta + k\left(\frac{\partial h}{\partial x}\right)^2\cos\theta - k\frac{\partial h}{\partial x}\sin\theta. \qquad (1)$$

The first and second terms on the right side are nonlinear in Eq. (1). Since $|\partial h/\partial x|$ is assumed to be small, the second term on the right side is neglected as

$$\lambda\frac{\partial h}{\partial t} = kh\frac{\partial^2 h}{\partial x^2}\cos\theta - k\frac{\partial h}{\partial x}\sin\theta. \qquad (2)$$

If we assume that h is constant h_o in the first term on the right side, Eq. (2) is written as

$$\lambda\frac{\partial h}{\partial t} = kh_0\frac{\partial^2 h}{\partial x^2}\cos\theta - k\frac{\partial h}{\partial x}\sin\theta. \qquad (3)$$

This is the linear diffusion equation. By linearizing the first term on the right side of Eq. (1), the nonlinear partial differential equation is expressed as

$$\lambda\frac{\partial h}{\partial t} = kh_o\frac{\partial^2 h}{\partial x^2}\cos\theta + k\left(\frac{\partial h}{\partial x}\right)^2 - k\frac{\partial h}{\partial x}\sin\theta \qquad (4)$$

Equation (4) is also linearized by the following two equations:

$$\lambda\frac{\partial H}{\partial t} = kh_o\frac{\partial^2 H}{\partial x^2}\cos\theta - k\frac{\partial H}{\partial x}\sin\theta \qquad (5)$$

$$h = \ln H \qquad (6)$$

Equation (5) is a linear partial differential equation. The function $\ln H$ is the solution of Eq. (4). Equations (5) and (6) are derived by the functional transformation in Chap 16. Usually the nonlinear partial differential equation is not easily solved. As an example, readers who are interested in solving nonlinear partial differential equations, refer to Mizumura (2002, 2003c, 2005a, 2006a, 2009a).

Problem 14.1. Show that Burgers equation (the one-dimensional Navier–Stokes equation)

$$\frac{\partial u}{\partial t} + u\frac{\partial u}{\partial x} = \nu\frac{\partial^2 u}{\partial x^2} \qquad (7)$$

is the diffusion type. In Eq. (4), ν is the kinematic viscosity.

14.2. Rayleigh Problem

We analyze the fluid motion when infinitely long plate abruptly starts to move with the velocity U_0. This is called a Rayleigh problem (Yih, 1969) and it is important to understand the effect of viscosity in fluid mechanics. The governing equation is the diffusion type and given by

$$\frac{\partial u}{\partial t} = \nu \frac{\partial^2 u}{\partial y^2}, \tag{14.4}$$

in which $u = $ the velocity in the x direction, $t = $ time, $\nu = $ the kinematic viscosity, and $y = $ the normal coordinate to the plate. The boundary conditions are

$$u = 0, \quad \text{at } t = 0, \tag{14.5}$$

$$u = U_0, \quad \text{at } y = 0 \ (t = \infty), \tag{14.6}$$

$$u = 0, \quad \text{at } y = \infty. \tag{14.7}$$

In general, the fundamental solution of Eq. (14.4) is obtained by the method of separation of variables. But since it is not easy to reach the final form, we use a similarity method. This method requires the restriction of the boundary conditions. The boundary conditions are $u = 0$ for $t = 0$ and $y = \infty$ from Eqs. (14.5) and (14.7), respectively. More boundary conditions are $u = U_o$ for $y = 0$ and $t = \infty$ from Eq. (14.6). Thus, a new variable such as y/t is considered for the boundary conditions of this problem. The two different boundary conditions become one, according to the new variable. Mathematical consideration of a similarity method (Bluman and Cole, 1974) shows $\eta = y/t^m \ (m > 0)$ as a new variable. Substituting $\eta = y/t^m \ (m > 0)$ into Eq. (14.4), we obtain

$$\frac{\partial u}{\partial t} = -\frac{m\eta}{t} \frac{du}{d\eta}, \tag{14.8}$$

$$\nu \frac{\partial^2 u}{\partial y^2} = \frac{\nu}{t^{2m}} \frac{d^2 u}{d\eta^2}. \tag{14.9}$$

If $m = 1/2$, after the substitution of Eqs. (14.8) and (14.9) into Eq. (14.4), the original independent variable t does not appear in Eq. (14.4). Considering the following procedure to become a simple form, we assume

$$\eta = \frac{y}{2\sqrt{\nu t}}. \tag{14.10}$$

Then, Eq. (14.4) becomes an ordinary differential equation.

$$\frac{d^2 u}{d\eta^2} + 2\eta \frac{du}{d\eta} = 0. \tag{14.11}$$

The independent variable is η in Eq. (14.11), although the independent variables are t and x in Eq. (14.4). Integrating Eq. (14.11) about η twice, we get

$$u = C_1 + C_2 \int_0^\eta e^{-\xi^2} d\xi, \tag{14.12}$$

in which C_1 and C_2 are integral constants. The boundary conditions about η are

$$u = 0, \quad \text{at } \eta = \infty, \tag{14.13}$$

$$u = U_0, \quad \text{at } \eta = 0. \tag{14.14}$$

Computing C_1 and C_2, we obtain

$$u = U_0[1 - erf(\eta)], \tag{14.15}$$

in which "*erf*" is the error function in Chap. 9. It is given by

$$erf(\eta) = \frac{2}{\sqrt{\pi}} \int_0^\eta e^{-\xi^2} d\xi. \tag{14.16}$$

Since two variables are combined as $\eta = y/(2\sqrt{\nu t})$, the method of the separation of variables cannot derive the solution of Eq. (14.4) easily.

Example 14.2. The horizontal wet soil sample in the cylinder is located. The cross-sectional area is constant and initial moisture content is $T_{t=0}$. When the x direction is horizontal and the moisture content T at $x = 0$ and $x = \ell$ are kept to zero, derive the distribution of moisture content in the sample.

Solution. The governing equation is

$$\frac{\partial T}{\partial t} = H^2 \frac{\partial^2 T}{\partial x^2}, \tag{1}$$

in which $x =$ the horizontal coordinate system, $t =$ time, and $H^2 =$ the transmissivity. The boundary conditions at $x = 0$ and ℓ are

$T = 0$. The initial condition is $T = T_{t=0}$. Assuming $T = e^{-mt}u(x)$ and m is a constant, we obtain

$$-mu = H^2 \frac{d^2 u}{dx^2}. \tag{2}$$

Defining $a^2 = m/H^2$ in Eq. (2), we have

$$u = A\sin ax + B\cos ax. \tag{3}$$

Since the boundary condition indicates $B = 0$ and $A \neq 0$, the equation $\sin a\ell = 0$ gives

$$a = \frac{r\pi}{\ell}, \quad \text{for } r = 1, 2, 3, \ldots. \tag{4}$$

Since m is the function of r, m is written as m_r. Then,

$$u_r = A_r \sin \frac{r\pi x}{\ell}, \quad m_r = \left(\frac{Hr\pi}{\ell}\right)^2. \tag{5}$$

If T is dependent on r, T is written as T_r.

$$T_r = A_r e^{-\left(\frac{Hr\pi}{\ell}\right)^2 t} \sin \frac{r\pi x}{\ell}. \tag{6}$$

The solution T is superposed by T_r as.

$$T = \sum_{r=1}^{\infty} T_r = \sum_{r=1}^{\infty} A_r e^{-\left(\frac{Hr\pi}{\ell}\right)^2 t} \sin \frac{r\pi x}{\ell}. \tag{7}$$

The initial condition shows

$$T_{t=0} = \sum_{r=1}^{\infty} A_r \sin \frac{r\pi x}{\ell}. \tag{8}$$

The coefficient A_r is

$$A_r = \frac{2}{\ell} \int_0^\ell T_{t=0} \sin \frac{r\pi x}{\ell} dx, \quad \text{for } r = 1, 2, \ldots. \tag{9}$$

Thus, when $T_{t=0}$ is constant, we get

$$A_r = \frac{2T_{t=0}}{r\pi}[1 - (-1)^r]. \tag{10}$$

The substitution of Eq. (10) into Eq. (7) gives the solution of this problem.

Problem 14.2. In the following partial differential equation:

$$\frac{\partial u}{\partial t} = H^2 \frac{\partial^2 u}{\partial x^2} - k(u - u_0), \tag{11}$$

in which H, k, and u_o are constants. Substituting $u(x,t) = u_0 + e^{-kt}v(x,t)$ into Eq. (11), show that Eq. (11) is

$$\frac{\partial v}{\partial t} = H^2 \frac{\partial^2 v}{\partial x^2}. \tag{12}$$

14.3. Flow near Oscillatory Plate

Consider the fluid motion near the oscillatory plate when the plate is infinitely long. The equation of motion is given by Eq. (14.4). The boundary conditions are

$$u = V_0 \sin \omega t, \quad \text{at } y = 0, \tag{14.17}$$

$$u = 0, \quad \text{at } y = \infty, \tag{14.18}$$

in which V_0 is constant. The solution is assumed to be

$$u(y, t) = \Im[v(y)e^{i\omega t}], \tag{14.19}$$

in which $\Im$ indicates the imaginary part. Substituting Eq. (14.19) into Eq. (14.4), we obtain

$$i\omega v = \nu \frac{d^2 v}{dy^2}. \tag{14.20}$$

When a is

$$a^2 = \frac{i\omega}{\nu} = \frac{\omega}{\nu} e^{i\left(\frac{\pi}{2} + 2k\pi\right)}, \quad \text{for } k = 0, 1, 2, \ldots, \tag{14.21}$$

the solution of Eq. (14.20) is

$$v = A e^{ay} + B e^{-ay}, \tag{14.22}$$

in which A and B are integral constants. As Eq. (14.21) indicates $a = \sqrt{\frac{\omega}{\nu}} e^{\frac{i\pi}{4}}$, Eq. (14.21) is given by

$$a = \sqrt{\frac{\omega}{\nu}} e^{\frac{i\pi}{4}} = \sqrt{\frac{\omega}{2\nu}} (1 + i). \tag{14.23}$$

Since $u = 0$ at $y = \infty$ in Eq. (14.18), $A = 0$ is derived in Eq. (14.22). Equation (14.17) shows

$$B = V_0. \tag{14.24}$$

Thus, we obtain

$$u(x, t) = \Im[V_0 e^{-ay} e^{i\omega t}] = V_0 e^{-\sqrt{\frac{\omega}{2\nu}} y} \sin\left(\omega t - \sqrt{\frac{\omega}{2\nu}} y\right). \tag{14.25}$$

The phase lag between Eqs. (14.17) and (14.25) is $-\sqrt{\frac{\omega}{2\nu}} y$.

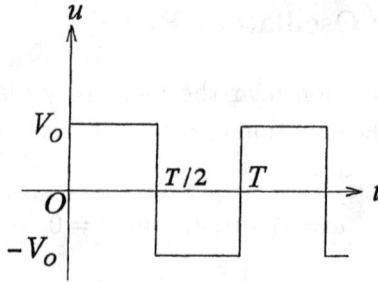

Figure 14.2.   Example 14.3.

Example 14.3. When the plate of infinite length at $y = 0$ moves in the function as shown in Fig. 14.2, find the fluid velocity near the plate.

Solution. The Fourier series of the function as shown in Fig. 14.2 is given by

$$u = \sum_{n=1}^{\infty} u_n = \sum_{n=1}^{\infty} V_n \sin \omega_n t, \tag{1}$$

in which

$$\omega_n = \frac{2\pi n}{T}, \quad u_n = V_n \sin \omega_n t. \tag{2}$$

The velocity u_n is derived from Eq. (14.25). Thus,

$$u_n = V_n e^{-\sqrt{\frac{\omega_n}{2\nu}}y} \sin\left(\omega_n t - \sqrt{\frac{\omega_n}{2\nu}}y\right). \tag{3}$$

The solution of u is

$$u = \sum_{n=1}^{\infty} u_n = \sum_{n=1}^{\infty} V_n e^{-\sqrt{\frac{\omega_n}{2\nu}}y} \sin\left(\omega_n t - \sqrt{\frac{\omega_n}{2\nu}}y\right). \tag{4}$$

Since the motion of the plate at $y = 0$ is expressed by the function in Fig. 14.2, we get

$$u = \sum_{n=1}^{\infty} V_n \sin \frac{2\pi n t}{T}. \tag{5}$$

Since the variable V_n is

$$\int_0^T u \sin \frac{2\pi r t}{T} dt = V_n \int_0^T \sin^2 \frac{2\pi r t}{T} dt, \tag{6}$$

defining $r = n$, we have

$$V_n = \frac{V_0}{\pi n}[1 - (-1)^n].$$

(7)

The substitution of Eq. (7) into Eq. (4) derives the solution.

Problem 14.3. Since the fluid motion induced by the oscillatory motion is dissipated fast and the viscous effect in the fluid flow does not propagate in a long distance, we can use the potential motion for the wave motion. Where does the amplitude of the fluid motion become 1% for the plate motion of 5 sec period by Eq. (14.25)? Assume $\nu = 0.01\,\mathrm{m^2/s}$.

14.4. Distribution of Suspended Load in Flow

The soil or sand which are transported in the flow in the suspended condition by turbulence is called suspended load. Since the distribution of suspended load is very similar to the diffusion or dispertion phenomena, the distribution of suspended load is analyzed by the diffusion equation (Graf, 1984). The continuity condition of suspended load is expressed by Eq. (14.2). Assuming that $v = 0$, $K_y = $ constant, $u = u_m = $ constant, and

$$\frac{\partial^2 C}{\partial x^2} \ll \frac{\partial^2 C}{\partial y^2},$$

(14.26)

we can simplify Eq. (14.2) as

$$\frac{\partial C}{\partial t} + u_m \frac{\partial C}{\partial x} = K_y \frac{\partial^2 C}{\partial y^2} + v_s \frac{\partial C}{\partial y},$$

(14.27)

in which $x = $ the flow direction and $y = $ the vertically upward direction to the bed. As shown in Fig. 14.3, if we assume that the bed is fixed for $x < 0$, movable for $x \geq 0$, and the water depth is h and the flow is steady and uniform, Eq. (14.27) is

$$u_m \frac{\partial C}{\partial x} = K_y \frac{\partial^2 C}{\partial y^2} + v_s \frac{\partial C}{\partial y}.$$

(14.28)

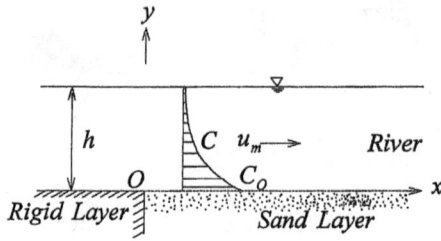

Figure 14.3.   Distribution of suspended load.

Referring to Fig. 14.3, the boundary conditions are

$$C = 0, \quad \text{at } x = 0, \tag{14.29}$$

$$C = 0, \quad \text{at } y = \infty, \tag{14.30}$$

$$C = C_0, \quad \text{at } y = 0. \tag{14.31}$$

Introducing the following assumption:

$$C(x, y) = f(y) + C'(x, y). \tag{14.32}$$

Substituting Eq. (14.32) into Eq. (14.28), we have the following two equations:

$$K_y \frac{d^2 f(y)}{dy^2} + V_s \frac{df(y)}{dy} = 0, \tag{14.33}$$

$$u_m \frac{\partial C'}{\partial x} = K_y \frac{\partial^2 C'}{\partial y^2} + V_s \frac{\partial C'}{\partial y}. \tag{14.34}$$

The boundary conditions for $f(y)$ are

$$f(y) = C_0, \quad \text{at } y = 0, \tag{14.35}$$

$$f(y) = 0, \quad \text{at } y = \infty. \tag{14.36}$$

Solving Eq. (14.33) with the boundary conditions Eqs. (14.35) and (14.36), we get

$$f(y) = C_0 e^{-\frac{v_s}{K_y} y}. \tag{14.37}$$

The concentration in the real flow is almost zero in the free water surface. But we use the boundary condition for Eq. (14.36) for convenience. The boundary conditions for $C'(x, y)$ are

$$C' = -C_0 e^{-\frac{v_s}{K_y} y}, \quad \text{at } x = 0, \tag{14.38}$$

$$C' = 0, \quad \text{at } y = n, \tag{14.39}$$

$$C' = 0, \quad \text{at } y = 0. \tag{14.40}$$

Applying the separation of variables to Eq. (14.34), we assume

$$C'(x, y) = X(x)Y(y). \tag{14.41}$$

Then, Eq. (14.34) is written as

$$\frac{u_m \dot{X}}{X} = \frac{K_y \ddot{Y} + v_s \dot{Y}}{Y} = -\mu, \tag{14.42}$$

in which "$\cdot$" shows differentiation and μ is a constant. Solving Eq. (14.42), we get

$$X(x) = e^{-\frac{\mu}{u_m} x}, \tag{14.43}$$

$$Y(y) = e^{-\frac{v_s}{2K_y} y}[C_1 \sin \beta y + C_2 \cos \beta y], \tag{14.44}$$

in which

$$\beta = \frac{1}{2}\sqrt{\frac{4\mu}{K_y} - \left(\frac{v_s}{K_y}\right)^2}, \quad \mu \geq \frac{v_s^2}{4K_y}. \tag{14.45}$$

Equation (14.40) indicates

$$C_2 = 0. \tag{14.46}$$

Since $C' = 0$ at $y = h$, we obtain

$$\beta = \frac{n\pi}{h}, \quad \text{for } n = 1, 2, 3, \ldots. \tag{14.47}$$

Since μ is dependent on n, we write μ_n. From Eqs. (14.45) and (14.47), we have

$$\mu_n = \frac{K_y}{4}\left[\left(\frac{v_s}{K_y}\right)^2 + \left(\frac{2n\pi}{h}\right)^2\right]. \tag{14.48}$$

Thus, C' is given by

$$C' = \sum_{n=1}^{\infty} A_n e^{-\frac{\mu_n}{u_m} x} e^{-\frac{v_s}{2K_y} y} \sin \frac{n\pi y}{h}. \tag{14.49}$$

The coefficient A_n in the Fourier sine series is obtained by Eq. (14.38) as follows:

$$-C_0 e^{-\frac{v_s}{2K_y}y} = \sum_{n=1}^{\infty} A_n \sin\frac{n\pi y}{h}. \qquad (14.50)$$

Thus, we have

$$A_r = -\frac{2}{h}C_0 \int_0^h e^{-\frac{v_s}{2K_y}y} \sin\frac{r\pi y}{h} dy$$

$$= -\frac{C_0 K_y}{\mu_r}\frac{2}{h}\frac{r\pi}{h}[1 - e^{-\frac{v_s h}{2K_y}}(-1)^r], \quad \text{for } r = 1,2,3,\ldots. \qquad (14.51)$$

Substituting Eq. (14.51) into Eq.(14.49), we obtain C'. The solution C consists of $f(y)$ and $C'(x, y)$. Thus, we get

$$C = C_0 e^{-\frac{v_s}{K_y}y} - \sum_{r=1}^{\infty} \frac{C_0 K_y}{\mu_r}\frac{2}{h}\frac{r\pi}{h}$$

$$\times [1 - e^{-\frac{v_s h}{2K_y}}(-1)^r]e^{-\frac{\mu_n}{u_m}x}e^{-\frac{v_s}{2K_y}y}\sin\frac{n\pi y}{h}. \qquad (14.52)$$

This is the solution of this problem.

Example 14.4. When the flow and the diffusion are steady in the concentration of suspended load, the concentration is dependent on y. Then, Eq. (14.2) is written as

$$\frac{d}{dy}\left(K_y\frac{dC}{dy}\right) + v_s\frac{dC}{dy} = 0. \qquad (1)$$

If the diffusion coefficient K_y in the y direction is given by

$$K_y = KU_*\frac{y(h-y)}{h}, \qquad (2)$$

show that the concentration of suspended load C is obtained by

$$\frac{C}{C_a} = \left(\frac{h-y}{y}\cdot\frac{a}{h-a}\right)^Z, \qquad (3)$$

in which $C = C_a$ at $y = a$, $C = 0$ at $y = h$, and $Z = \frac{v_s}{KU_*}$.

Solution. Integrating Eq. (1) with the condition of $C = 0$ at $y = h$, we obtain

$$\frac{KU_*y}{h}(h-y) + v_sC = 0. \qquad (4)$$

Integrating Eq. (4) with the condition $C = C_a$ at $y = a$, we have

$$\frac{C}{C_a} = \left(\frac{h-y}{y} \frac{a}{h-a} \right)^Z. \tag{5}$$

This is the concentration of suspended load and called a Rouse equation in sediment transport.

Problem 14.4. Obtain the solution when $f(h) = 0$ instead of Eq. (14.36). This indicates that the concentration of suspended load is zero at the free water surface $y = h$.

14.5. Well Hydraulics

Consider an aquifer with the following characteristics: (1) horizontal, (2) confined between impermeable layers on top and bottom, (3) infinite in horizontal extent, (4) constant thickness, (5) homogeneous and isotropic, and (6) Darcy law condition. Furthermore, assume for simplicity: (1) a single pumping, (2) a constant pumping rate, (3) negligible well diameter relative to the aquifer's horizontal dimensions, (4) well penetration through the entire aquifer depth, and (5) uniform initial piezometric (hydraulic) head throughout the aquifer. The horizontal motion of confined groundwater is given

$$\frac{\partial^2 h}{\partial x^2} + \frac{\partial^2 h}{\partial y^2} = \frac{S}{T} \frac{\partial h}{\partial t}, \tag{14.53}$$

in which $h =$ the piezometric head, $S =$ the strativity, x and $y =$ the horizontal and vertical coordinate system, respectively, $t =$ time, and $T =$ the transmissivity. Since the flow is symmetric about the origin, the piezometric head is also point symmetry. Equation (14.53) in the Cartesian coordinate system $x - y$ is transformed to the cylindrical coordinate system $r - \theta$. Then, Eq. (14.53) is transformed to

$$\frac{\partial^2 h}{\partial r^2} + \frac{1}{r} \frac{\partial h}{\partial r} + \frac{1}{r^2} \frac{\partial^2 h}{\partial \theta^2} = \frac{S}{T} \frac{\partial h}{\partial t}. \tag{14.54}$$

Since the piezometric head h is independent of θ, Eq. (14.54) is written as

$$\frac{\partial^2 h}{\partial r^2} + \frac{1}{r} \frac{\partial h}{\partial r} = \frac{S}{T} \frac{\partial h}{\partial t}. \tag{14.55}$$

The boundary conditions are

$$h = h_0, \quad \text{at } t = 0, \tag{14.56}$$

$$h = h_0, \quad \text{at } r = \infty. \tag{14.57}$$

Referring to Sec. 14.2, we transform the independent variables t and r to a new variable ξ as (Bluman and Cole, 1974)

$$\xi = \frac{1}{2}\sqrt{\frac{S}{T}}\frac{r}{\sqrt{t}}. \tag{14.58}$$

Substituting Eq. (14.58) into Eq. (14.55), we obtain the following ordinary differential equation:

$$\frac{d^2h}{d\xi^2} + \left(\frac{1}{\xi} + 2\xi\right)\frac{dh}{d\xi} = 0. \tag{14.59}$$

Integrating Eq. (14.59) twice, we have

$$h = h_0 - C_1 \int_\xi^\infty \frac{e^{-\zeta^2}}{\zeta}d\zeta. \tag{14.60}$$

An integral constant C_1 is determined from the condition of the constant pumping discharge Q as

$$Q = 2\pi r Dk \left.\frac{dh}{dr}\right|_{r=0} = 2\pi T C_1. \tag{14.61}$$

Then, we get

$$C_1 = \frac{Q}{2\pi T}. \tag{14.62}$$

Substituting $\eta = \zeta^2$ into Eq. (14.60), we gain

$$h_0 - h = \frac{Q}{2\pi T}\int_\eta^\infty \frac{e^{-\eta}}{2\eta}d\eta = \frac{Q}{4\pi T}W(\eta). \tag{14.63}$$

The function $W(\eta)$ is called a Theis well function. The function $W(\eta)$ is expressed by

$$W(\eta) = -\gamma - \ln\eta + \eta - \frac{\eta^2}{2\cdot 2!} + \frac{\eta^3}{3\cdot 3!} - \cdots, \tag{14.64}$$

in which $\gamma =$ a Euler number. The Euler number is defined by

$$\gamma = \lim_{n \to \infty} \left(1 + \frac{1}{2} + \frac{1}{3} + \cdots + \frac{1}{n} - \log n \right) = \int_0^\infty e^{-t} \ln \left(\frac{1}{t} \right) dt.$$

(14.65)

Equation (14.65) is transformed to

$$\int_\eta^\infty \frac{e^{-t}}{t} dt = -\gamma - \lim_{t \to 0} e^{-t} \ln t - \int_0^\eta \frac{1}{t} \left(1 - t + \frac{t^2}{2!} - \frac{t^3}{3!} + \cdots \right) dt,$$

(14.66)

in which

$$\lim_{t \to 0} (e^{-t} - 1) \ln t = 0$$

(14.67)

and

$$\int_\eta^\infty \frac{e^{-t}}{t} dt = -\gamma - \ln \eta + \eta - \frac{\eta^2}{2 \cdot 2!} + \frac{\eta^3}{3 \cdot 3!} - \cdots .$$

(14.68)

The left side in Eq. (14.68) corresponds to $W(\eta)$.

Example 14.5. Derive Eq. (14.59).

Solution.

$$\frac{\partial h}{\partial r} = \frac{dh}{d\xi} \cdot \frac{\partial \xi}{\partial r} = \frac{dh}{d\xi} \cdot \frac{1}{2} \sqrt{\frac{S}{T}} \cdot \frac{1}{\sqrt{t}},$$

(1)

$$\frac{\partial^2 h}{\partial r^2} = \frac{d^2 h}{d\xi^2} \cdot \left(\frac{\partial \xi}{\partial r} \right)^2 = \frac{d^2 h}{d\xi^2} \cdot \frac{1}{4} \frac{S}{T} \cdot \frac{1}{t},$$

(2)

$$\frac{\partial h}{\partial t} = \frac{dh}{d\xi} \cdot \left(-\frac{1}{4} \right) \sqrt{\frac{S}{T}} \cdot \frac{r}{t^{1.5}}.$$

(3)

Substituting Eqs. (1), (2), and (3) into Eq. (14.55), we obtain Eq. (14.59).

Problem 14.5. Solve the following partial differential equation:

$$\frac{\partial \phi}{\partial t} = \frac{\partial^2 \phi}{\partial x^2}$$

(4)

with the following boundary conditions:

$$\phi = 0, \quad \text{at } x = 0, \tag{5}$$

$$\phi = 0, \quad \text{at } t = \infty, \tag{6}$$

$$\phi = 1, \quad \text{at } x = \infty, \tag{7}$$

$$\phi = 1, \quad \text{at } t = 0. \tag{8}$$

Solution.

$$\phi = \frac{2}{\sqrt{\pi}} \int_0^\xi e^{-\frac{\xi^2}{2}} d\xi, \tag{9}$$

in which $\xi = \frac{x}{2\sqrt{t}}$.

14.6. Water Content Distribution in Ground

The infiltration as the vertical motion of water content in the ground is expressed by

$$\frac{\partial v}{\partial t} = H^2 \frac{\partial^2 v}{\partial x^2}, \tag{14.69}$$

in which H^2 = the transmissivity and $F(x)$ = the initial water content. Since the solution of Eq. (14.69) decreases as t increases, we assume

$$v = e^{-a^2 t} u(x). \tag{14.70}$$

Substituting Eq. (14.70) into Eq. (14.69), we get

$$-a^2 u = H^2 \frac{d^2 u}{dx^2}. \tag{14.71}$$

Defining α by the following equation:

$$a^2 / H^2 = \alpha^2. \tag{14.72}$$

We assume the special solution of Eq. (14.71) as

$$u = \cos \alpha (x - \beta), \tag{14.73}$$

in which β = an integral constant. Using Eq. (14.70), we have

$$v_1 = e^{-\alpha^2 H^2 t} \cos \alpha (x - \beta). \tag{14.74}$$

Multiplying $F(\beta)/\pi$ to Eq. (14.74) and defining v_2, we have

$$v_2 = \frac{F(\beta)}{\pi} e^{-\alpha^2 H^2 t} \cos \alpha(x - \beta). \tag{14.75}$$

The double integrals of v_2 about α and β is given by

$$v = \frac{1}{\pi} \int_0^\infty \int_{-\infty}^\infty F(\beta) e^{-\alpha^2 H^2 t} \cos \alpha(x - \beta) d\alpha\, d\beta, \tag{14.76}$$

in which $\alpha > 0$ and $-\infty < \beta < \infty$. For $t = 0$, we have

$$v(x, t = 0) = \frac{1}{\pi} \int_0^\infty \int_{-\infty}^\infty F(\beta) \cos \alpha(x - \beta) d\alpha\, d\beta. \tag{14.77}$$

Equation (14.77) is the Fourier integral of $F(x)$ (Eq. (10.47)). The following integral is computed as

$$\int_0^\infty e^{-y^2} \cos 2by\, dy = \frac{\sqrt{\pi} e^{-b^2}}{2}. \tag{14.78}$$

Thus, Eq. (14.76) is written as

$$v(x, t) = \frac{1}{2H\sqrt{\pi t}} \int_{-\infty}^\infty F(\beta) e^{-\frac{(x-\beta)^2}{4H^2 t}} d\beta. \tag{14.79}$$

This is the solution of Eq. (14.69) for the initial water content $F(x)$.

Example 14.6. Show that if $F(\beta) = 1$ in Eq. (14.79), $v(x, t) = 1$.

Solution. Defining

$$\frac{x - \beta}{2H\sqrt{t}} = u. \tag{1}$$

Differentiating Eq. (1), we get

$$-d\beta = 2H\sqrt{t}du. \tag{2}$$

Equation (14.79) is

$$v(x, t) = \frac{1}{2H\sqrt{\pi t}} \int_{-\infty}^\infty 2H\sqrt{t} e^{-u^2} du = 1, \tag{3}$$

in which

$$1 = \frac{2}{\sqrt{\pi}} \int_0^\infty e^{-\xi^2} d\xi. \tag{4}$$

The range of integral of u is from ∞ to $-\infty$ in Eq. (3), since $-\infty < \beta < \infty$.

Problem 14.6. Show that when the two-dimensional heat conduction becomes steady, the governing equation becomes the potential equation.

14.7. Modeling of Coastal Changes

Let us consider the one-line theory to explain coastal changes presented by Pelnard (Mizumura, 1992b). The continuity equation of sand is given by

$$\frac{\partial X_0}{\partial t} + \frac{1}{(1-\lambda)h_i}\frac{\partial Q_y}{\partial y} = 0, \qquad (14.80)$$

in which t = time, X_0 = the position of the shoreline, y = the coordinate along the shoreline, λ = void ratio of sand, h_i = the critical water depth for sand moving, and Q_y = the longshore discharge of sand. When the water depth is shallower than h_i, sand moves and the shore changes. The longshore discharge of sand Q_y is assumed to be

$$Q_y = F\sin\alpha_b \cos\alpha_b, \qquad (14.81)$$

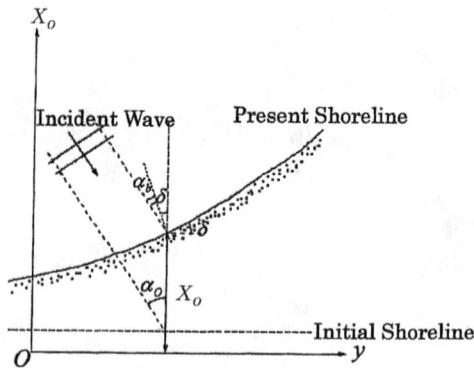

Figure 14.4.   One-line theory.

in which $F = $ a constant and $\alpha_b = $ an angle of incidence at the breaking point. The angle of incidence is measured from the normal to the shoreline. Referring to Fig. 14.4, we get the following relationship:

$$\alpha_b = \alpha_0 - \delta = \alpha_0 - \tan^{-1}\frac{\partial X_0}{\partial y}, \tag{14.82}$$

in which $\alpha_0 = $ an angle of incidence for the initial shoreline and $\delta = $ an angle between the present and the initial shorelines. Substituting Eq. (14.82) into Eq. (14.81) and assuming that α_0, α_b, and $\partial X_0/\partial y$ are very small, we get the Taylor series as

$$\sin\alpha_b = \sin\left(\alpha_0 - \tan^{-1}\frac{\partial X_0}{\partial y}\right) \cong \sin\alpha_0 - \cos\alpha_0 \tan^{-1}\frac{\partial X_0}{\partial y}. \tag{14.83}$$

If we assume

$$\sin\alpha_0 \cong \tan\alpha_0, \quad \cos\alpha_0 \cong 1, \quad \tan^{-1}\frac{\partial X_0}{\partial y} \cong \frac{\partial X_0}{\partial y}, \quad \cos\alpha_b \cong 1, \tag{14.84}$$

we have

$$\sin\alpha_b \cong \tan\alpha_0 - \frac{\partial X_0}{\partial y}. \tag{14.85}$$

Thus, the longshore discharge of sand is given by

$$Q_y = F\left(\tan\alpha_0 - \frac{\partial X_0}{\partial y}\right). \tag{14.86}$$

Substituting Eq. (14.86) into Eq. (14.80), we obtain

$$\frac{\partial X_0}{\partial t} - \frac{F}{(1-\lambda)h_i}\frac{\partial^2 X_0}{\partial y^2} = 0. \tag{14.87}$$

This is the diffusion equation. This is applicable to estimate shoreline changes after the construction of a groyne or breakwater. The boundary condition of the construction of a groyne at $y = 0$ and $t = 0$ is given by

$$X_0 = 0, \quad \text{at } t = 0, \tag{14.88}$$

$$\frac{\partial X_0}{\partial y} = \tan\alpha_0, \quad \text{at } y = 0, \tag{14.89}$$

$$X_0 = 0, \quad \text{at } y = \pm\infty. \tag{14.90}$$

Defining in Eq. (14.87)

$$A_0 = \frac{F}{(1-\lambda)h_i},\qquad(14.91)$$

we use the Laplace transform of X_0 about t in Chap. 11. Then, Eq. (14.87) is

$$sx_0 - A_0\frac{d^2x_0}{dy^2} = 0,\qquad(14.92)$$

in which $x_0 = L\{X_0\}$. Since the Laplace transform of a constant C_0 is C_0/s, assuming that $X_0 = C_0$ at $y = 0$, the solution of Eq. (14.92) is

$$x_0 = \frac{C_0}{s}e^{-qy},\qquad(14.93)$$

in which $q = \sqrt{s/A_0}$. The Laplace transform of $\partial X_0/\partial y|_{y=0}$ about t is

$$\frac{dx_0}{dy}\bigg|_{y=0} = L\left\{\frac{\partial X_0}{\partial y}\bigg|_{y=0}\right\} = \frac{\tan\alpha_0}{s}.\qquad(14.94)$$

From Eq. (14.93), we get

$$\frac{dx_0}{dy}\bigg|_{y=0} = -\frac{C_0}{\sqrt{sA_0}}.\qquad(14.95)$$

The comparison between Eqs. (14.94) and (14.95) shows

$$C_0 = -\frac{\sqrt{A_0}\tan\alpha_0}{\sqrt{s}}.\qquad(14.96)$$

Substituting Eq. (14.96) into Eq. (14.93), we get

$$x_0 = -\frac{\sqrt{A_0}\tan\alpha_0}{s^{3/2}}e^{-qy}.\qquad(14.97)$$

Since $q = \sqrt{s/A_0}$, let us consider the inverse Laplace transform of

$$\frac{e^{-k\sqrt{s}}}{s^{3/2}},\quad\text{for } k \geq 0.\qquad(14.98)$$

The inverse Laplace transform (Churchill, 1958) is

$$L^{-1}\left\{\frac{e^{-k\sqrt{s}}}{s^{3/2}}\right\} = 2\sqrt{\frac{t}{\pi}}\exp\left(-\frac{k^2}{4t}\right) - k\,erfc\left(\frac{k}{2\sqrt{t}}\right).\qquad(14.99)$$

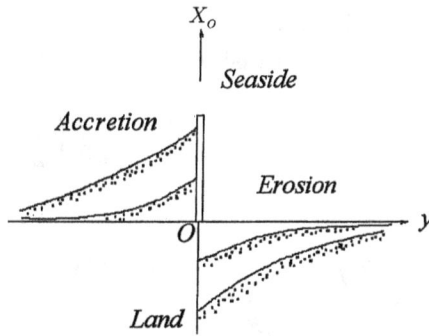

Figure 14.5.   Shoreline changes by one-line theory.

If $k = y/\sqrt{A_0}$, Eq. (14.97) is

$$X_0 = -\sqrt{A_0} \tan \alpha_0 \left[2\sqrt{\frac{t}{\pi}} \exp\left(-\frac{y^2}{4tA_0}\right) - \frac{y}{\sqrt{A_0}} erfc\left(\frac{y}{2\sqrt{A_0 t}}\right)\right],$$

$$(14.100)$$

in which

$$erfc\, X = \frac{2}{\sqrt{\pi}} \int_X^\infty e^{-p^2} dp. \qquad (14.101)$$

Equation (14.100) is described in Fig. 14.5. As explained, the linear problem which has the boundary conditions can be solved by using the Laplace transform. When we use the Laplace transform, we treat the problem of the inverse Laplace transform. This is not easy. Therefore, the Laplace transform is not widely used in the field of hydraulic engineering.

Example 14.7. We used the boundary conditions Eqs. (14.88), (14.89), and (14.90) for the solution of Eq. (14.87). Instead of Eq. (14.89), if we use

$$X_0 = C_0, \quad \text{at } y = 0, \qquad (1)$$

derive the solution of Eq. (14.87).

Solution. The inverse Laplace transform is Eq. (14.93) for this condition. The inverse Laplace transform (Churchill, 1958) is

$$X_0 = L^{-1} \left\{ \frac{C_0}{s} e^{-qy} \right\} = C_0 \left[1 - erf \left(\frac{y}{2\sqrt{Dt}} \right) \right], \tag{2}$$

in which

$$erf\ X = \frac{2}{\sqrt{\pi}} \int_0^X e^{-p^2}\, dp. \tag{3}$$

Equation (2) is the solution of this problem.

Problem 14.7. Assuming $X_0 = f[\theta(y,t)]$, solve the following diffusion equation:

$$\frac{\partial X_0}{\partial t} + u_0 \frac{\partial X_0}{\partial y} = A_0 \frac{\partial^2 X_0}{\partial y^2}, \tag{4}$$

in which u_0 and A_0 are constants.

Chapter 15

Solution of Nonlinear Equations

The energy equation (Bernoulli equation) is fundamental in hydraulic engineering and it is a nonlinear algebraic equation. When the nonlinear wave equation is numerically solved, the method of characteristics is often used. Then, the algebraic equations that are obtained by the method of characteristics form nonlinear equations and are numerically solved. When a nonlinear equation is quadratic, we can use the root formula of a quadratic equation. But when the order of the algebraic equation is more than two, the iteration method is often used in engineering. The algebraic equation of whose order is more than two has more than two solutions. We must be careful of the selection of the solutions. Usually we check the physical condition of the solution and find one of them. A nonlinear solution often contains unnecessary solutions that are not physically realistic.

15.1. Cubic Equation

A cubic equation is generally written as

$$Z^3 + A_2 Z^2 + A_1 Z + A_0 = 0, \tag{15.1}$$

in which A_2, A_1 and A_0 are constants. Substituting the following variable Z into Eq. (15.1):

$$Z = x - \frac{A_2}{3} \tag{15.2}$$

and eliminating the term of Z^2, we have

$$r_1 = \frac{A_2^2}{3} - A_1, \tag{15.3}$$

$$r_2 = -A_0 - \frac{2}{27} A_2^3 + \frac{A_1 A_2}{3}. \tag{15.4}$$

Then, Eq. (15.1) becomes

$$x^3 - r_1 x - r_2 = 0. \tag{15.5}$$

273

Equation (15.5) has one real solution. According to the coefficients r_1 and r_2 in Eq. (15.5), it has one or three real solutions. When the number of the solutions is one, $27r_2^2 > 4r_1^3$ is established. When the number of the solutions is three, $27r_2^2 < 4r_1^3$ is formed. As an example, consider the steady open channel flow when the specific energy E is constant.

$$E = \frac{v^2}{2g} + h, \tag{15.6}$$

in which $v = $ the velocity, $g = $ the gravitational acceleration, and $h = $ the water depth. If the discharge per unit width is q, Eq. (15.6) is rewritten by

$$E = \frac{q^2}{2gh^2} + h. \tag{15.7}$$

Since the specific energy takes a minimum $3h_c/2$ for the critical depth h_c, if the specific energy is equal to $2h_c$, we can write

$$\frac{q^2}{2gh^2} + h = 2h_c. \tag{15.8}$$

Equation (15.8) has two roots for subcritical and supercritical flow.

Since the critical depth is equal to $h_c = (q^2/g)^{1/3}$, we have from Eq. (15.8),

$$h^3 - 2h_ch^2 + \frac{h_c^3}{2} = 0. \tag{15.9}$$

In this case, since $A_0 = h_c^3/2$, $A_1 = 0$, and $A_2 = -2h_c$, we have

$$r_1 = \frac{4h_c^2}{3}, \tag{15.10}$$

$$r_2 = -\frac{h_c^3}{2} + \frac{2}{27} \cdot 8h_c^3 = \frac{5}{54}h_c^3. \tag{15.11}$$

Thus, since

$$27r_2^2 < 4r_1^3 \tag{15.12}$$

is formed, Eq. (15.8) has three roots. Since Eq. (15.9) is positive for $h = 0$, Eq. (15.9) has two positive roots and a negative root or three negative roots. Since Eq. (15.9) has an extreme at $h = 0$, Eq. (15.9) has two positive roots. The root formula for the cubic equation gives

$$x_1 = \frac{2}{\sqrt{3}} r_1^{1/2} \cos \frac{\phi}{3}, \qquad (15.13)$$

$$x_2 = -\frac{2}{\sqrt{3}} r_1^{1/2} \cos \frac{\pi - \phi}{3}, \qquad (15.14)$$

$$x_3 = -\frac{2}{\sqrt{3}} r_1^{1/2} \cos \frac{\pi + \phi}{3}, \qquad (15.15)$$

in which ϕ is defined as

$$\cos \phi = \left(\frac{3}{r_1}\right)^{3/2} \frac{r_2}{2}. \qquad (15.16)$$

If $27r_2^2 > 4r_1^3$, the cubic equation has one root. For $r_1 > 0$ and $r_2 > 0$,

$$x_0 = \frac{2}{\sqrt{3}} r_1^{1/2} \cosh \frac{\phi}{3}, \quad \cos \phi = \left(\frac{3}{r_1}\right)^{1.5} \frac{r_2}{2}. \qquad (15.17)$$

For $r_1 < 0$ and $r_2 > 0$,

$$x_0 = \frac{2}{\sqrt{3}} (-r_1)^{1/2} \sinh \frac{\phi}{3}, \quad \sinh \phi = \left(\frac{3}{-r_1}\right)^{1.5} \frac{r_2}{2}. \qquad (15.18)$$

For $r_2 < 0$, the substitution of $X = -x$ into Eq. (15.5) shows the cubic equation of X corresponds to the previous two cases.

Example 15.1. From Eq. (15.8) calculate the water depth h when $q = 1\,\mathrm{m^2/s}$.

Solution. In Eq. (15.9), $h_c = (q^2/g)^{1/3} = 0.467$. Equation (15.9) becomes

$$h^3 - 2 \times 0.467h^2 + \frac{0.467^3}{2} = 0. \qquad (1)$$

Thus, we get $r_1 = 0.291$ and $r_2 = 9.43 \times 10^{-3}$. Equation (15.16) indicates

$$\phi = 81°. \qquad (2)$$

Therefore, Eqs. (15.13), (15.14) and (15.15) show

$$x_1 = 0.555, \quad \Rightarrow \quad h_1 = 0.866, \tag{3}$$

$$x_2 = -0.522, \quad \Rightarrow \quad h_2 = -0.211, \tag{4}$$

$$x_3 = -0.0326, \quad \Rightarrow \quad h_3 = 0.279, \tag{5}$$

in which h_1 and h_3 indicate the subcritical and supercritical water depths, respectively. If we use the iteration method, it takes much more time and we must be careful whether it is the subcritical or supercritical water depth. In this case, $A_0 = 0.051, A_1 = 0$, and $A_2 = -0.934$ are obtained in Eq. (15.1).

Problem 15.1. Obtain the water depth h in Eq. (15.8) for $q = 5\,\text{m}^2/\text{s}$.

15.2. Quintic Equation

The algebraic equation of which order is more than four does not have a root formula. But in hydraulic engineering we need to solve a quintic equation to get a uniform depth when Manning's formula is used. The Manning's formula for mean velocity in pipes or open channels is given as

$$v = \frac{1}{n} R^{2/3} S_0^{1/2}, \tag{15.19}$$

in which $n = $ Manning's roughness coefficient, $S_0 = $ the bottom slope, and $R = $ the hydraulic radius. When the discharge Q is given, consider the water depth h in the open channel of rectangular cross section of which width is B. The continuity equation shows

$$Q = Bhv. \tag{15.20}$$

The hydraulic radius R is defined by

$$R = \frac{Bh}{B + 2h}. \tag{15.21}$$

Substituting Eqs. (15.19) and (15.21) into Eq. (15.20), we have

$$Q = \frac{Bh}{n} \left(\frac{Bh}{B + 2h} \right)^{2/3} S_0^{1/2}. \tag{15.22}$$

Rearranging Eq. (15.22), we get

$$\left(\frac{Qn}{B\sqrt{S_0}}\right)^3 = h^3 \left(\frac{Bh}{B+2h}\right)^2. \qquad (15.23)$$

This leads to the following quintic equation about h:

$$h^5 = \frac{1}{B^2}\left(\frac{Qn}{B\sqrt{S_0}}\right)^3 (4h^2 + 4Bh + B^2). \qquad (15.24)$$

This is generally solved by the iteration method. Assume h on the right side of Eq. (15.24) as $1\,\mathrm{m}$ (the water depth in natural rivers is almost $1\,\mathrm{m}$). A new water depth h is obtained from the left side of Eq. (15.24). Substituting this obtained water depth to the right side, we get a new water depth from the left side. These procedures must be continued until the water depth becomes constant and converges.

Example 15.2. Calculate the uniform water depth h from Eq. (15.24) when discharge $Q = 10\,\mathrm{m}^3/\mathrm{s}$, the width of open channel $B = 3\,\mathrm{m}$, the bottom slope $S_0 = 0.01$, and the Manning's roughness coefficient $n = 0.03$.

Solution. Equation (15.24) with the numerical values of Q, B, S_0, and n leads to the following quintic equation:

$$h^5 = 0.111(4h^2 + 12h + 9). \qquad (1)$$

Substituting $h = 1\,\mathrm{m}$ into the right side of Eq. (1), we get a new water depth $h = 1.227\,\mathrm{m}$ from the left side. Substituting this water depth $h = 1.227\,\mathrm{m}$ into the right side, we get $h = 1.27\,\mathrm{m}$ again. Stillmore, we get $1.278\,\mathrm{m}, \ldots, 1.28\,\mathrm{m}$. The final value is convergent. This is the solution of Eq. (1).

Problem 15.2. When discharge $Q = 20\,\mathrm{m}^3/\mathrm{s}$, the channel width $B = 10\,\mathrm{m}$, the bottom slope $S_0 = 0.001$, and the Manning's roughtness coefficient $n = 0.04$, calculate the water depth h from Eq. (15.24).

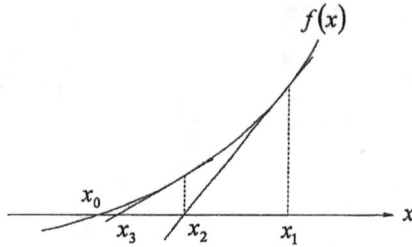

Figure 15.1.   Newton method.

15.3. Newton Method

When a function is convex in the downward, as shown in Fig. 15.1, the function intersects at x_0 on the x axis. The method to compute the intersection point x_0 is the Newton method in Chap. 2. Select a point x_1 which is much larger than x_0. This is an initial value. Then, a tangential line of $f(x)$ at x_1 is written as

$$y - f(x_1) = f'(x_1)(x - x_1), \qquad (15.25)$$

in which "$'$" $= d/dx$. An intersection point x_2 between the tangential line Eq. (15.25) and the x axis is obtained by substituting $y = 0$ in Eq. (15.25). It is given by

$$x_2 = x_1 - \frac{f(x_1)}{f'(x_1)}. \qquad (15.26)$$

Stillmore, obtain a tangential line of $f(x)$ at x_2 and an intersection point x_3 between the tangential line and the x axis is obtained in the same way. These procedures immediately approach the intersection point x_0 between $f(x)$ and the x axis. As an example, let us consider an intersection point between the logarithmic law and the viscous sub-layer in viscous flow. The logarithmic law is formed in the whole cross-section except being near the wall. At the wall the effect of viscosity is very important and the velocity distribution is strongly affected by the viscosity. Thus, the intersection point is given by a point of the same velocity between the logarithmic law and the viscous sub-layer as

$$\frac{U_*^2 y}{\nu} = U_* \left[2.5 \ln \frac{U_* y}{\nu} + 5.5 \right], \qquad (15.27)$$

in which $U_* =$ shear velocity, $\nu =$ the kinematic viscosity, and $y =$ the coordinate system normal to the wall. Substituting $X = U_* y / \nu$ into Eq. (15.27), we get

$$X = 2.5 \ln X + 5.5. \tag{15.28}$$

Defining $f(X)$ in the following equation:

$$f(X) = X - 2.5 \ln X - 5.5. \tag{15.29}$$

The derivative of Eq. (15.29) is

$$f'(X) = 1 - \frac{2.5}{X}. \tag{15.30}$$

Substituting Eqs. (15.29) and (15.30) into Eq. (15.26), we obtain

$$X_2 = X_1 - \frac{X_1 - 2.5 \ln X_1 - 5.5}{1 - 2.5/X_1}. \tag{15.31}$$

Substituting a large value for X_1 in Eq. (15.31) as the initial value, we have X_2 from Eq. (15.31). A new value X_3 is computed from Eq. (15.31) substituting X_2 on the right side instead of X_1. The difference between X_2 and X_3 decreases and the procedures must be continued until the difference becomes very small and negligible. As the selection of the initial value, the range of a root of the behavior of the function must be investigated beforehand.

Example 15.3. Derive X from Eq. (15.31). Assume the initial value of $X_1 = 20$.

Solution. Substituting $X_1 = 20$ into Eq. (15.31), we get $X_2 = 11.99$. Stillmore, substituting $X_1 = 11.99$ into Eq. (15.31), we obtain $X_2 = 11.64$. This is the correct value. The convergence is very fast. Equation (15.28) has another root near $X \cong 0$. The derivative $f'(X)$ is zero at $X = 2.5$ and $f(X = 2.5) < 0$. Since $f(X) \to \infty$ as $X \to 0$ and $f(X) \to \infty$ as $X \to +\infty$, $f(X) = 0$ has two roots.

Problem 15.3. Solve Example 15.1 by the Newton method.

15.4. Solution of Differential Equation and Secant Method

The water depth or water surface profile of the open channel flow (Chow, 1959) is given by

$$\frac{dh}{dx} = \frac{S_0 - S_f}{1 - F^2},$$ (15.32)

in which h = the water depth, S_0 = the bottom slope, S_f = the energy slope, and F = the Froude number. If the channel width is very large, the Froude number is defined by

$$F = \frac{v}{\sqrt{gh}}$$ (15.33)

and the energy slope is also obtained from the Manning's formula as

$$S_f = \frac{v^2 n^2}{h^{4/3}},$$ (15.34)

in which v = the velocity and n = the Manning's roughness coefficient. Since the general river flow is subcritical, substituting Eq. (15.33) and (15.34) into Eq. (15.32) and referring to Fig. 15.2, we get

$$h_j = h_{j+1} - \frac{\Delta x}{2}\left[S_{0,j+1} + S_{0,j} - S_{f,j+1} - S_{f,j}\right] \div \left[1 - \frac{F_{j+1}^2 + F_j^2}{2}\right],$$ (15.35)

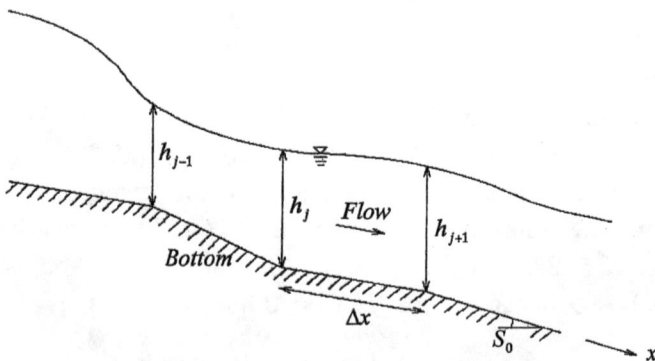

Figure 15.2.   Open channel flow.

in which Δx = space increment, j = nodal point in space, and $\div$ indicates division. The most downstream water depth h_{j+1} is given as the downstream boundary condition. In computations in the fields, the boundary condition must be given at any place in the computational domain. The selection of the boundary condition is more important than solving nonlinear differential equations. The water depth h_j at j step is computed from Eq. (15.35). Since $S_{f,j}$ and F_j in Eq. (15.35) contain an unknown variable h_j, first assume $h_j = h_{j+1}$ for the computations of $S_{f,j}$ and F_j and obtain a new water depth h_j. Using this h_j, we compute a new h_j in the same procedure. These procedures must be continued until the difference between the previous h_j and the new h_j is smaller than a very small value. Then, this h_j satisfies Eq. (15.35). This computation usually converges very fast, because the assumed initial value is near the convergent value.

A function $f(x)$ is not always continuous and differentiable in engineering. The Newton method cannot be applicable for a discontinuous function. If there exists a root of $f(x) = 0$ between a and b as shown in Fig. 15.3, then we have

$$f(a)f(b) < 0, \quad \text{for } a < b. \tag{15.36}$$

We assume that $f(a) < 0$ and $f(b) > 0$. Compute $f(x)$ at $(a+b)/2$ and assume

$$f\left(\frac{a+b}{2}\right) < 0. \tag{15.37}$$

Then, the root must be between $(a+b)/2$ and b. Next, compute $f(x)$ at $\frac{a+3b}{4}$. Whether $f(x)$ is positive or negative indicates the existence

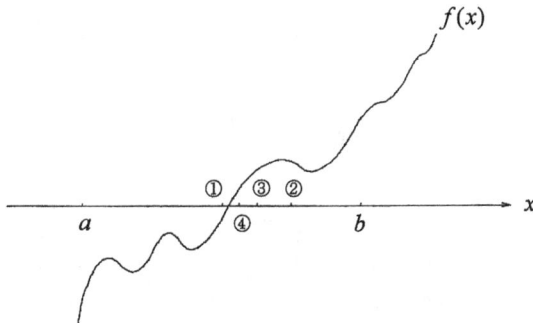

Figure 15.3. Secant method.

range of the root. The point in Fig. 15.3 approaches the root as ①, ②, ③, ④, ... The continuous computations make the length of an existence range of the root smaller. Finally, we get the approximate root when the length of the existence range of the root becomes very small. If the original length of the existence range is $b - a$, the length of the existence range of the root is $(\frac{1}{2})^{20}(b - a) \cong 9.5 \times 10^{-7}(b - a)$ after 20 times of these procedures. This procedure is applicable to the computation of a root of a discontinuous and continuous functions. This is called a secant method.

Example 15.4. The discharge $q = 5\,\mathrm{m}^2/\mathrm{s}$ is flowing in a rectangular channel. When the water depth is $5\,\mathrm{m}$ at the downstream end, compute the water depth $1\,\mathrm{km}$ upstream. Assume that $S_0 = 1/500, n = 0.03$, and $\Delta x = 1\,\mathrm{km}$.

Solution. Substituting the numerical values into Eq. (15.35), we get

$$h_1 = h_2 - \frac{1000}{2} \left[\frac{2}{500} - \frac{(5 \times 0.03)^2}{h_1^{10/3}} - \frac{(5 \times 0.03)^2}{h_2^{10/3}} \right]$$

$$\div \left[1 - \frac{1}{2} \left(\frac{5^2}{9.8 h_1^3} + \frac{5^2}{9.8 h_2^3} \right) \right]. \tag{1}$$

Substituting $h_2 = 5.00\,\mathrm{m}$ on the right side of Eq. (1), we have

$$h_1 = 5 - 500 \left[0.004 - \frac{0.0225}{h_1^{10/3}} - 1.05 \times 10^{-4} \right]$$

$$\div \left[1 - \frac{1}{2} \left(\frac{2.55}{h_1^3} + 0.0204 \right) \right]$$

$$= 5 - 500 \left[0.003895 - \frac{0.0225}{h_1^{10/3}} \right] \div \left[0.9898 - \frac{2.55}{2 h_1^3} \right]. \tag{2}$$

By substituting $h_1 = 4.50\,\mathrm{m}$ on the right side, the left side gives $3.08\,\mathrm{m}$. Sequently, we obtain $3.22\,\mathrm{m}, 3.19\,\mathrm{m}, 3.20\,\mathrm{m}$, and $3.20\,\mathrm{m}$. The computations converge and we get $h_1 = 3.20\,\mathrm{m}$. Usually, the space increment Δx must be $100\,\mathrm{m}$ or $200\,\mathrm{m}$.

Problem 15.4. When there is a turbulent flow in a smooth pipe, the coefficient of Darcy–Weisbach is given by

$$\frac{1}{\sqrt{f}} = 2.03 \log_{10} \left(Re \sqrt{f} \right) - 0.8. \tag{3}$$

If $Re = 100,000$, derive f in Eq. (3). The parameters are $Re = vD/\nu$, $D =$ the pipe diameter, $v =$ the cross-sectional average velocity, and $\nu =$ the knematic viscosity.

15.5. Hardy–Cross Method

When many pipes form networks, it is called the networks of pipes. Figure 15.4 constructs the networks of pipes. This has three circuits and ten pipes. The number of pipes in each circuit are $n_1 = 4$, $n_2 = 4$, and $n_3 = 4$. Let us consider the discharge in each pipe. The continuity equations of water and energy are nonlinear equations. Assume discharges in each circuit j and the continuity equation derives discharge Q'_{ji} in each pipe. When the energy loss in each pipe is $k_{ji}Q'^m_{ji}$, the total energy loss in each circuit is 0, according to a Kirchhoff theorem. If this discharge Q'_{ji} is not a true value, they do not satisfy the Kirchhoff theorem. We add a corrective discharge ΔQ_j to Q'_{ji} in each pipe in order that the correct discharge in each pipe satisfies the Kirchhoff theorem in each circuit. The Kirchhoff theorem is written as

$$\sum_{i=1}^{n_j}{}^* k_{ji}(Q'_{ji} + \Delta Q_j)^m = 0, \tag{15.38}$$

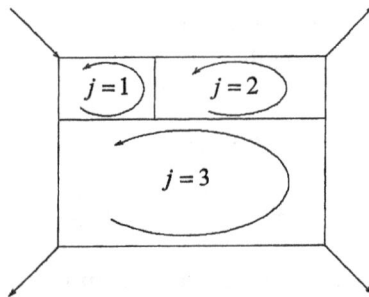

Figure 15.4.   Hardy–Cross method.

in which n_j is the number of pipes in a circuit j. The symbol $\sum^*$ indicates the sum of positive discharge if the flow direction is counterclockwise and negative discharge if the flow direction is clockwise. The positive and negative directions of a circuit is defined to be counterclockwise and clockwise, respectively. The exponent m in the mean velocity of Eq. (15.38) is 1.85 or 2. This is dependent on the used velocity formula in the pipe flow. When a Chezy, the Manning, or Darcy–Weisbach formulas are used, $m = 2$ is defined. When a Hagen–Williams formula is used, $m = 1.85$ is defined. The corrective discharge ΔQ_j is much smaller than the assumed discharge Q'_{ji}. Then, we get

$$|Q'_{ji}| >> |\Delta Q_j|. \tag{15.39}$$

Thus, neglecting higher terms of ΔQ_j in Eq. (15.38), we obtain the corrective discharge as

$$\Delta Q_j = -\frac{\sum^{*n_j}_{i=1} k_{ji} Q'^m_{ji}}{m \sum^{n_j}_{i=1} k_{ji} |Q'_{ji}|^{m-1}}. \tag{15.40}$$

The sum in the numerator of Eq. (15.40) shows $\sum^*$ and the sum in the denominator of Eq. (15.40) is $\sum$. Since Q'_{ji} in the denominator of Eq. (15.40) is positive, we use the absolute value. Since ΔQ_j in Eq. (15.40) is the difference between the correct discharge and the assumed discharge in the circuit j, the correct discharge is obtained by the sum of the assumed discharge Q'_{ji} and corrective discharge ΔQ_j. As long as the absolute corrective discharge $|\Delta Q_j|$ in Eq. (15.40) does not monotonically decrease, the assumed discharge must be initialized again. This method is not applicable when the assumed discharge is not close to a true discharge. This is called a Hardy–Cross method (Streeter, 1958). The value of k_{ji} is calculated from the Darcy–Weisbach equation (Daily and Harleman, 1966) as

$$k_{ji} = \frac{f_{ji} \ell_{ji}}{2g D_{ji}} \left(\frac{4}{\pi D^2_{ji}}\right)^2, \tag{15.41}$$

in which f_{ji} = the friction loss coefficient in the pipe i of the circuit j due to the Darcy–Weisbach equation, ℓ_{ji} = the length of pipe i and the circuit j, and D_{ji} = the diameter of the pipe i and the circuit j.

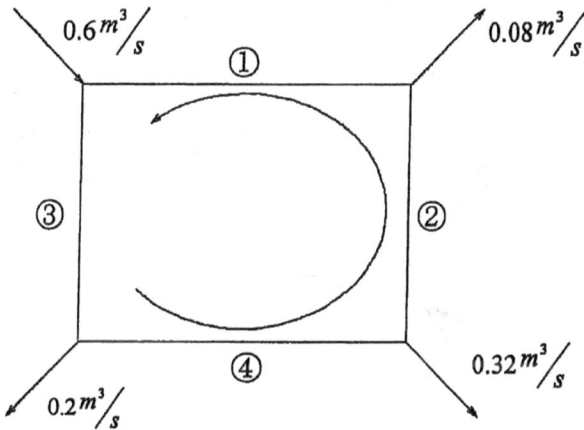

Figure 15.5. Example 15.5.

Example 15.5. There is a circuit as shown in Fig. 15.5. Calculate discharges in each pipe. Assume that $m = 2$, $k_1 = 160$, $k_2 = 270$, $k_3 = 40$, and $k_4 = 100$. The unit is s^2/m^5.

Solution. Assume the discharges in the pipe 1, 2, 3, and 4 are -0.24 m^3/s, -0.16 m^3/s, 0.36 m^3/s, and 0.16 m^3/s, respectively. Since $j = 1$, we have

$$\sum_{i=1}^{4}{}^{*} k_{1i} Q_{1i}'^2 = -9.216 - 6.912 + 5.184 + 2.56 = -8.384, \quad (1)$$

$$\sum_{i=1}^{4} |k_{1i} Q_{1i}'| = 38.4 + 43.2 + 14.4 + 16 = 112. \quad (2)$$

From Eq. (15.40) we get the corrective discharge as

$$\Delta Q_1 = -\frac{8.384}{2 \times 112} = -0.0374 \ (\text{m}^3/\text{s}). \quad (3)$$

The correct discharge is $Q_{1i} = Q_{1i}' + \Delta Q_1$. Thus, we have

$$Q_{11} = -0.203 \ (\text{m}^3/\text{s}), \quad Q_{12} = -0.123 \ (\text{m}^3/\text{s}),$$
$$Q_{13} = 0.3974 \ (\text{m}^3/\text{s}), \quad Q_{14} = 0.1974 \ (\text{m}^3/\text{s}). \quad (4)$$

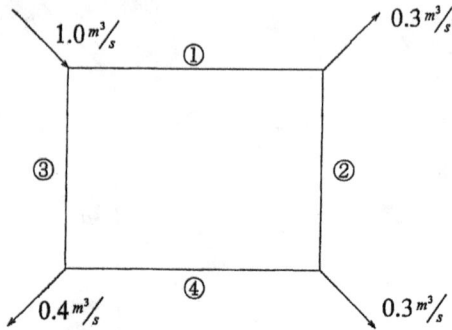

Figure 15.6.   Problem 15.5.

Equation (4) indicates

$$\sum_{i=1}^{4}{}^{*} k_{1i}Q_{1i}'^{2} = -6.59 - 4.08 + 6.32 + 3.90 = -0.45, \qquad (5)$$

$$\sum_{i=1}^{4} |k_{1i}Q_{1i}| = 32.48 + 33.21 + 15.90 + 19.74 = 101.33. \qquad (6)$$

Therefore, we have

$$\Delta Q_1 = -\frac{0.45}{2 \times 101.33} = -2.22 \times 10^{-3}. \qquad (7)$$

Since this is very small, this computation converges. The discharges in each pipe are

$$Q_{11} = -0.201 \ (\text{m}^3/\text{s}), \quad Q_{12} = -0.121 \ (\text{m}^3/\text{s}),$$

$$Q_{13} = 0.40 \ (\text{m}^3/\text{s}), \quad Q_{14} = 0.20 \ (\text{m}^3/\text{s}). \qquad (8)$$

Problem 15.5. Compute discharges in each pipe of the circuit as shown in Fig. 15.6. The k values of the pipe 1, 2, 3, and 4 are $15.5 \ \text{s}^2/\text{m}^5$.

15.6. Method of Characteristics

We consider flood analysis using Eqs. (12.107) and (12.108). Referring to Fig. 15.7, Eq. (12.107) is formulated in the finite difference

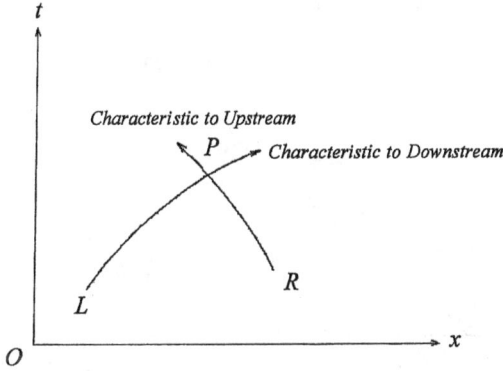

Figure 15.7.   Method of characteristics.

scheme (Liggett, 1994; Chiu *et al.*, 1976) as

$$\frac{x_P - x_L}{t_P - t_L} = \frac{v_P + v_L}{2} + \sqrt{g\frac{h_P + h_L}{2}}, \qquad (15.42)$$

$$\frac{x_P - x_R}{t_P - t_R} = \frac{v_P + v_R}{2} - \sqrt{g\frac{h_P + h_R}{2}}. \qquad (15.43)$$

Equation (12.108) is also formulated in the finite difference scheme as

$$\frac{v_P - v_L}{t_P - t_L} + \sqrt{\frac{2g}{h_P + h_L}}\frac{h_P - h_L}{t_P - t_L} = g\left(S_0 - \frac{S_{fP} + S_{fL}}{2}\right), \qquad (15.44)$$

$$\frac{v_P - v_R}{t_P - t_R} - \sqrt{\frac{2g}{h_P + h_R}}\frac{h_P - h_R}{t_P - t_R} = g\left(S_0 - \frac{S_{fP} + S_{fR}}{2}\right). \qquad (15.45)$$

Rewriting Eqs. (15.42) and (15.43), we have

$$x_P - x_L = A_L(t_P - t_L), \qquad (15.46)$$

$$x_P - x_R = A_R(t_P - t_R), \qquad (15.47)$$

in which

$$A_L = \frac{v_P + v_L}{2} + \sqrt{g\frac{h_P + h_L}{2}}, \qquad (15.48)$$

$$A_R = \frac{v_P + v_R}{2} - \sqrt{g\frac{h_P + h_R}{2}}. \qquad (15.49)$$

Equations (15.44) and (15.45) are written as

$$v_P - v_L + B_L(h_P - h_L) = D_L(t_P - t_L), \qquad (15.50)$$

$$v_P - v_R - B_R(h_P - h_R) = D_R(t_P - t_R), \qquad (15.51)$$

in which

$$B_L = \sqrt{\frac{2g}{h_P + h_L}}, \quad B_R = \sqrt{\frac{2g}{h_P + h_R}}, \qquad (15.52)$$

$$D_L = g\left(S_0 - \frac{S_{fP} + S_{fL}}{2}\right), \quad D_R = g\left(S_0 - \frac{S_{fP} + S_{fR}}{2}\right). \qquad (15.53)$$

From Eqs. (15.46) and (15.47), we get

$$t_P = \frac{x_R - x_L + A_L t_L - A_R t_R}{A_L - A_R}, \qquad (15.54)$$

$$x_P = x_L + A_L\frac{x_R - x_L - A_R t_R + A_R t_L}{A_L - A_R}. \qquad (15.55)$$

From Eqs. (15.48) and (15.49), we obtain

$$h_P = \frac{D_L(t_P - t_L) - D_R(t_P - t_R) + v_L - v_R + B_L h_L + B_R h_R}{B_L + B_R}, \qquad (15.56)$$

$$v_P = v_L + D_L(t_P - t_L)$$
$$- B_L\frac{D_L(t_P - t_L) - D_R(t_P - t_R) + v_L - v_R + B_R h_R - B_R h_L}{B_L + B_R}. \qquad (15.57)$$

Compute A_L, A_R, B_L, B_R, D_L, and D_R with h_P and v_P which are assumed beforehand. Usually we use h and v at the point near the point P on the $x - t$ plane as the values of h_P and v_P. The computation must be continued until the values of h_P and v_P converge (Stoker, 1957).

Example 15.6. When $t_L = t_R = 0$, $x_L = 0$, $x_R = 20\,\text{m}$, $h_L = 2\,\text{m}$, $h_R = 1.5\,\text{m}$, $v_L = 0.6\,\text{m/s}$, $v_R = 0.3\,\text{m/s}$, $n = 0.03$, and $S_0 = 0.001$, solve Eqs. (15.42) $\sim$ Eq. (15.45) and obtain x_P, t_P, h_P, and v_P.

Solution.

$$v_P = \frac{v_L + v_R}{2} = 0.45 \text{ m/s}, \quad h_P = \frac{h_L + h_R}{2} = 1.75 \text{ m}. \quad (1)$$

Then, since $S_{fP} = 4.79 \times 10^{-4}$ indicates $A_L = 4.81$ m/s, $A_R = -3.62$ m/s, $B_L = 2.29$ 1/s, $B_R = 2.46$ 1/s, $D_L = 8.32 \times 10^{-3}$ m/s^2, and $D_R = 9.15 \times 10^{-3}$ m/s^2, we obtain $t_P = 2.32$ s, $x_P = 11.16$ m, $h_P = 1.80$ m, and $v_P = 1.08$ m/s.

Using new x_P, t_P, h_P, and v_P, we get $A_L = 5.15$ m/s, $A_R = -3.33$ m/s, $B_L = 2.27$ 1/s, $B_R = 2.44$ 1/s, $D_L = 6.82 \times 10^{-3}$ m/s^2, $D_R = 7.22 \times 10^{-3}$ m/s^2, $t_P = 2.36$ s, $x_P = 12.15$ m, $h_P = 1.80$ m, and $v_P = 1.07$ m/s.

The value of h_P becomes almost constant. But v_P is still varying for this computation. Continue the same procedure.

$A_L = 5.15$ m/s, $A_R = -3.33$ m/s, $B_L = 2.27$ 1/s, $B_R = 2.44$ 1/s, $D_L = 6.82 \times 10^{-3}$ m/s^2, $D_R = 7.22 \times 10^{-3}$ m/s^2, $t_P = 2.36$ s, $x_P = 12.15$ m, $h_P = 1.80$ m, and $v_P = 1.07$ m/s.

Since h_P and v_P do not change, finally $h_P = 1.80$ m, $v_P = 1.07$ m/s, $t_P = 2.36$ s, and $x_P = 12.15$ m are obtained.

Problem 15.6. When $n = 0.05$, re-compute Example 15.6.

15.7. Lax–Wendroff Scheme

This is the finite difference scheme (Chap. 2) for solving a nonlinear partial differential equation. This scheme produces artificial viscosity in the computation and the artificial viscosity stabilizes the computation. The partial differential equation

$$\frac{\partial u}{\partial t} = A(u)\frac{\partial u}{\partial x} \quad (15.58)$$

is solved by the Lax–Wendroff scheme (Ames, 1969). The Taylor series of $u(x, t)$ about t is given by

$$u(x, t + \Delta t) = u(x, t) + \Delta t\frac{\partial u}{\partial t} + \frac{(\Delta t)^2}{2!}\frac{\partial^2 u}{\partial t^2} + \cdots. \quad (15.59)$$

Since

$$\frac{\partial^2 u}{\partial t^2} = A(u)^2\frac{\partial^2 u}{\partial x^2} + A(u)\frac{\partial A}{\partial u}\left(\frac{\partial u}{\partial x}\right)^2, \quad (15.60)$$

we get

$$u(x, t + \Delta t) = u(x, t) + \Delta t \, A(u) \frac{\partial u}{\partial x} + \frac{(\Delta t)^2}{2!} \left[A(u)^2 \frac{\partial^2 u}{\partial x^2} \right.$$

$$\left. + A(u) \frac{\partial A}{\partial u} \left(\frac{\partial u}{\partial x} \right)^2 \right] + \cdots . \tag{15.61}$$

The differentiation about x is approximated by the following finite difference scheme:

$$u(x, t + \Delta t) \cong u(x, t) + \Delta t A(u(x, t)) \frac{u(x + \Delta x, t) - u(x - \Delta x, t)}{2\Delta x}$$

$$+ \frac{(\Delta t)^2}{2!} \left\{ A(u(x, t)) \frac{u(x + \Delta x, t) - 2u(x, t) + u(x - \Delta x, t)}{(\Delta x)^2} \right.$$

$$\left. + A(u(x, t)) \frac{\partial A(u(x, t))}{\partial u} \left[\frac{u(x + \Delta x, t) - u(x - \Delta x, t)}{2\Delta x} \right]^2 \right\}. $$

$$\tag{15.62}$$

Equation (15.62) is called the Lax–Wendroff scheme. This is often used for the computation of nonlinear partial differential equations of the first-order.

Example 15.7. When $A(u) = A = const$ in Eq. (15.58), derive Eq. (15.62).

Solution. Eq. (15.62) becomes

$$u(x, t + \Delta t) \cong u(x, t) + \Delta t A \frac{u(x + \Delta x, t) - u(x - \Delta x, t)}{2\Delta x}$$

$$+ \frac{(\Delta t)^2}{2!} \left[A^2 \frac{u(x + \Delta x, t) - 2u(x, t) + u(x - \Delta x, t)}{(\Delta x)^2} \right]. $$

$$\tag{15.63}$$

Problem 15.7. When $A = const$ and $u(x, t = 0) = x$, obtain $u(\Delta x, t)$.

Chapter 16

Linearization Methods

Since the flow motion is expressed by the Euler equation of motion or Navier–Stokes equations which are nonlinear, the resultant fluid motion is also nonlinear. When we use the potential equation which is linear, the boundary conditions at the free water surface are nonlinear. The conditions at the free water surface are the kinematic and dynamic boundary conditions. The nonlinear partial differential equation with linear boundary conditions or the linear partial differential equation with nonlinear boundary conditions cannot be analytically solved. The development of fluid mechanics is battles against nonlinearity. The many methods of linearization were found to get analytical solutions from the Euler equation of motion or Navier–Stokes equations in fluid mechanics.

16.1. Fundamentals in Linearization

When the exponents of dependent variables are not 1 or 0, a partial or ordinary differential equations are called nonlinear. We studied linear mathematics in school. But it is not available to nonlinear mathematics. We must use a new method to solve nonlinear differential equations. There are several elementary linearization techniques. When x is much smaller than 1, we have the following nonlinear equation about x:

$$(1 + x)^n \cong 1 + nx, \tag{16.1}$$

in which n = a real. The left side is nonlinear and the right side is linear. This approximation makes Eq. (16.1) linear. When there are two positive dependent variables x and y, the following equation is formed:

$$xy \cong ax + by, \tag{16.2}$$

in which a and b are constant coefficients. The left side is nonlinear and the right side is linear. When a dependent variable h and a partial derivative $\partial h/\partial x$ are positive, we have

$$h\frac{\partial h}{\partial x} \cong a'h + b'\frac{\partial h}{\partial x}, \tag{16.3}$$

in which a' and b' are constant coefficients. The nonlinear term on the left side is linearized. When $x \gg y > 0$ and $n \neq 1$, applying Eq. (16.1), we have

$$(x + y)^n \cong x^n \left(1 + \frac{ny}{x}\right). \tag{16.4}$$

The nonlinearity of y is linearized in Eq. (16.4).

Example 16.1. When $x \gg y$ and $n \neq 1$, approximate Eq. (16.4).

Solution.

$$(x+y)^n = x^n \left[1 + \binom{n}{1}\frac{y}{x} + \binom{n}{2}\left(\frac{y}{x}\right)^2 + \cdots + \binom{n}{n}\left(\frac{y}{x}\right)^n\right]$$
$$\cong x^n \left(1 + \frac{ny}{x}\right). \tag{1}$$

Problem 16.1. When $h = h_0 + h'$, $v = v_0 + v'$, $h_0 \gg h' > 0$, $v_0 \gg v' > 0$, $S_f = n^2 v^2/h^{4/3}$, and n is constant, derive the following equation:

$$S_f \cong \frac{n^2 v_0^2}{h_0^{4/3}}\left(1 + \frac{2v'}{v_0}\right)\left(1 - \frac{4h'}{3h_0}\right). \tag{2}$$

in which $h =$ water depth, $v =$ velocity, $S_f =$ energy slope, and $n =$ Manning's roughness coefficient.

16.2. Stokes Wave

The motion of surface waves is derived from the potential equation. The velocity potential as the surface wave implicitly has a time-varying term in the solution. Thus, we have

$$\frac{\partial^2 \phi}{\partial x^2} + \frac{\partial^2 \phi}{\partial y^2} = 0, \tag{16.5}$$

in which $\phi =$ the velocity potential, $x =$ the horizontal coordinate system on the still water surface, and $y =$ the vertically upward coordinate system. Assuming that the water depth is constant h, we have the boundary condition at the bottom as

$$\frac{\partial \phi}{\partial y} = 0, \quad \text{at } y = -h. \tag{16.6}$$

The kinematic boundary condition at the free water surface is

$$\frac{\partial \eta}{\partial t} + \frac{\partial \phi}{\partial x}\frac{\partial \eta}{\partial x} = \frac{\partial \phi}{\partial y}, \quad \text{at } y = \eta. \tag{16.7}$$

The dynamic boundary condition at the free water surface is

$$\frac{\partial \phi}{\partial t} + g\eta + \frac{1}{2}\left[\left(\frac{\partial \phi}{\partial x}\right)^2 + \left(\frac{\partial \phi}{\partial y}\right)^2\right] = 0, \quad \text{at } y = \eta, \tag{16.8}$$

in which $\eta =$ the displacement of the free water surface from the still water surface. We obtained the solution of Eq. (16.5) in Chap. 13 by linearizing Eqs. (16.7) and (16.8) when the wave has an infinitesimally small amplitude. But if the incident wave reaches shallower zone, the assumption of the infinitesimally small amplitude wave is not formed. Then, Stokes in 1847 (Debnath, 1994) found a new computational technique, introducing the moving coordinate system with the celerity C, under this assumption waves do not move. The velocity potential and stream function satisfy the potential equation. Therefore, the velocity potential and stream function are given by

$$\frac{\phi}{C} = -x + \beta \cosh k(h + y) \sin kx, \tag{16.9}$$

$$\frac{\psi}{C} = -y + \beta \sinh k(h + y) \cos kx, \tag{16.10}$$

in which $\beta =$ a small parameter, $k =$ an angular frequency, $2\pi/L$, and $L =$ the wave length. Equation (16.9) satisfies the boundary condition at the bottom, Eq. (16.6). The streamline which corresponds to the bottom is obtained by substituting $y = -h$ in Eq. (16.10) as

$$\psi = Ch. \tag{16.11}$$

Assuming that $\psi = 0$ at $y = \eta$, we get the water surface profile from Eq. (16.10). Thus, the water surface profile η is given by

$$\eta = \beta \sinh k(h + \eta) \cos kx. \tag{16.12}$$

This is the nonlinear equation about η. The iteration method gives η at x. But herein, when β is very small, we assume a form of the function η as follows:

$$\eta = a_1\beta + a_2\beta^2 + a_3\beta^3, \tag{16.13}$$

in which a_1, a_2, and a_3 are determined later. Assuming $h \gg |\eta|$, the hyperbolic sine function sinh in Eq. (16.12) is expanded by the Taylor series as

$$\sinh k(h + \eta) = \sinh kh + k\eta \cosh kh + \frac{k^2\eta^2}{2}\sinh kh$$

$$+ \frac{k^3\eta^3}{6}\cosh kh + \cdots. \tag{16.14}$$

Substituting Eq. (16.14) into Eq. (16.12), we get the water surface profile η. Substituting Eq. (16.13) into η, we obtain

$$a_1\beta + a_2\beta^2 + a_3\beta^3$$

$$= \beta\Bigg[\sinh kh + k(a_1\beta + a_2\beta^2 + a_3\beta^3)\cosh kh$$

$$+ \frac{k^2}{2}(a_1\beta + a_2\beta^2 + a_3\beta^3)^2 \sinh kh$$

$$+ \frac{k^3}{6}(a_1\beta + a_2\beta^2 + a_3\beta^3)^3 \cosh kh + \cdots \Bigg] \cos kx. \tag{16.15}$$

Comparing the power of β in Eq. (16.15), we obtain

$$a_1 = \sinh kh \cos kx, \tag{16.16}$$

$$a_2 = ka_1 \cosh kh \cos kx, \tag{16.17}$$

$$a_3 = ka_2 \cosh kh \cos kx + \frac{k^2}{2}a_1 \sinh kh \cos kx. \tag{16.18}$$

Substituting Eq. (16.16) into Eq. (16.17), we have

$$a_2 = k \cosh kh \cos kx \cdot \sinh kh \cos kx$$

$$= k \cosh kh \sinh kh \cdot \frac{1 + \cos 2kx}{2}. \tag{16.19}$$

Substituting Eqs. (16.16) and (16.19) into Eq. (16.18), we get

$$a_3 = k^2 \cosh^2 kh \sinh kh \cos kx \cdot \frac{1 + \cos 2kx}{2} + \frac{k^2}{2} \sinh^2 kh \cos^2 kx$$

$$= k^2 \cosh^2 kh \sinh kh \cdot \frac{\cos 3kx + 3 \cos kx}{4}$$

$$+ k^2 \sinh^2 kh \cdot \frac{1 + \cos 2kx}{4}. \tag{16.20}$$

Substituting Eqs. (16.16), (16.19), and (16.20) into Eq. (16.13), we obtain the water surface profile as

$$\eta = \beta \sinh kh \cos kx + \beta^2 k \cosh kh \sinh kh \cdot \frac{1 + \cos 2kx}{2}$$

$$+ \beta^3 k^2 \cosh^2 kh \sinh kh \cdot \frac{\cos 3kx + 3 \cos kx}{4}$$

$$+ \beta^3 k^2 \sinh^2 kh \cdot \frac{1 + \cos 2kx}{4}. \tag{16.21}$$

Equation (16.21) converges for a small parameter β. To determine the small parameter β, the wave height H is used. The wave height H is equal to the difference of $\eta_{\max}$ and $\eta_{\min}$. That is, from Eq. (16.12), we have

$$\eta_{\max} - \eta_{\min} = 2\beta \sinh kh = H. \tag{16.22}$$

Therefore, we have

$$\beta = \frac{H}{2 \sinh kh}. \tag{16.23}$$

Thus, Eq. (16.21) is rewritten as

$$\eta = \frac{1}{8} kH^2 \coth kh + \frac{H}{2} \cos kx + \frac{1}{8} kH^2 \coth kh \cdot \cos 2kx$$

$$+ \frac{1}{32} k^2 H^3 \left(\coth^2 kh + \frac{1}{2} \right) \cos 3kx + \cdots. \tag{16.24}$$

Therefore, the water surface level increases

$$\delta = \frac{1}{L} \int_0^L \eta dx = \frac{1}{8} kH^2 \coth kh, \tag{16.25}$$

in which $L =$ the wave length. If δ is not negligibly small, the water depth must be modified as $h + \delta$. The existence of $\cos 2kx$ shows that the peak and trough are steeper and flatter than the sine wave, respectively. This is called the Stokes wave.

Example 16.2. For the deep sea waves, Eq. (16.10) is expressed by

$$\frac{\psi}{C} = -y + \beta e^{ky} \cos kx. \tag{1}$$

When $\eta = a_1\beta + a_2\beta^2$, obtain a_1 and a_2.

Solution. As the water surface profile, substituting $\psi = 0$ and $y = \eta$ into Eq. (1), we have

$$\eta = \beta e^{k\eta} \cos kx. \tag{2}$$

Substituting $\eta = a_1\beta + a_2\beta^2$ into Eq. (2), we get

$$a_1\beta + a_2\beta^2 = \beta e^{k(a_1\beta + a_2\beta^2)} \cos kx$$

$$= \beta \cos kx \left[1 + k(a_1\beta + a_2\beta^2) + \frac{k^2(a_1\beta + a_2\beta^2)^2}{2!} + \cdots \right]. \tag{3}$$

The comparison of Eq. (3) indicates

$$a_1 = \cos kx, \tag{4}$$

$$a_2 = \cos kx \cdot ka_1 = k \cdot \frac{1 + \cos 2kx}{2}. \tag{5}$$

Problem 16.2. Obtain a_1, a_2, and a_3 in Eq. (16.13) in Example 16.1.

16.3. KdV Equation

We call the long wave if its wave length L is very large in comparison with the water depth h. When the wave height is very small in comparison with the water depth h, we call the wave of infinitesimally small amplitude. Consider a wave of finite height and whose wave length is not long. This wave is called a Boussinesq wave. The analysis of the Boussinesq wave (Wadachi, 1992) is related to the perturbation method in Chap. 19. The celerity of wave of infinitesimally small amplitude in Chap. 13 is given by

$$C = \sqrt{\frac{g}{k} \tanh kh},$$ (16.26)

in which k = the angular frequency. The Maclaurin series of $\tanh kh$ is

$$\tanh kh = kh - \frac{(kh)^3}{3} + \frac{2}{15}(kh)^5 - \frac{17}{315}(kh)^7 + - \cdots.$$ (16.27)

Using $C = \omega/k$ and expanding kh in Eq. (16.26), we get

$$\omega^2 = gk \tanh kh = C_0^2 k^2 - \frac{1}{3} C_0^2 h^2 k^4 + \cdots,$$ (16.28)

in which $\omega = 2\pi/T$, an angular number of oscillation, T = wave period, and $C_0 = \sqrt{gh}$. Thus, ωt is rewritten as

$$\omega t = \sqrt{gk \tanh kh} t = \left[C_0^2 k^2 - \frac{1}{3} C_0^2 h^2 k^4 + \cdots \right]$$

$$\cong C_0 kt - \frac{1}{6} C_0 h^2 k^3 t + \cdots.$$ (16.29)

Thus, we have

$$kx - \omega t = k(x - C_0 t) + \frac{1}{6} C_0 h^2 k^3 t + \cdots$$

$$= k \left[x - C_0 \left(1 - \frac{1}{6} h^2 k^2 \right) t + \cdots \right].$$ (16.30)

Since the celerity is C_0 and the phase difference is $h^2 k^2 t/6$, the difference ϵ is expressed by

$$\epsilon \sim O(h^2 k^2),$$ (16.31)

in which O indicates an order. If ϵ is a small parameter, we have

$$\epsilon^{1/2} \sim kh.$$ (16.32)

Since $h \ll L$, we introduce new variables as

$$\xi = \epsilon^{1/2}(x - C_0 t), \quad \tau = \epsilon^{3/2} t.$$ (16.33)

We expand η and ϕ in the following forms:

$$\eta = \epsilon \eta^{(1)}(\xi, \tau) + \epsilon^2 \eta^{(2)}(\xi, \tau) + \epsilon^3 \eta^{(3)}(\xi, \tau) + \cdots$$ (16.34)

and

$$\phi = \epsilon^{1/2}\phi^{(1)}(\xi, y, \tau) + \epsilon^{3/2}\phi^{(2)}(\xi, y, \tau) + \epsilon^{5/2}\phi^{(3)}(\xi, y, \tau) + \cdots .$$
$$(16.35)$$

Because

$$\partial\phi/\partial x = \epsilon^{1/2}\partial\phi/\partial\xi, \tag{16.36}$$

the expansion of ϕ starts from $\epsilon^{1/2}$ in Eq. (16.35). Next, since Eqs. (16.5), (16.6), (16.7), and (16.8) are transformed by Eq. (16.33), by replacing x and t by ξ and τ and by substituting Eqs. (16.34) and (16.35) into Eqs. (16.5) and (16.6), they are arranged in the power of ϵ. Then, from Eq. (16.5), we get

$$\frac{\partial^2\phi^{(1)}}{\partial y^2} = 0, \quad \frac{\partial^2\phi^{(n)}}{\partial y^2} + \frac{\partial^2\phi^{(n-1)}}{\partial\xi^2} = 0, \quad \text{for } n \geq 2. \tag{16.37}$$

From Eq. (16.6), we get

$$\frac{\partial\phi^{(n)}}{\partial y} = 0, \quad \text{at } y = -h, \ n \geq 1. \tag{16.38}$$

From Eqs. (16.37) and (16.38), we obtain

$$\phi^{(1)} = \phi^{(1)}(\xi, \eta), \tag{16.39}$$

$$\phi^{(2)} = -\frac{1}{2}(y+h)^2\frac{\partial^2\phi^{(1)}}{\partial\xi^2} + a(\xi, \tau), \tag{16.40}$$

$$\phi^{(3)} = \frac{1}{24}(y+h)^4\frac{\partial^4\phi^{(1)}}{\partial\xi^4} - \frac{1}{2}(y+h)^2\frac{\partial^2 a(\xi, \tau)}{\partial\xi^2} + b(\xi, \tau), \tag{16.41}$$

in which a and b are integral functions. $\phi(\xi, \eta, \tau)$ is expanded by the Maclaurin series at $y = 0$ to obtain the boundary condition at $y = \eta$. That is,

$$\phi(\xi, \eta, \tau) = \phi(\xi, 0, \tau) + \frac{\partial\phi(\xi, 0, \tau)}{\partial y}\eta + \frac{1}{2}\frac{\partial^2\phi(\xi, 0, \tau)}{\partial y^2}\eta^2 + \cdots .$$
$$(16.42)$$

The derivative $\frac{\partial\phi}{\partial t}$ is given by

$$\frac{\partial\phi}{\partial t} = \frac{\partial\phi}{\partial\xi}\cdot\frac{\partial\xi}{\partial t} + \frac{\partial\phi}{\partial\tau}\cdot\frac{\partial\tau}{\partial t} = \epsilon^{1/2}\left(-C_0\frac{\partial\phi}{\partial\xi} + \epsilon\frac{\partial\phi}{\partial\tau}\right) \cong -\epsilon^{1/2}C_0\frac{\partial\phi}{\partial\xi}.$$
$$(16.43)$$

Using Eqs. (16.41) and (16.42), we put that the coefficients are zero for the same power of ϵ in Eq. (16.8).

$$-C_0 \frac{\partial \phi^{(1)}}{\partial \xi} + g\eta^{(1)} = 0, \quad \text{at } y = 0, \tag{16.44}$$

$$\frac{\partial \phi^{(1)}}{\partial \tau} - C_0 \frac{\partial \phi^{(2)}}{\partial \xi} + \frac{1}{2} \left(\frac{\partial \phi^{(1)}}{\partial \xi} \right)^2 + g\eta^{(2)} = 0, \quad \text{at } y = 0. \tag{16.45}$$

In the same way, Eq. (16.7) indicates

$$\frac{\partial \phi^{(2)}}{\partial y} = -C_0 \frac{\partial \eta^{(1)}}{\partial \xi}, \quad \text{at } y = 0, \tag{16.46}$$

$$\frac{\partial \phi^{(3)}}{\partial y} + \eta^{(1)} \frac{\partial^2 \phi^{(2)}}{\partial y^2} = \frac{\partial \eta^{(1)}}{\partial \tau} - C_0 \frac{\partial \eta^{(2)}}{\partial \xi} + \frac{\partial \phi^{(1)}}{\partial \xi} \frac{\partial \eta^{(1)}}{\partial \xi}, \quad \text{at } y = 0. \tag{16.47}$$

From Eq. (16.43), we get

$$\frac{\partial \phi^{(1)}}{\partial \xi} \frac{\partial \eta^{(1)}}{\partial \xi} = \frac{C_0}{h} \eta^{(1)} \frac{\partial \eta^{(1)}}{\partial \xi}. \tag{16.48}$$

From Eqs. (16.41) and (16.44), we have

$$\frac{\partial \phi^{(3)}(\xi, 0, \tau)}{\partial y} = \frac{1}{6} h^3 \frac{\partial^4 \phi^{(1)}}{\partial \xi^4} - h \frac{\partial^2 a(\xi, \tau)}{\partial \xi^2}$$

$$= \frac{1}{6} C_0 h^2 \frac{\partial^3 \eta^{(1)}}{\partial \xi^3} - h \frac{\partial^2 a(\xi, \tau)}{\partial \xi^2}. \tag{16.49}$$

From Eqs. (16.40) and (16.44), we obtain

$$\eta^{(1)} \frac{\partial^2 \phi^{(2)}(\xi, 0, \tau)}{\partial y^2} = -\eta^{(1)} \frac{\partial^2 \phi^{(1)}}{\partial \xi^2} = -\frac{C_0}{h} \eta^{(1)} \frac{\partial \eta^{(1)}}{\partial \xi}. \tag{16.50}$$

From Eqs. (16.39), (16.40), (16.44), and (16.45), we have

$$-C_0 \frac{\partial \eta^{(2)}}{\partial \xi} = \frac{C_0}{g} \left[\frac{\partial^2 \phi^{(1)}}{\partial \xi \partial \eta} - C_0 \frac{\partial^2 \phi^{(2)}}{\partial \xi^2} + \frac{\partial \phi^{(1)}}{\partial \xi} \frac{\partial^2 \phi^{(1)}}{\partial \xi^2} \right]$$

$$= \frac{\partial \eta^{(1)}}{\partial \tau} - \frac{1}{2} C_0 h^2 \frac{\partial^3 \eta^{(1)}}{\partial \xi^3} - h \frac{\partial^2 a(\xi, \eta)}{\partial \xi^2} + \frac{C_0}{h} \eta^{(1)} \frac{\partial^2 \eta^{(1)}}{\partial \xi}. \tag{16.51}$$

Substituting Eqs. (16.48), (16.49), (16.50), and (16.51) into Eq. (16.47) and arranging it, we obtain

$$\frac{\partial \eta^{(1)}}{\partial \tau} + \frac{3}{2}\frac{C_0}{h}\eta^{(1)}\frac{\partial \eta^{(1)}}{\partial \xi} + \frac{C_0 h^2}{6}\frac{\partial^3 \eta^{(1)}}{\partial \xi^3} = 0. \tag{16.52}$$

This equation is called a KdV equation. Equation (16.52) was derived by Korteweg and de Vries in the Netherlands in 1895. The KdV equation is generally written as

$$\frac{\partial \eta}{\partial t} + C_1 \eta \frac{\partial \eta}{\partial \xi} + C_2 \frac{\partial^3 \eta}{\partial \xi^3} = 0, \tag{16.53}$$

in which C_1 and C_2 are constant coefficients. The coefficients in the KdV equation are arbitrarily selected. For example,

$$t \to (6h/C_0)t, \quad \xi \to hx \quad \text{and} \quad \eta \to 2hu/3 \tag{16.54}$$

are used in Eq. (16.52) is transformed to

$$\frac{\partial u}{\partial t} + 6u\frac{\partial u}{\partial x} + \frac{\partial^3 u}{\partial x^3} = 0. \tag{16.55}$$

This is also called the KdV equation. The second term in Eq. (16.55) is nonlinear and it makes the front of the wave form steeper as the wave propagates. The third term in Eq. (16.55) is a dissipation term and it makes the wave form milder as the wave propagates. The two terms have opposite effects to maintain the wave form for a long time. For the progressing wave, $\xi = x - Ct$ shows the wave form as

$$u(x, t) = u(x - Ct) = u(\xi). \tag{16.56}$$

Substituting Eq. (16.56) into Eq. (16.55), we obtain

$$-C\frac{du}{d\xi} + 6u\frac{du}{d\xi} + \frac{d^3 u}{d\xi^3} = 0. \tag{16.57}$$

Integrating Eq. (16.57) twice, we have

$$\frac{1}{2}\left(\frac{du}{d\xi}\right)^2 + u^3 - \frac{1}{2}Cu^2 + Au + B = 0, \tag{16.58}$$

in which A and B are integral constants. Equation (16.58) is similar to an energy integral in mechanics. If $du/d\xi$ is velocity in Eq. (16.58), the potential energy of mass 1 is given by

$$V(u) = u^3 - \frac{1}{2}Cu^2 + Au + B. \tag{16.59}$$

$$V(u)$$

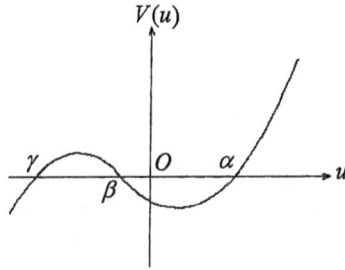

Figure 16.1. Computation of KdV equation, Eq. (16.59).

Equation (16.58) indicates the energy conservation equation. If roots of $V(u)$ are α, β, and $\gamma (\alpha > \beta > \gamma)$, $V(u)$ is expressed by

$$V(u) = (u - \alpha)(u - \beta)(u - \gamma). \qquad (16.60)$$

As Fig. 16.1 shows, in order for Eq. (16.58) to be meaningful, $V(u)$ must be negative, because $\frac{1}{2}\left(\frac{du}{d\xi}\right)^2$ is always positive. The domains of negative $V(u)$ are $u \leq \gamma$ and $\beta \leq u \leq \alpha$ as shown in Fig. 16.1. The oscillation solution to express wave phenomena corresponds to $\beta \leq u \leq \alpha$. Substituting Eq. (16.60) into Eq. (16.58) and integrating it, we have

$$\int_\alpha^u \frac{du}{\sqrt{(\alpha - u)(u - \beta)(u - \gamma)}} = \pm \int_{\xi_0}^\xi \sqrt{2}d\xi, \quad \text{for } \beta < u < \alpha.$$
$$(16.61)$$

The integral on the left side of Eq. (16.61) is the Jacobi's elliptic integral $sn(z, k)$ in Chap. 9. The left side is written as

$$\frac{2}{\sqrt{\alpha - \gamma}} sn^{-1}\left(\sqrt{\frac{\alpha - u}{\alpha - \beta}}, k\right) = \mp\sqrt{2}(\xi - \xi_0), \qquad (16.62)$$

in which $z = \sqrt{\frac{\alpha - u}{\alpha - \beta}}$, $k = \sqrt{\frac{\alpha - \beta}{\alpha - \gamma}}$, and $k = $ the modulus of sn function. Thus, Eq. (16.62) is rewritten as

$$\frac{\alpha - u}{\alpha - \beta} = sn^2\left(\sqrt{\frac{\alpha - \gamma}{2}}(\xi - \xi_0), k\right). \qquad (16.63)$$

Since

$$sn^2(z, k) + cn^2(z, k) = 1, \qquad (16.64)$$

Eq. (16.63) is written as

$$u(x,t) = \beta + (\alpha - \beta)cn^2\left(\sqrt{\frac{\alpha-\gamma}{2}}(x-Ct-\xi_0),k\right). \qquad (16.65)$$

The progressing wave of this type ($C > 0$) is called a cnoidal wave. The cnoidal wave is applicable in the shallower domain than the domain in which the Stokes wave is applicable. Let us consider a wave which is applicable in the shallower domain than the domain in which the cnoidal wave is applicable.

When $\xi \to \infty$ in Eq. (16.57), we assume that η and its derivative are zero. Then, we obtain $A = B = 0$. Since $\beta = \gamma = 0$ and $\alpha = C/2$, Eq. (16.61) is integrable by the elementary function as follows:

$$u(x,t) = \frac{C}{2}sech^2\left(\frac{\sqrt{C}}{2}(x-Ct-\xi_0)\right), \quad \text{for } C > 0, \qquad (16.66)$$

in which sech $x = 1/\cosh x$. This wave has one peak and we call it a solitary wave (Toda, 1983). In Eq. (16.66), C is the celerity, $C/2$ is the wave height, $2/\sqrt{C}$ is the wave width. This indicates that if the celerity is large, the wave height is large and the wave width becomes short. The person who found the solitary wave was Scott Russel in 1834 in England. Although the KdV equation is nonlinear, it has two analytical solutions. The persons who numerically computed the KdV equation with $C_1 = 1$ and $C_2 = 0.000484$ in Eq. (16.53) from the initial condition $\eta = \cos \pi \xi$ were Zabusky and Kruskal in 1965 in USA (Toda, 1983). They found that the waves move independently and are expressed by the solitary wave. They collide with each other and reappear when the original wave form reappears. Each wave is reserved and acts as a particle. Thus, a solitary wave is called a soliton. To solve the KdV equation, Hirota found an inverse scattering method. This is the applicable method of the Fourier transform to nonlinear systems.

Example 16.3. Using

$$\tanh^{-1} x = \frac{1}{2}\ln\left|\frac{1+x}{1-x}\right|. \qquad (1)$$

Derive Eq. (16.66).

Solution. If in Eq. (16.61), we define $\alpha = C/2$ and $\beta = \gamma = 0$, we get

$$\int_{C/2}^{u} \frac{d\eta}{\sqrt{C/2 - u \cdot u}} = \frac{1}{\sqrt{C/2}} \ln \left| \frac{\sqrt{C/2 - u} - \sqrt{C/2}}{\sqrt{C/2 - u} + \sqrt{C/2}} \right|$$

$$= \sqrt{\frac{2}{C}} \ln \left| \frac{X - 1}{X + 1} \right| = \pm\sqrt{2}(\xi - \xi_0), \qquad (2)$$

in which $X = \sqrt{1 - 2u/C}$. Rewriting Eq. (2) by a function tanh, we have

$$1 - \frac{2u}{C} = \tanh^2 \left[\frac{\sqrt{C}}{2}(\xi - \xi_0) \right], \qquad (3)$$

in which $1 - \tanh^2 X = \text{sech}^2 X$. Thus, we get

$$u = \frac{C}{2} \text{sech}^2 \left[\frac{\sqrt{C}}{2}(x - Ct - \xi_0) \right]. \qquad (4)$$

Problem 16.3. Derive Eq. (16.62) from Eq. (16.61).

16.4. Stokes Equation

Since the nonlinearity in the Navier–Stokes equations is not weak, it is very difficult for us to obtain analytical solutions of the Navier–Stokes equations. When a sphere of radius a is fixed in the uniform flow of velocity U, Stokes (1851) approximately solved the Navier–Stokes equations. When the flow is very small, nonlinear terms are negligible in the Navier–Stokes equations. When the flow is line symmetry about the progressing direction, we introduce a Stokes stream function ψ. By using the Stokes stream function, the Navier–Stokes equations in the cylindrical coordinate system are written as

$$\left[\frac{\partial^2}{\partial r^2} + \frac{\sin \theta}{r^2} \frac{\partial}{\partial \theta} \left(\frac{1}{\sin \theta} \frac{\partial}{\partial \theta} \right) \right]^2 \psi = 0. \qquad (16.67)$$

The flow is uniform at the long distance from the sphere. The Stokes stream function for the uniform flow is

$$\psi = \frac{1}{2} U r^2 \sin^2 \theta. \qquad (16.68)$$

The form of Eq. (16.68) indicates the solution of Eq. (16.67) as follows:

$$\psi = f(r)\sin^2\theta. \tag{16.69}$$

Substituting Eq. (16.69) into Eq. (16.67), we obtain

$$\left[\frac{\partial^2}{\partial r^2} + \frac{\sin\theta}{r^2}\frac{\partial}{\partial\theta}\left(\frac{1}{\sin\theta}\frac{\partial}{\partial\theta}\right)\right]\left[\left(\frac{d^2f(r)}{dr^2} - \frac{2f(r)}{r^2}\right)\sin^2\theta\right] = 0. \tag{16.70}$$

Rearranging Eq. (16.70), we get

$$\left(\frac{d^2}{dr^2} - \frac{2}{r^2}\right)\cdot\left(\frac{d^2}{dr^2} - \frac{2}{r^2}\right)f(r) = 0. \tag{16.71}$$

If we assume that the form of $f(r)$ is the polynomial of r^n, the exponent n satisfies the following equation:

$$[(n-2)(n-3) - 2][n(n-1) - 2] = 0. \tag{16.72}$$

The solution of Eq. (16.72) is

$$n = -1, \quad 1, \quad 2, \quad 4. \tag{16.73}$$

Thus, the general solution of Eq. (16.71) is

$$f(r) = \frac{A}{r} + Br + Cr^2 + Dr^4, \tag{16.74}$$

in which A, B, C, and D are integral constants. Equation (16.68) shows

$$C = \frac{1}{2}U, \quad D = 0. \tag{16.75}$$

Thus, since the function $f(r)$ is obtained,

$$\psi = \left(\frac{A}{r} + Br + \frac{1}{2}Ur^2\right)\sin^2\theta. \tag{16.76}$$

The velocity components in the r and θ directions are given by

$$v_r = \frac{1}{r\sin\theta}\frac{1}{r}\frac{\partial\psi}{\partial\theta} = -U\cos\theta + 2\left(\frac{A}{r^3} + \frac{B}{r}\right)\cos\theta, \tag{16.77}$$

$$v_\theta = -\frac{1}{r\sin\theta}\frac{\partial\psi}{\partial r} = U\sin\theta + \left(\frac{A}{r^3} - \frac{B}{r}\right)\sin\theta. \tag{16.78}$$

The boundary conditions $v_r = v_\theta = 0$ on the surface of the sphere $r = a$ indicate

$$v_r = 0 \Rightarrow -U + 2\left(\frac{A}{a^3} + \frac{B}{a}\right) = 0, \tag{16.79}$$

$$v_\theta = 0 \Rightarrow U + \frac{A}{a^3} - \frac{B}{a} = 0. \tag{16.80}$$

The integral constants A and B are obtained by Eqs. (16.79) and (16.80).

$$A = -\frac{1}{4}Ua^3, \quad B = \frac{3}{4}Ua. \tag{16.81}$$

Substituting A and B into Eq. (16.76), the stream function ψ is

$$\psi = \frac{1}{2}U\left(r^2 + \frac{3}{2}ar - \frac{1}{2}\frac{a^3}{r}\right)\sin^2\theta = \frac{1}{4}U(r-a)^2(2r-a)\sin^2\theta. \tag{16.82}$$

The substitution of $\pi - \theta$ instead of θ into Eq. (16.82) results in the same equation as Eq. (16.82). This indicates that the flow is symmetric about $\theta = \pi/2$. That is, the flow in front of the sphere is the same as it is in the lee. This shows that this flow does not induce the drag force due to pressure changes. The drag force is induced by the friction resistance. Usually, the viscous flow in front of the sphere is completely different from it in the lee. The flow in the lee forms a wake. The flow analysis without the nonlinear terms is called a Stokes flow. When the fluid is at rest and the sphere moves with the velocity U, the term indicates the uniform flow in Eq. (16.82) is deleted. Then,

$$\psi = \frac{1}{4}\left(3ar - \frac{a^3}{r}\right)\sin^2\theta. \tag{16.83}$$

If D is the friction force between the sphere and the fluid, the work that the sphere exerts on the fluid is DU. This is equal to dissipated energy due to viscosity. When the vorticity is ζ, the energy dissipation is $\mu\zeta^2$. The vorticity is given by

$$\zeta = \frac{1}{r}\frac{\partial(rv_\theta)}{\partial r} - \frac{1}{r}\frac{\partial v_r}{\partial \theta} = \frac{3a}{2r^2}U\sin\theta. \tag{16.84}$$

Integrating the vorticity in the fluid, we get

$$DU = \mu \int_0^\infty dr \int_0^\pi \frac{9a^2}{4r^4} U^2 \sin^2 \theta \cdot 2\pi r^2 \sin \theta \, d\theta = 6\pi \mu U^2 a. \quad (16.85)$$

Thus, the drag force D is

$$D = 6\pi \mu U a. \quad (16.86)$$

Equation (16.86) is called a Stokes' drag law or equation. The definition of the drag force is given by

$$D = \frac{1}{2} C_D \rho \pi a^2 U^2. \quad (16.87)$$

When the Reynolds number $Re = 2aU/\nu$ is less than 1, the drag coefficient C_D is obtained from Eqs. (16.86) and (16.87) as follows:

$$C_D = \frac{24}{Re}, \quad \text{for } Re < 1. \quad (16.88)$$

If we solve nonlinear differential equations by assuming the neglection of the nonlinear terms in nonlinear differential equations, we call it a Stokes approximation. This is the first step of the nonlinear analysis. But when we apply the Stokes approximation to the uniform flow past a circular cylinder, the flow does not become the uniform flow at a long distance from the circular cylinder. This is called a Stokes paradox. Usually when we solve nonlinear differential equations, we first assume that the nonlinear terms are zero. Then, to improve the method of the solution, instead of neglecting the nonlinear terms, we use the following assumption:

$$u \frac{\partial u}{\partial x} \cong U \frac{\partial u}{\partial x}, \quad (16.89)$$

in which $U = $ the uniform velocity. Using this method Eq. (16.89), Oseen obtained the drag coefficient. The drag coefficient is called an Oseen equation. It is written by

$$C_D = \frac{24}{Re} \left(1 + \frac{3}{16} Re \right). \quad (16.90)$$

This approximation does not generate the Stokes paradox. Goldstein in 1929 solved the Navier–Stokes equations using an Oseen

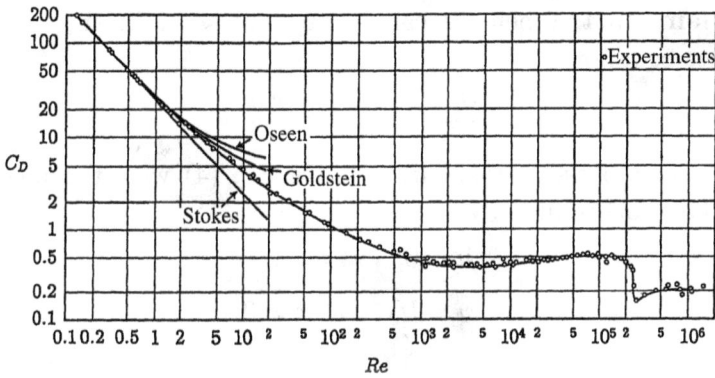

Figure 16.2. Drag coefficient of sphere.

approximation. This is

$$C_D = \frac{24}{Re}\left(1 + \frac{3}{16}Re - \frac{16}{1280}Re^2 + \frac{71}{20480}Re^3 - \cdots\right). \qquad (16.91)$$

The drag coefficient C_D is given in Fig. 16.2.

Example 16.4. Derive Eq. (16.72).

Solution. Equation (16.71) is rearranged as

$$\frac{d^4 f}{dr^4} - \frac{2}{r^2}\frac{d^2 f}{dr^2} - \frac{d^2}{dr^2}\left(\frac{2f}{r^2}\right) + \frac{4f}{r^4} = 0. \qquad (1)$$

Substituting $f(r) = r^n$ into Eq. (1), we get

$$n(n-1)(n-2)(n-3)r^{n-4}$$
$$- 2n(n-1)r^{n-4} - 2[n(n-1) - 4n + 6]r^{n-4} + 4r^{n-4} = 0. \qquad (2)$$

Since Eq. (2) is always equal to zero, we have

$$n(n-1)(n-2)(n-3) - 2(2n^2 - 6n + 6) + 4$$
$$= [(n-2)(n-3) - 2][n(n-1) - 2]. \qquad (3)$$

The condition that Eq. (3) becomes zero is Eq. (16.72).

Problem 16.4. When we can assume $|u| \ll 1$ in the following Burgers equation:

$$\frac{\partial u}{\partial t} + u\frac{\partial u}{\partial x} = \nu\frac{\partial^2 u}{\partial x^2}. \tag{4}$$

Using the Stokes approximation, solve Eq. (4) with the following boundary conditions:

$$u = 0, \quad \text{at } t = 0, \tag{5}$$

$$u = U_0, \quad \text{at } x = 0, \tag{6}$$

$$u = 0, \quad \text{at } x = \infty. \tag{7}$$

16.5. Burgers Equation

To understand the turbulent flow, Burgers investigated the one-dimensional Navier–Stokes equation. It is written as

$$\frac{\partial u}{\partial t} + u\frac{\partial u}{\partial x} = \nu\frac{\partial^2 u}{\partial x^2}. \tag{16.92}$$

Equation (16.92) has the fundamental property of the turbulent flow. The left side has an unsteady term and a nonlinear term and the right side has a diffusion term. If the nonlinear term is linear, Eq. (16.92) is easily solved. The Stokes approximation of Eq. (16.92) indicates

$$\frac{\partial u}{\partial t} = \nu\frac{\partial^2 u}{\partial x^2}. \tag{16.93}$$

The Oseen approximation of Eq. (16.92) with $u \cong u_0$ in the convection term indicates

$$\frac{\partial u}{\partial t} + u_0\frac{\partial u}{\partial x} = \nu\frac{\partial^2 u}{\partial x^2}, \tag{16.94}$$

in which $u_0 = $ a constant velocity. But since Eqs. (16.93) and (16.94) become linear, the two solutions cannot express turbulent phenomena. The turbulent phenomena produce many vortices and energy cascades which are generated by the nonlinear terms. To solve Eq. (16.92), a functional transform (Ames, 1965, 1972) is applied to Eq. (16.92). It is

$$u = f[\theta(x, t)]. \tag{16.95}$$

Then, we have

$$\frac{\partial u}{\partial t} = \frac{df}{d\theta}\frac{\partial \theta}{\partial t}, \tag{16.96}$$

$$\frac{\partial u}{\partial x} = \frac{df}{d\theta}\frac{\partial \theta}{\partial x}, \tag{16.97}$$

$$\frac{\partial^2 u}{\partial x^2} = \frac{d^2 f}{d\theta^2}\left(\frac{\partial \theta}{\partial x}\right)^2 + \frac{df}{d\theta}\frac{\partial^2 \theta}{\partial x^2}. \tag{16.98}$$

Substituting Eqs. (16.95), (16.96), (16.97), and (16.98) into Eq. (16.92), we derive

$$\frac{df}{d\theta}\frac{\partial \theta}{\partial t} + f\frac{df}{d\theta}\frac{\partial \theta}{\partial x} = \nu\left[\frac{d^2 f}{d\theta^2}\left(\frac{\partial \theta}{\partial x}\right)^2 + \frac{df}{d\theta}\frac{\partial^2 \theta}{\partial x^2}\right]. \tag{16.99}$$

If in Eq. (16.99) we assume the following relationships:

$$f\frac{df}{d\theta}\frac{\partial \theta}{\partial x} = \nu\frac{d^2 f}{d\theta^2}\left(\frac{\partial \theta}{\partial x}\right)^2, \tag{16.100}$$

$$\frac{df}{d\theta}\frac{\partial \theta}{\partial t} = \nu\frac{df}{d\theta}\frac{\partial^2 \theta}{\partial x^2}. \tag{16.101}$$

Equation (16.100) is rearranged as

$$f\frac{df}{d\theta} = \nu\frac{d^2 f}{d\theta^2}\frac{\partial \theta}{\partial x}, \tag{16.102}$$

in which $\partial\theta/\partial x \neq 0$. Integrating Eq. (16.102) about θ, we get

$$\frac{f^2}{2} = \nu\frac{df}{d\theta}\frac{\partial \theta}{\partial x}. \tag{16.103}$$

Thus, integrating Eq. (16.103) about θ, we obtain

$$f = -\frac{2\nu}{\theta}\frac{\partial \theta}{\partial x} = -2\nu\frac{\partial}{\partial x}(\ln\theta). \tag{16.104}$$

If we assume $df/d\theta \neq 0$, Eq. (16.101) becomes

$$\frac{\partial \theta}{\partial t} = \nu\frac{\partial^2 \theta}{\partial x^2}. \tag{16.105}$$

Equation (16.105) is the linear diffusion equation and the solution θ is easily obtained. Substituting θ into Eq. (16.104), we have the solution of the nonlinear partial differential equation, Eq. (16.92).

The transform of Eq. (16.104) is called a Cole–Hopf transform. In this way, the functional transform sometimes makes the nonlinear partial differential equation to the linear partial differential equation. Assuming $\nu = 0$ in Eq. (16.92), we have

$$\frac{\partial u}{\partial t} + u \frac{\partial u}{\partial x} = 0. \tag{16.106}$$

The solution is

$$u = F(x - ut), \tag{16.107}$$

in which $F(\cdot) = $ an arbitrary function. The solution of

$$\frac{\partial u}{\partial t} + f(u) \frac{\partial u}{\partial x} = 0 \tag{16.108}$$

is given by

$$u = F[x - f(u)t]. \tag{16.109}$$

The solution of Eq. (16.94) is also easily obtained. Substituting Eq. (16.95) into Eq. (16.94), we get

$$\frac{df}{d\theta} \frac{\partial \theta}{\partial t} + u_0 \frac{df}{d\theta} \frac{\partial \theta}{\partial x} = \frac{d^2 f}{d\theta^2} \left(\frac{\partial \theta}{\partial x} \right)^2 + \frac{df}{d\theta} \frac{\partial^2 \theta}{\partial x^2}. \tag{16.110}$$

If we assume the following two equations:

$$\frac{\partial \theta}{\partial t} = \frac{\partial^2 \theta}{\partial x^2}, \tag{16.111}$$

$$u_0 \frac{df}{d\theta} = \frac{d^2 f}{d\theta^2} \frac{\partial \theta}{\partial x}. \tag{16.112}$$

Then, we can solve Eq. (16.112) and determine $f(\theta)$ as

$$f(\theta) = \exp\left(\frac{u_0 \theta}{\theta_x} \right). \tag{16.113}$$

Substituting the solution $\theta(x, t)$ of Eq. (16.111) into Eq. (16.113), we get the solution of Eq. (16.94).

Example 16.5. Linearize the following nonlinear partial differential equation:

$$\frac{\partial u}{\partial t} + u^2 \frac{\partial u}{\partial x} = \frac{\partial^2 u}{\partial x^2}. \tag{1}$$

Solution. Substituting $u = f[\theta(x, t)]$ into Eq. (1), we have

$$\frac{df}{d\theta} \cdot \frac{\partial \theta}{\partial t} + f^2 \frac{df}{d\theta} \cdot \frac{\partial \theta}{\partial x} = \frac{d^2 f}{d\theta^2} \left(\frac{\partial \theta}{\partial x} \right)^2 + \frac{df}{d\theta} \cdot \frac{\partial^2 \theta}{\partial x^2}. \tag{2}$$

Assuming that the second term on the left side is equal to the first term on the right side, we get

$$f^2 \frac{df}{d\theta} = \frac{d^2 f}{d\theta^2} \frac{\partial \theta}{\partial x}. \tag{3}$$

Integrating Eq. (3) about θ, we obtain

$$\frac{\partial \theta}{\partial x} = -\frac{2\theta f^2}{3}. \tag{4}$$

Equation (4) indicates

$$f = \pm \sqrt{-\frac{3}{2\theta} \cdot \frac{\partial \theta}{\partial x}}. \tag{5}$$

From the remaining terms in Eq. (2), we obtain

$$\frac{\partial \theta}{\partial t} = \frac{\partial^2 \theta}{\partial x^2}. \tag{6}$$

The solution of Eq. (6) is easily obtained and the substitution of θ in Eq. (6) into Eq. (5) shows the solution of Eq. (1).

Problem 16.5. Linearize the following partial differential equation:

$$\frac{\partial u}{\partial t} + \frac{\partial u}{\partial x} + u \frac{\partial u}{\partial x} = \frac{\partial^2 u}{\partial x^2}. \tag{7}$$

16.6. Series Solution

Let us consider the following ordinary differential equation (Yoshida, 1950):

$$p_0(x) \frac{d^n y}{dx^n} + p_1(x) \frac{d^{n-1} y}{dx^{n-1}} + \cdots + p_{n-1}(x) \frac{dy}{dx} + p_n(x) y = q(x). \tag{16.114}$$

If $p_0(x) \neq 0$, $p_k(x)/p_0(x)$ at $x = x_0$ is expanded by the Taylor series as

$$\frac{p_k(x)}{p_0(x)} = \sum_{j=0}^{\infty} C_{kj}(x - x_0)^j, \quad \text{for } k = 1, 2, \ldots, n. \tag{16.115}$$

Then, we obtain the following sequential equations:

$$y = y_1, \frac{dy_1}{dx} = y_2, \frac{dy_2}{dx} = y_3, \dots, \frac{dy_{n-1}}{dx} = y_n,$$

$$\frac{dy_n}{dx} = \frac{q(x)}{p_0(x)} - \frac{p_n(x)}{p_0(x)}y_1 - \frac{p_{n-1}(x)}{p_0(x)}y_2 - \dots - + \frac{p_1(x)}{p_0(x)}y_{n-1}. \quad (16.116)$$

Equation (16.116) are equal to Eq. (16.114). A point $x = x_0$ is a normal point of Eq. (16.114), since $p_0(x_0) \neq 0$. When $p_0(x_0) = 0$, a point $x = x_0$ is a singular point of Eq. (16.114). When Eq. (16.115) is rearranged as

$$\frac{p_k(x)}{p_0(x)} = \frac{1}{(x-x_0)^k}\sum_{j=0}^{\infty}C_{kj}(x-x_0)^j, \quad \text{for } k = 1, 2, \dots, n. \quad (16.117)$$

Then, since the limiting value

$$\lim_{x \to x_0}(x - x_0)^k \frac{p_k(x)}{p_0(x)} = C_{k0} \quad (16.118)$$

is finite, we call this point a definite singular point. If the expansion as Eq. (16.118) is impossible, the point of $x = x_0$ is called an indefinite singular point. The method to obtain a solution near a definite singular point is called a Frobenius method (Yoshida, 1950). As an example of this method, there are the solutions of Hermite, Bessel or Legandre differential equations. Rewriting Eqs. (16.116) in general form, we get the following sequential equations:

$$\frac{dy_i}{dx} = f_i(x, y_1, y_2, \dots, y_n), \quad \text{for } i = 1, 2, \dots, n. \quad (16.119)$$

The boundary conditions are

$$y_i(a) = b_i, \quad \text{for } i = 1, 2, \dots, n. \quad (16.120)$$

Defining new variables $\bar{x}$ and $\bar{y}$ as

$$x = \bar{x} + a, \quad y_i = \bar{y}_i + b_i, \quad \text{for } i = 1, 2, \dots, n \quad (16.121)$$

and changing the variables x and y_i to $\bar{x}$ and $\bar{y}_i$, respectively, we have the solution which satisfies $\bar{y}_i(0) = 0$. To obtain the solution which

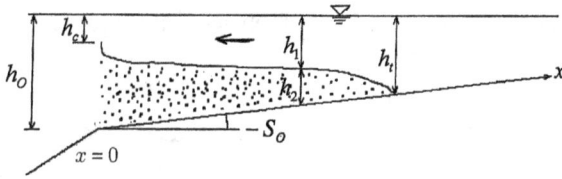

Figure 16.3. Saline water intrusion.

satisfies $\bar{y}_i(0) = 0$ $(i = 1, 2, \ldots, n)$, we have the following formal expression:

$$\bar{y}_i(x) = C_i^{(1)}\bar{x} + C_i^{(2)}\bar{x}^2 + \cdots + C_i^{(m)}\bar{x}^m + \cdots . \qquad (16.122)$$

Substituting Eq. (16.122) into Eq. (16.114) and comparing the same power of $\bar{x}$ in Eq. (16.114), we get coefficients $C_i^{(1)}$, $C_i^{(2)}$, ... We get a formal solution similar to Eq. (16.117). Equation (16.122) is called the formal solution of Eq. (16.114). When the formal solution converges, the formal solution approximates an analytical solution. Let us apply this method to the analysis of a saline wedge. When saline layer intrudes into the pure water as shown in Fig. 16.3, the saline wedge moves under the layer of pure water until an equilibrium condition is attained. Based on the physical condition of the saline wedge, the saline wedge attains an equilibrium condition and becomes steady for any bottom slope when the flow in the upper layer is steady and the depth-averaged velocity in the lower layer is zero. The length of the saline wedge is large for the small bottom slope. The limiting length of the saline wedge is caused by the flow on the horizontal bottom and the saline wedge reaches most upstream. When the velocity in the upper layer and the bottom slope are zero, the length of the saline wedge attains infinity. The solution of the shape of the saline wedge is easily obtained by applying the Saint Venant equations. Let us consider a river near its mouth and the saline wedge intrudes into the pure water layer. Under the some physical condition, two layers of the upper and lower are formed. The upper and lower layers consist of pure and saline water, respectively. The pure water in the upper layer flows over the saline water and the saline water is at rest in depth-averaged motion. The momentum equation of the upper layer is written as (Mizumura, 1992b)

$$\frac{dh_1}{dx} = \frac{f_i}{2} \frac{h_0 - S_0 x}{h_2(h_1^3 - h_c^3)},$$ (16.123)

in which h_1 = the thickness of the pure water layer, x = coordinate system in Fig. 16.3, f_i = an internal friction coefficient between the pure water layer and saline water layer, S_0 = the bottom slope, and h_c = internal critical depth. With the following transformation:

$$\eta = \frac{h_1}{h_0}, \quad \xi = \frac{x}{h_0}, \quad \eta_c = \frac{h_c}{h_0}, \quad h_0 + S_0 x = h_1 + h_2.$$ (16.124)

Equation (16.123) is written as

$$\frac{d\xi}{d\eta} = \frac{2}{f_i} \frac{(1 + S_0\xi - \eta)(\eta^3 - \eta_c^3)}{1 + S_0\xi}.$$ (16.125)

The boundary condition for Eq. (16.125) is

$$\eta = \eta_c, \quad \text{at } \xi = 0.$$ (16.126)

Equation (16.126) indicates that the pure water depth is critical at the river mouth. If $S_0 = 0$, Eq. (16.125) is easily integrable and its result is

$$\frac{\eta^4}{4} - \eta_c^3\eta - \frac{\eta^5}{5} + \frac{\eta_c^3}{2}\eta^2 + \frac{3}{4}\eta_c^4 - \frac{3}{10}\eta_c^5 = \frac{f_i}{2}\xi.$$ (16.127)

The existence of S_0 induces the nonlinearity in Eq. (16.125). By considering Eq. (16.126), the formal solution of Eq. (16.125) is assumed to be

$$\xi = \sum_{n=1}^{\infty} a_n(\eta - \eta_c)^{n+1}.$$ (16.128)

Substituting Eq. (16.128) into Eq. (16.125) and comparing the power of $\eta - \eta_c$, we get sequently

$$a_1 = \frac{3\eta_c^2(1 - \eta_c)}{f_i},$$ (16.129)

$$a_2 = \frac{2\eta_c(1 - 2\eta_c)}{f_i},$$ (16.130)

$$a_3 = \frac{1}{2f_i}\left[1 - 4\eta_c + \frac{9S_0\eta_c^5(1 - \eta_c)}{f_i}\right],$$ (16.131)

$$a_4 = \frac{2}{5f_i}\left[-1 + \frac{6S_0\eta_c^4(4 - 5\eta_c)}{f_i}\right],$$ (16.132)

$$\vdots$$

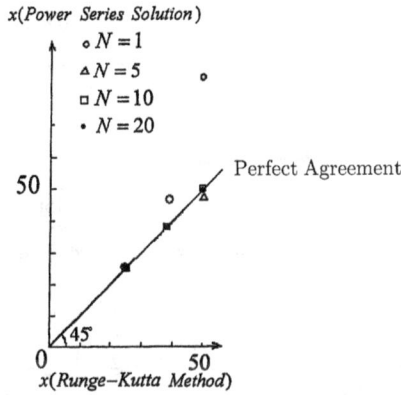

Figure 16.4. Comparison of formal solution with the Runge–Kutta method.

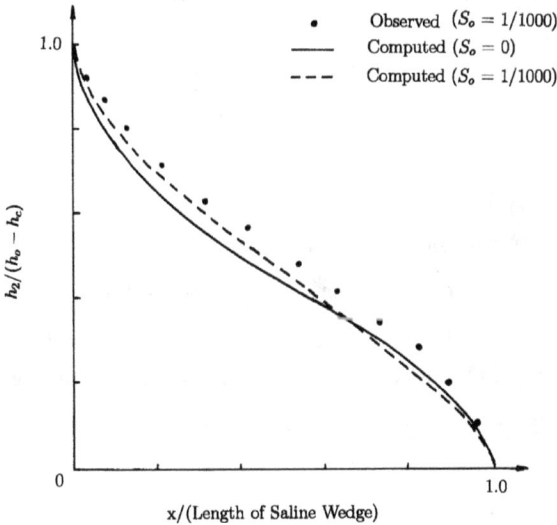

Figure 16.5. Comparison of formal solution with experiments ($F_1 = 0.087$).

To verify the formal solution, the Runge–Kutta method is used for numerical computation of Eq. (16.125). The comparison of $x (= h_0\xi)$ as the function of η is represented in Fig. 16.4. In the numerical computation, we used $S_0 = -0.001$, $q_1 = 1\,\mathrm{m}^2/\mathrm{s}$, $g_2 = 0.01\,g$, $f_i = 0.2/(\sqrt{Re_1} \cdot F_1)$, and $h_0 = 5\,\mathrm{m}$. Symbols Re_1 and F_1 indicate the Reynolds number and the Froude number of the upper layer, respectively. This result shows that Eq. (16.128) converges. Figure 16.5 shows the effect of the

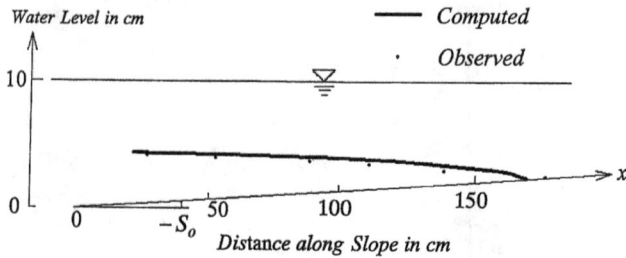

Figure 16.6.   Comparison of formal solution with experiment.

bottom slope on the form of the saline wedge. Figure 16.6 compares the formal solution with the experimental data. The number of used terms in the formal solution is 20.

Example 16.6. Solve the following ordinary differential equation:

$$\frac{d^2y}{dx^2} + xy = 0 \tag{1}$$

with the following bounadry conditions:

$$\frac{dy(0)}{dx} = 0, \quad y(0) = 1. \tag{2}$$

Solution. Defining the formal solution as

$$y(x) = C_0 + C_1 x + C_2 x^2 + C_3 x^3 + \cdots + C_n x^n + \cdots . \tag{3}$$

The boundary conditions indicate $C_0 = 1$ and $C_1 = 0$. Substituting Eq. (3) into Eq. (1) and comparing the power of x, we get

$$C_2 = 0, \quad C_3 = -\frac{1}{3!}, \quad C_4 = C_5 = 0, \quad C_6 = \frac{4}{6!},$$

$$C_7 = C_8 = 0, \quad C_9 = -\frac{4 \times 7}{9!}, \dots. \tag{4}$$

Therefore, the solution is

$$y(x) = 1 - \frac{1}{3!}x^3 + \frac{1 \times 4}{6!}x^6 - \frac{1 \times 4 \times 7}{9!}x^9 + \cdots . \tag{5}$$

Referring to Eq. (1.67), we have

$$\rho = \lim_{n \to \infty} \frac{x^3}{(3n+3)(3n+2)(3n+1)} = 0 \tag{6}$$

Equation (6) indicates the convergence of Eq. (5).

Problem 16.6. Obtain the formal solution of

$$\frac{dy}{dx} = 1 + x + x^2 + y. \tag{7}$$

The boundary condition is $y(0) = C$. The solution is $y(x) = (C+4)$ $e^x - 4 - 3x - x^2$.

16.7. Approximate Solution of Nonlinear Ordinary Differential Equation

Consider the approximate solution of the following nonlinear ordinary differential equation:

$$a_1 y + a_2 y^2 + a_3 y^3 + \cdots = k\phi(t), \quad a_1 \neq 0, \tag{16.133}$$

in which $t =$ an independent variable, $y =$ an dependent variable, $k =$ a constant, $\phi(t) =$ a function of t, $a_i =$ a function of D, and $D =$ a differential operator d/dt. It is defined by

$$D = \frac{d}{dt}. \tag{16.134}$$

As the formal solution of Eq. (16.133), we assume

$$y = A_1 k + A_2 k^2 + A_3 k^3 + \cdots . \tag{16.135}$$

Substituting Eq. (16.135) into Eq. (16.133) and equating the power of k, we obtain the functions A_i for $i = 1, 2, 3, \ldots$.

$$A_1 = \frac{\phi(t)}{a_1}, \tag{16.136}$$

$$A_2 = -\frac{a_2 A_1^2}{a_1}, \tag{16.137}$$

$$A_3 = -\frac{1}{a_1} [2a_2 A_1 A_2 + a_3 A_1^3], \tag{16.138}$$

$$A_4 = -\frac{1}{a_1} [a_2 (A_2^2 + 2A_1 A_3) + 3a_3 A_1^2 A_2 + a_4 A_1^4], \tag{16.139}$$

$$A_5 = -\frac{1}{a_1} [2a_2 (A_1 A_4 + A_2 A_3) + 3a_3 (A_1 A_2^2 + A_1^2 A_3) + 4a_4 A_1^3 A_2 + a_5 A_1^5], \tag{16.140}$$

$$A_6 = -\frac{1}{a_1}[a_2(A_3^2 + 2A_1A_5 + 2A_2A_4)$$
$$+ a_3(A_2^3 + 3A_1^2A_4 + 6A_1A_2A_3)$$
$$+ 2a_4(2A_1^2A_2^2 + 2A_1^3A_3) + 5a_5A_1^4A_2 + a_6A_1^6]. \quad (16.141)$$
$$\vdots$$

The formal solution of the nonlinear differential equation of special type is mechanically obtained. This is called a reversion method (Pipes and Harrill, 1971).

Example 16.7. Solve the following ordinary differential equation:

$$\frac{dy}{dt} - y^2 = 1. \quad (1)$$

The boundary condition is $y = 0$ at $t = 0$.

Solution.

$$a_1 = \frac{d}{dt} = D, \quad (2)$$

$$a_2 = -1, \quad k\phi(t) = 1, \quad k = 1. \quad (3)$$

Equation (16.136) shows

$$A_1 = \frac{1}{a_1} \quad \text{or} \quad DA_1 = 1. \quad (4)$$

This indicates

$$\frac{dA_1}{dt} = 1 \quad \text{or} \quad A_1 = t + C_1. \quad (5)$$

The boundary condition shows an integral constant $C_1 = 0$. Thus, we have $A_1 = t$. The function A_2 is derived from Eq. (16.137) as

$$A_2 = -\frac{a_2A_1^2}{a_1} = \frac{A_1^2}{D}. \quad (6)$$

Thus, we get

$$DA_2 = t^2. \quad (7)$$

The boundary condition indicates

$$A_2 = \frac{t^3}{3}. \quad (8)$$

The function A_3 is derived from Eq. (16.138),

$$A_3 = -\frac{1}{a_1}\left[-\frac{2t^4}{3}\right] = \frac{1}{D}\cdot\left(\frac{2t^4}{3}\right). \tag{9}$$

Thus, we have

$$\frac{dA_3}{dt} = \frac{2t^4}{3}. \tag{10}$$

The boundary condition gives

$$A_3 = \frac{2t^5}{15}. \tag{11}$$

Therefore, the approximate solution until A_3 term is

$$y = t + \frac{t^3}{3} + \frac{2t^5}{15}. \tag{12}$$

Equation (12) is a Maclurin series of $\tan t$. The analytical solution of Eq. (1) is given by

$$y = \tan t. \tag{13}$$

Problem 16.7. Solve the following nonlinear ordinary differential equation:

$$\frac{dy}{dt} + y^2 = t \tag{14}$$

with the boundary condition $y = 1$ at $t = 0$.

Chapter 17

Method of Boundary-Layer Theory

We, human beings, were interested in the computation of a drag force in fluid. First, the theory of perfect fluid was used to obtain the drag force. But due to the difference of the flow of real fluid from the flow of perfect fluid, the Navier–Stokes equations were derived independently by Navier in France in 1822, and Stokes in England in 1845. Since the Navier–Stokes equations are nonlinear partial differential equations, it was also very difficult to compute the drag force by using the Navier–Stokes equations. In 1904, Prandtl presented a method to simplify the Navier–Stokes equations. This was based on the physical consideration that the range of viscous effect near the body is very thin and restricted, and flow outside the range is irrotational. Therefore, the representative length in the normal direction to the body surface is very small. This simplifies the Navier–Stokes equations. In this chapter we study the theory of the laminar boundary-layer and apply it to jets, wakes, river flows, and sea flows. Mathematically this method is applied to a singular perturbation method in Chap. 19.

17.1. Boundary-Layer Equation

Let us consider the thin plate of which surface is located in the two-dimensional uniform flow as shown in Fig. 17.1. The plate is located parallel to the flow direction. The x and y directions are parallel and normal to the plate surface, respectively. The origin is located at the leading edge of the plate. The Navier–Stokes equations are

$$\frac{\partial u}{\partial x} + \frac{\partial v}{\partial y} = 0, \tag{17.1}$$

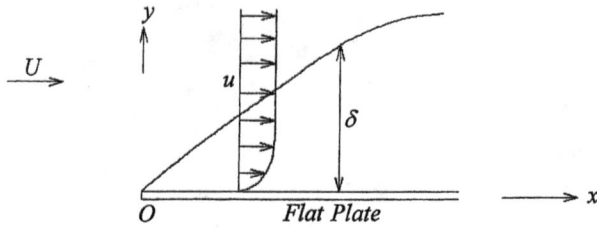

Figure 17.1.   Laminar boundary layer.

$$u\frac{\partial u}{\partial x} + v\frac{\partial u}{\partial y} = -\frac{1}{\rho}\frac{\partial p}{\partial x} + \nu\left(\frac{\partial^2 u}{\partial x^2} + \frac{\partial^2 u}{\partial y^2}\right), \qquad (17.2)$$

$$u\frac{\partial v}{\partial x} + v\frac{\partial v}{\partial y} = -\frac{1}{\rho}\frac{\partial p}{\partial y} + \nu\left(\frac{\partial^2 v}{\partial x^2} + \frac{\partial^2 v}{\partial y^2}\right), \qquad (17.3)$$

in which $x =$ the direction of the main flow, $y =$ normal to the plate, $u =$ the velocity in the x direction, $v =$ the velocity in the y direction, $p =$ the pressure, $\rho =$ the water density, and $\nu =$ the dynamic viscosity. The boundary conditions are

$$u = v = 0, \quad \text{at } y = 0. \qquad (17.4)$$

Equation (17.4) indicates that the velocities in the x and y directions are zero at the plate surface. When the Reynolds number Re is very large, the irrotational flow occupies the whole domain except for that near the plate surface. The effect of the viscosity is restricted in the thin layer near the plate surface. The velocity is uniform in the irrotational flow, but it decreases to zero at the plate surface, due to the effect of the viscosity. This domain is defined by the boundary-layer by Prandtl. Thus, the effect of the viscosity must be considered in the thin boundary-layer. The fact that the boundary-layer is very thin makes the Navier–Stokes equations simple. The thickness of the boundary-layer is defined by δ. Assume that the representative lengths in the x and y directions are ℓ and δ, respectively. The non-dimensional lengths in the x and y directions are X and Y, respectively. The non-dimensional coordinate system is defined by

$$x = \ell X, \quad y = \delta Y, \qquad (17.5)$$

in which ℓ corresponds to the length of the plate. Considering Eq. (17.4) and integrating Eq. (17.1) about y from 0 to δ, we get

$$v|_{y=\delta} = -\int_0^\delta \frac{\partial u}{\partial x} dy = -\frac{\delta}{\ell} \int_0^1 \frac{\partial u}{\partial X} dY. \qquad (17.6)$$

Since from Eq. (17.5) X and Y change from 0 to 1, the order of v is almost $(\delta/\ell)u$. According to Eq. (17.5), Eq. (17.2) are transformed to

$$\frac{1}{\ell}\left(u\frac{\partial u}{\partial X} + \frac{\ell}{\delta}v\frac{\partial u}{\partial Y} \right) = -\frac{1}{\rho\ell}\frac{\partial p}{\partial X} + \nu\left(\frac{1}{\ell^2}\frac{\partial^2 u}{\partial X^2} + \frac{1}{\delta^2}\frac{\partial^2 u}{\partial Y^2} \right). \qquad (17.7)$$

On the left side of Eq. (17.7), the orders of two terms are the same, because the order of v is almost $(\delta/\ell)u$. In the parentheses on the right side of Eq. (17.7), the first term is neglected in comparison with the second term, because δ^2/ℓ^2 is much smaller than 1. Since the order of $u/\ell \cdot \partial u/\partial X$ of the first term on the left side is the same as $\nu/\delta^2 \cdot \partial^2 u/\partial Y^2$ of the third term on the right side. Thus, since the representative velocity of u is U, the order of δ is given by

$$\delta^2 \sim \frac{\nu\ell}{U}. \qquad (17.8)$$

Therefore, we obtain

$$\frac{\delta}{\ell} \sim \frac{1}{\sqrt{U\ell/\nu}} = \frac{1}{\sqrt{Re}}, \qquad (17.9)$$

in which $Re =$ the Reynolds number. Equation (17.9) shows that δ/ℓ is proportional to $Re^{-1/2}$. Next, Eq. (17.3) is transformed to

$$\frac{1}{\ell}\left(u\frac{\partial v}{\partial X} + \frac{\ell}{\delta}v\frac{\partial v}{\partial Y} \right) = -\frac{1}{\rho\delta}\frac{\partial p}{\partial Y} + \nu\left(\frac{1}{\ell^2}\frac{\partial^2 v}{\partial X^2} + \frac{1}{\delta^2}\frac{\partial^2 v}{\partial Y^2} \right). \qquad (17.10)$$

The order of the terms on the left side is $u/\ell \cdot v$. Thus, the order is almost $U^2\delta/\ell^2$. Since $\nu \sim \delta^2$ from Eq. (17.8), the largest term on the right side is the first term on the right side. Since the order of δ is very small, $\partial p/\partial Y$ is almost zero. This indicates that the pressure p is constant in the y direction. This indicates that the pressure p is constant within the boundary-layer. The pressure outside the boundary-layer is the same as the pressure on the plate surface. Equation (17.2) is simplified to the following equation in the boundary-layer as follows:

$$u\frac{\partial u}{\partial x} + v\frac{\partial u}{\partial y} = -\frac{1}{\rho}\frac{\partial p}{\partial x} + \nu\frac{\partial^2 u}{\partial y^2}. \tag{17.11}$$

This equation and $\partial p/\partial y = 0$ from Eq. (17.3) are called Prandtl's boundary-layer equation or the boundary-layer equations (Shames, 1962; Schlichting, 1966).

Example 17.1. When the flow is steady and uniform, consider the following similarity transform of the indepentent variables x and y to a new variable η (Bluman and Cole, 1974):

$$\eta = \sqrt{\frac{U}{\nu x}} \cdot y. \tag{1}$$

The stream function is defined as

$$\psi = \sqrt{\nu U x} \cdot f(\eta). \tag{2}$$

Substituting Eqs. (1) and (2) into Eq. (17.11), we can derive the following equation:

$$2\frac{d^3 f}{d\eta^3} + f\frac{d^2 f}{d\eta^2} = 0. \tag{3}$$

Equation (3) is called a Blasius differential equation.

Solution. Rewriting velocities u and v in the x and y directions, respectively we obtain

$$u = \frac{\partial \psi}{\partial y} = \frac{\partial \psi}{\partial \eta} \cdot \frac{\partial \eta}{\partial y} = U\frac{df}{d\eta}, \tag{4}$$

$$v = -\frac{\partial \psi}{\partial x} = -\frac{\partial \psi}{\partial \eta} \cdot \frac{\partial \eta}{\partial x} - \frac{\partial \psi}{\partial x} = \frac{1}{2}\sqrt{\frac{U\nu}{x}}\left(\eta\frac{df}{d\eta} - f\right). \tag{5}$$

The differentiations of u and v with Eqs. (1), (4), and (5) give

$$\frac{\partial u}{\partial x} = \frac{\partial u}{\partial \eta} \cdot \frac{\partial \eta}{\partial x} = -U\frac{d^2 f}{d\eta^2} \cdot \frac{\eta}{2x}, \tag{6}$$

$$\frac{\partial u}{\partial y} = \frac{\partial u}{\partial \eta} \cdot \frac{\partial \eta}{\partial y} = U\sqrt{\frac{U}{\nu x}} \cdot \frac{d^2 f}{d\eta^2}, \tag{7}$$

$$\frac{\partial^2 u}{\partial y^2} = \frac{\partial}{\partial \eta}\left(\frac{\partial u}{\partial y}\right) \cdot \frac{\partial \eta}{\partial y} = \frac{U^2}{\nu x} \cdot \frac{d^3 f}{d\eta^3}. \tag{8}$$

Substituting Eqs. (6), (7), and (8) into Eq. (17.11), we obtain

$$2\frac{d^3 f}{d\eta^3} + f\frac{d^2 f}{d\eta^2} = 0. \tag{9}$$

Because the flow is steady and uniform, $\frac{df}{dx} = 0$. This is the nonlinear ordinary differential equation. Blasius, who was Prandtl's student, solved this nonlinear ordinary differential equation by using the series expansion method in Chap. 16 and the porturbation method in Chap. 19.

Problem 17.1. If we assume that the velocity outside the boundary-layer is $U(x)$, the boundary-layer equation, Eq. (17.11), is written as

$$u\frac{\partial u}{\partial x} + v\frac{\partial u}{\partial y} = -\frac{1}{\rho}\frac{\partial p}{\partial x} + \nu\frac{\partial^2 u}{\partial y^2}. \tag{10}$$

Since the momentum equations outside the boundary-layer is the Euler equations of motion in the x direction, we have

$$U\frac{\partial U}{\partial x} = -\frac{1}{\rho}\frac{\partial p}{\partial x}. \tag{11}$$

The substitution of Eq. (11) into Eq. (10) shows

$$u\frac{\partial u}{\partial x} + v\frac{\partial u}{\partial y} = U\frac{\partial U}{\partial x} + \nu\frac{\partial^2 u}{\partial y^2}. \tag{12}$$

When the velocity is given by

$$U(x) = u_1 x^m \quad (u_1: \text{constant}), \tag{13}$$

new variables are defined as

$$\eta = y\sqrt{\frac{m+1}{2} \cdot \frac{u_1}{\nu}} \cdot x^{(m-1)/2}, \tag{14}$$

$$\psi = \sqrt{\frac{2}{m+1}}\sqrt{\nu u_1} \cdot x^{(m+1)/2} f(\eta). \tag{15}$$

When velocities are

$$u = \frac{\partial \psi}{\partial x} = Uf'(\eta), \tag{16}$$

$$v = -\frac{\partial \psi}{\partial y} = -\sqrt{\frac{m+1}{2}\nu u_1 x^{m-1}}\left(f + \frac{m-1}{m+1}\eta f'\right), \tag{17}$$

derive the following equation:

$$f''' + ff'' + \beta(1 - f'^2) = 0, \tag{18}$$

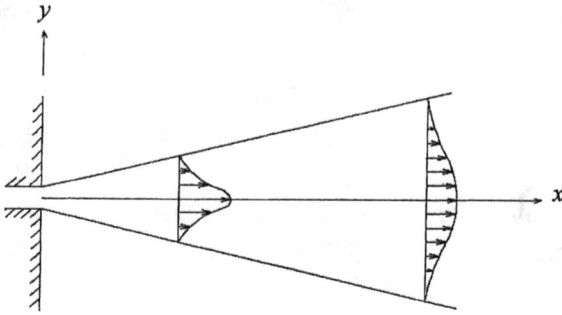

Figure 17.2.   Free jet.

in which

$$\beta = \frac{2m}{m+1}. \tag{19}$$

For the special cases of flow, the analytic or approximate solutions of Eq. (18) were obtained (Schlichting, 1966).

17.2. Two-Dimensional Jet

When fluid emits into the same fluid through a nozzle or orifice, the emitted fluid entrains the surrounding fluid and forms a free jet as shown in Fig. 17.2. This free jet is hydrodynamically very unstable and easily becomes turbulent. In the two-dimensional jet, $Re_c = 4.0$ is critical. The Reynolds number Re_c is based on the velocity at the free jet axis, the width of the free jet, and dynamic viscosity. When the fluid density of the free jet is different from the surrounding fluid, the gravitational and buoyancy forces must be considered. These are called a gravity current or plume. In the fluid field of a free jet and wake, there are no walls to influence the fluid motion. In the flow of the free jet and wake, the width of the flow is not infinite, the width of the flow is very thin in comparison with the distance from the origin, and velocity difference in the lateral direction is large. When the velocity in the flow direction is u and velocity in the lateral direction is v, the x and y directions indicate the flow axis and the lateral directions in the two dimensions, respectively. The boundary-layer equation is

$$u\frac{\partial u}{\partial x} + v\frac{\partial u}{\partial y} = \frac{1}{\rho}\frac{\partial \tau}{\partial y}. \qquad (17.12)$$

Since the wall or boundary do not exist in these flows, a laminar friction force is negligible in comparison with a turbulent friction force. Thus, the shear stress τ in Eq. (17.12) indicates the Reynolds stress. We assume that the pressure is constant through the fluid. In the free turbulent flow, the Reynolds stress is given by Prandtl as follows (Schlichting, 1966):

$$\tau = \rho\epsilon\frac{\partial u}{\partial y}, \qquad (17.13)$$

in which ϵ = the eddy viscosity. The eddy viscosity is also given by

$$\epsilon = k_1 b(x)(u_{max}(x) - u_{min}(x)), \qquad (17.14)$$

in which k_1 = a constant, $b(x)$ = the width of the free jet, $u_{max}(x)$ = the maximum velocity, and $u_{min}(x)$ = the minimum velocity. Usually the minimum velocity is defined 0. Then, the solution of the free jet was obtained by Goertler in 1942 (Schlichting, 1966). Substituting Eq. (17.13) into Eq. (17.12), we get

$$u\frac{\partial u}{\partial x} + v\frac{\partial u}{\partial y} = \epsilon\frac{\partial^2 u}{\partial y^2}. \qquad (17.15)$$

Assuming $u_{min} = 0$ in Eq. (17.14), we have the eddy viscosity as

$$\epsilon = k_1 b(x)u_{max}(x). \qquad (17.16)$$

If the maximum velocity at $x = s$ on the x axis is U_s and the width of the jet is b_s, the maximum velocity $u_{max}(x)$ on the x axis and the width of the jet $b(x)$ are respectively given by

$$u_{max}(x) = U_s f_1\left(\frac{x}{s}\right), \qquad (17.17)$$

$$b(x) = b_s f_2\left(\frac{x}{s}\right) = b_s\left(\frac{x}{s}\right)^m, \qquad (17.18)$$

in which f_1 and f_2 are arbitrary functions of x/s. Defining the non-dimensional length η as

$$\eta = \frac{y}{b(x)}, \qquad (17.19)$$

we assume that the velocity distribution in the lateral direction is similar about η. Then, the stream function $\Psi(x,y)$ is defined by

$$\Psi(x,y) = f(x)\psi(\eta), \tag{17.20}$$

$$f(x) = b_s U_s \left(\frac{x}{s}\right)^p, \tag{17.21}$$

in which $\psi(\eta)$ is an arbitrary function of η. Thus, the velocities u and v are

$$u = \frac{\partial \Psi}{\partial y} = \frac{f(x)}{b(x)}\frac{d\psi(\eta)}{d\eta} = U_s \left(\frac{x}{s}\right)^{p-m}\frac{d\psi(\eta)}{d\eta}, \tag{17.22}$$

$$v = -\frac{\partial \Psi}{\partial x} = -\frac{df(x)}{dx}\psi(\eta) + \frac{f(x)}{b(x)}\frac{db(x)}{dx}\eta\frac{d\psi(\eta)}{d\eta}, \tag{17.23}$$

In the same way, the partial derivatives of u and the eddy viscosity ϵ are

$$\frac{\partial u}{\partial x} = \frac{f'(x)b(x) - f(x)b'(x)}{b^2(x)}\psi'(x) - \frac{f(x)b'(x)}{b^2(x)}\eta\psi''(\eta), \tag{17.24}$$

$$\frac{\partial u}{\partial y} = \frac{f(x)}{b^2(x)}\psi''(\eta), \tag{17.25}$$

$$\frac{\partial^2 u}{\partial y^2} = \frac{f(x)}{b^3(x)}\psi'''(\eta), \tag{17.26}$$

$$\epsilon = k_1 u(x,0)b(x) = k_1 f(x)\psi'(0), \tag{17.27}$$

in which "$'$" indicates $d/d\eta$ and $u_{\max}(x) = u(x,0)$. The exponents m in Eq. (17.18) and p in Eq. (17.21) are determined by the following two conditions:

(i)   The momentum flux J is independent of the x direction. That is

$$J = \rho \int_{-\infty}^{\infty} u^2(x,y)\,dy = \text{const.} \tag{17.28}$$

(ii)   The inertia term on the left side in Eq. (17.15) is the same order of the turbulent friction term on the right side.

Thus, Eq. (17.28) is

$$J = \rho b_s U_s^2 \left(\frac{x}{s}\right)^{2p-m}\int_{-\infty}^{\infty} [\psi'(\eta)]^2 d\eta = \text{const.} \tag{17.29}$$

In order that J is independent of x, we get

$$2p - m = 0. \tag{17.30}$$

On the other hand, from Eqs. (17.24), (17.26), and (17.27), $u\partial u/\partial x$, $\partial^2 u/\partial y^2$, and ϵ are approximately written by

$$u\frac{\partial u}{\partial x} \sim x^{2p-2m-1}, \tag{17.31}$$

$$\frac{\partial^2 u}{\partial y^2} \sim x^{p-3m}, \tag{17.32}$$

$$\epsilon \sim x^p. \tag{17.33}$$

The symbol "$\sim$" indicates proportionality. Applying Eqs. (17.24), (17.26), and (17.27) into Eq. (17.15), the condition (ii) shows the relationship of the exponents of x,

$$2p - 2m - 1 = p + (p - 3m). \tag{17.34}$$

From Eqs. (17.30) and (17.34), we get

$$m = 1, \quad p = \frac{m}{2} = \frac{1}{2}. \tag{17.35}$$

Therefore, we obtain

$$b(x) \sim x, \tag{17.36}$$

$$u_{\max}(x) \sim x^{-1/2}, \tag{17.37}$$

$$f(x) \sim x^{1/2}, \tag{17.38}$$

$$\epsilon \sim x^{1/2}. \tag{17.39}$$

Equation (17.35) shows that the two-dimensional free jet expands linearly and the maximum velocity on the x axis decreases in the power of $-1/2$. Substituting Eqs. (17.22) to (17.27) into Eq. (17.15) with the usage of Eq. (17.35), we have

$$\frac{2\epsilon_s s}{U_s b_s^2}\psi''' + \psi\psi'' + (\psi')^2 = 0, \tag{17.40}$$

in which

$$\epsilon_s = k_1 b_s U_s. \tag{17.41}$$

The boundary conditions of the nonlinear ordinary differential equation are $\partial u/\partial y = 0$ and $v = 0$ at $y = 0$ and $u \to 0$ for $y \to \infty$.

These are given by Eqs. (17.23), (17.25), and (17.22) as follows:

$$\psi = 0, \quad \psi'' = 0, \quad \text{at } \eta = 0, \tag{17.42}$$

$$\psi' = 0, \quad \text{at } \eta \to \infty. \tag{17.43}$$

The boundary conditions are Eqs. (17.42) and (17.43) to solve Eq. (17.40). Assuming the indefinite width of the free jet b in Eq. (17.43) which is proportional to the distance from the origin as

$$\sigma = \frac{b_s}{s} = \frac{b}{x} = 4k_1, \tag{17.44}$$

we get

$$\frac{2\epsilon_s s}{U_s b_s^2} = \frac{1}{2}. \tag{17.45}$$

Assuming $\psi = 1$ as $\eta \to \infty$, we can integrate Eq. (17.40) as

$$\psi^2 + \psi' = 1. \tag{17.46}$$

Using Eq. (17.42), we integrate Eq. (17.46) once more, then we get

$$\psi(\eta) = \tanh \eta. \tag{17.47}$$

Equations (17.22), (17.23), and (17.47) derive the velocities u and v as

$$u = \frac{\sqrt{3}}{2} \sqrt{\frac{J}{\rho \sigma x}} (1 - \tanh^2 \eta), \tag{17.48}$$

$$v = \frac{\sqrt{3}}{4} \sqrt{\frac{J\sigma}{\rho x}} [2\eta(1 - \tanh^2 \eta) - \tanh \eta], \tag{17.49}$$

in which

$$\eta = \frac{y}{\sigma x}. \tag{17.50}$$

Example 17.2. High density fluid including suspended load moves down on a slope in a sea, lake, or river. We call this phenomenon a gravity current. The momentum equation of the gravity current (Turner, 1973) on the slope is given by

$$\rho_0 \frac{d}{dx} \int_0^{\delta_s} u^2 \, dy = g \sin \theta \int_0^{\delta_s} \Delta \rho \, dy - \tau_b, \tag{1}$$

in which ρ_0 = the water density, x = the flow direction, δ_s = the thickness of the gravity current, u = the velocity in the gravity current, θ = the angle of the slope, $\Delta\rho$ = the density difference between the surrounding pure water and the gravity current, and τ_b = the shear stress on the slope. We assume the following forms of the velocity in the gravity current, the thickness of the gravity current, and the density difference between the surrounding pure water and the gravity current:

$$u = u_{\max}(x)F(\eta) = u_0 \left(\frac{x}{x_0}\right)^\alpha F(\eta), \tag{2}$$

$$\delta_s(x) = \delta_0 \left(\frac{x}{x_0}\right)^\beta, \tag{3}$$

$$\Delta\rho = \Delta\rho_{\max}(x)H(\eta) = \rho_0 \left(\frac{x}{x_0}\right)^\gamma H(\eta), \tag{4}$$

in which α, β, and γ = constants, $F(\eta)$ and $H(\eta)$ = arbitrary functions, $\eta = y/\delta_s$, and subscript 0 indicates the base point. The resistance law on the slope is explained by

$$\tau_b = 0.0564\, \rho_s u_{\max}^2(x) \left[\frac{\nu}{u_{\max}(x)\delta_s(x)}\right]^{1/4}. \tag{5}$$

Then, obtain the coefficients α, β, and γ in Eqs. (2), (3), and (4).

Solution. Substituting Eqs. (2), (3), (4), and (5) into Eq. (1) and rearranging Eq. (1) about the exponent of x/x_0, we get

$$\alpha + 5\beta = 4, \quad \gamma = 2\alpha - 1. \tag{6}$$

If the gravity current is assumed to be the jet, the coefficient $\beta = 1$ from Eq. (17.30). Thus,

$$\alpha = -1, \quad \gamma = -3. \tag{7}$$

The application of the gravity current in the groundwater is the problem of leachate from landfills (Mizumura, 2003b). The leachate includes chloride ion and equi-concentration curves of chloride ion form the shape of the gravity current.

Problem 17.2. Derive Eq. (17.46) from Eqs. (17.42) and (17.43).

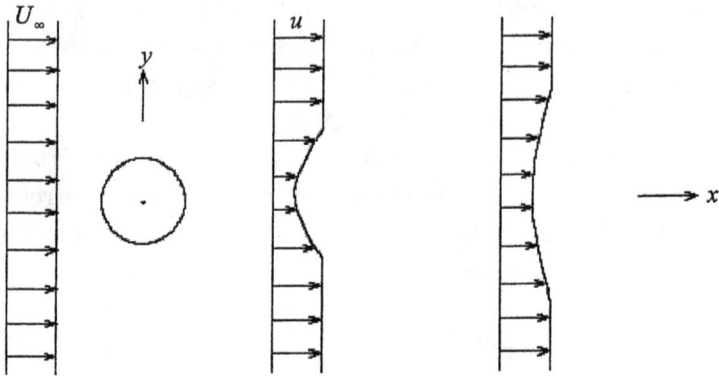

Figure 17.3.   Wake.

17.3. Two-Dimensional Wake

When a blunt body is located in the flow U_∞, the boundary-layer develops along the blunt body from the leading edge of the blunt body in Fig. 17.3, and separates from the blunt body, because the pressure cannot increase in the lee of the blunt body. The separated boundary-layer consists of various vortices and forms a wake in the lee of the blunt body. The lateral motion in the wake is restricted in comparison with the longitudinal motion. Thus, we can apply the boundary-layer equations to the analysis of the wake (Schlichting, 1966). In the velocity field in the wake, we can estimate the longitudinal velocity in the wake $u(y) \cong U_\infty$. Neglecting $v\partial u/\partial y$ in Eq. (17.15), we have

$$U_\infty \frac{\partial u_1}{\partial x} = \epsilon \frac{\partial^2 u_1}{\partial y^2}, \tag{17.51}$$

in which u_1 is defined as

$$u_1 = U_\infty - u. \tag{17.52}$$

The residual velocity u_1 indicates the deviation of the velocity from the velocity of the uniform flow. The symbol ϵ is the eddy viscosity. From Eq. (17.16) we get

$$\epsilon = k_1 u_{1\max}(x)b(x). \tag{17.53}$$

Similar to the free jet, the similarity assumption indicates that the non-dimensional length η is given by

$$\eta = \frac{y}{b(x)}, \tag{17.54}$$

in which y = the lateral distance from the center line of the wake and $b(x)$ = the width of the wake. The stream function $\Psi(x, y)$ is defined as

$$\Psi(x, y) = f(x)\psi(\eta), \tag{17.55}$$

in which $f(x)$ and $\psi(\eta)$ are arbitrary functions. Let us assume that the width of the wake $b(x)$ and an arbitrary function in the stream function $f(x)$ are the following power functions of x:

$$b(x) = \sigma x^m, \tag{17.56}$$

in which σ = an arbitrary constant. The function $f(x)$ is

$$f(x) = x^p. \tag{17.57}$$

The functional forms of $b(x)$, $f(x)$, and $\psi(\eta)$ are determined by the following conditions:

(i) $b(x)$, $f(x)$, and $\psi(\eta)$ satisfy Eqs. (17.51) and (17.53).
(ii) $b(x)$, $f(x)$, and $\psi(\eta)$ satisfy the relationship that the drag D is equal to the momentum flux.

Then, we have

$$D = \rho \int_{-\infty}^{\infty} u(U_\infty - u)dy \cong \rho U_\infty \int_{-\infty}^{\infty} u_1 dy, \tag{17.58}$$

in which u_1 is much less than U_∞ and approximated by $u(y) = U_\infty - u_1 \cong U_\infty$. From Eqs. (17.55), (17.56), and (17.57), we obtain

$$u_1 = u_1(x, y) = \frac{\partial \Psi}{\partial y} = f(x)\frac{d\psi}{d\eta} \cdot \frac{\partial \eta}{\partial y} = \frac{x^{p-m}}{\sigma}\psi'(\eta), \tag{17.59}$$

$$\frac{\partial u_1}{\partial x} = [(p - m)\psi'(\eta) - m\eta\psi''(\eta)]\frac{x^{p-m-1}}{\sigma}, \tag{17.60}$$

$$\frac{\partial u_1}{\partial y} = \frac{x^{p-2m}}{\sigma^2}\psi''(\eta), \tag{17.61}$$

$$\frac{\partial^2 u_1}{\partial y^2} = \frac{x^{p-3m}}{\sigma^3}\psi'''(\eta), \tag{17.62}$$

in which "$'$" $= d/d\eta$. On the other hand, we have

$$\epsilon = k_1 u_1(x,0)b(x) = k_1 x^p \psi'(0). \tag{17.63}$$

Substituting these equations into Eqs. (17.51) and (17.53), we get

$$[(p-m)\psi'(\eta) - m\eta\psi''(\eta)]x^{p-m-1} = \frac{k_1}{\sigma^2 U_\infty}\psi'(0)\psi'''(\eta)x^{2p-3m}. \tag{17.64}$$

From Eq. (17.58), we have

$$\int_{-\infty}^{\infty} u_1(x,y)\,dy = b(x)\int_{-\infty}^{\infty} u_1(x,\eta)d\eta = x^p \int_{-\infty}^{\infty} \psi'(\eta)d\eta = \text{const.} \tag{17.65}$$

From the condition that the powers of x in both sides of Eqs. (17.64) and (17.65) are the same, we obtain

$$p = 0, \quad m = \frac{1}{2}. \tag{17.66}$$

In order that the coefficient $[k_1/(\sigma^2 U_\infty)]$ on the right side of Eq. (17.64) becomes 1, the arbitrary constant σ in Eq. (17.56) is defined as

$$\sigma = \sqrt{\frac{k_1\psi'(0)}{U_\infty}}. \tag{17.67}$$

Then, the function $\psi(\eta)$ satisfies the following ordinary differential equation:

$$\psi'''(\eta) + \frac{\eta}{2}\psi''(\eta) + \frac{1}{2}\psi'(\eta) = 0. \tag{17.68}$$

The boundary conditions are $\partial u_1/\partial y = 0$ at $y = 0$ and $u_1 \to 0$ as $y \to \infty$. That is, we have

$$\psi''(\eta) = 0, \quad \text{at } \eta = 0, \tag{17.69}$$

$$\psi'(\eta) = 0, \quad \text{as } \eta \to \infty. \tag{17.70}$$

Using Eqs. (17.69) and (17.70), the integral of $\psi'(\eta)$ in Eq. (17.68) is obtained as

$$\psi'(\eta) = Ae^{-\eta^2/4}. \tag{17.71}$$

A constant A is derived from Eq. (17.58) as follows:

$$A = \frac{\frac{1}{2}C_D U_\infty d}{\int_{-\infty}^{\infty} e^{-\eta^2/4} d\eta} = \frac{C_D U_\infty d}{4\sqrt{\pi}}, \tag{17.72}$$

in which $d =$ the width of the body and $C_D =$ the drag coefficient in the definition $D = C_D d(\rho U_\infty^2/2)$. Therefore, the velocity distribution u_1 of the difference from the uniform flow U_∞ in the wake is derived from Eqs. (17.59), (17.71), and (17.72) as follows:

$$u_1(x, y) = U_\infty C_D \frac{1}{4\sqrt{\pi}} \sqrt{\frac{U_\infty d}{\epsilon_0}} \left(\frac{x}{d}\right)^{-1/2} e^{-\eta^2/4}. \tag{17.73}$$

Thus, the velocity distribution u is

$$u = U_\infty - u_1(x, y). \tag{17.74}$$

Example 17.3. There is a wall jet as shown in Fig. 17.4. Analyze it using the boundary-layer equation.

Solution. Integrating the boundary-layer equation of Eq. (17.11) under the condition of the constant pressure p from y to ∞ and

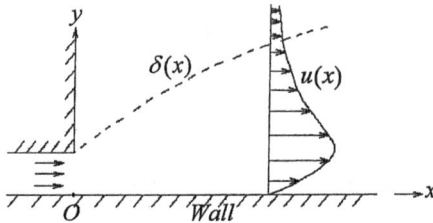

Figure 17.4.   Example 17.3.

noting $\partial u/\partial y|_{y=\infty} = 0$, we get

$$\int_y^\infty u \frac{\partial u}{\partial x} dy + \int_y^\infty v \frac{\partial u}{\partial y} dy = -\nu \frac{\partial u}{\partial y}. \tag{1}$$

Multiplying u to Eq. (1) and integrating from 0 to ∞, we obtain

$$\int_0^\infty \left(u \frac{\partial}{\partial x} \int_y^\infty \frac{u^2}{2} dy \right) dy$$

$$+ \int_0^\infty \left(u \int_y^\infty v \frac{\partial u}{\partial y} dy \right) dy + \frac{\nu}{2} \int_0^\infty \frac{\partial}{\partial y} u^2 \, dy = 0. \tag{2}$$

Since $u = 0$ at $y = 0$ and ∞, the third term in Eq. (2) is zero. Applying the partial integral to the second term and using the relationship $\partial v/\partial y = -\partial u/\partial x$, we obtain

$$\int_0^\infty \left(u \int_y^\infty v \frac{\partial u}{\partial y} dy \right) dy = \int_0^\infty u \left([vu]_y^\infty - \int_y^\infty u \frac{\partial v}{\partial y} dy \right) dy$$

$$= \int_0^\infty u \left(-vu + \int_y^\infty u \frac{\partial u}{\partial x} dy \right) dy$$

$$= \int_0^\infty u \left(-vu + \frac{\partial}{\partial x} \int_y^\infty \frac{u^2}{2} dy \right) dy. \tag{3}$$

Thus, Eq. (2) becomes

$$\int_0^\infty \left(u \frac{\partial}{\partial x} \int_y^\infty u^2 \, dy \right) dy - \int_0^\infty u^2 v \, dy = 0, \tag{4}$$

in which

$$\frac{d}{dx} \int_0^\infty \left(u \int_y^\infty u^2 \, dy \right) dy = \int_0^\infty \left(\frac{\partial u}{\partial x} \int_y^\infty u^2 \, dy \right) dy$$

$$+ \int_0^\infty \left(u \frac{\partial}{\partial x} \int_y^\infty u^2 \, dy \right) dy. \tag{5}$$

Since the continuity equation is

$$\frac{\partial u}{\partial x} = -\frac{\partial v}{\partial y}, \tag{6}$$

the first term on the right side of Eq. (5) is transformed to

$$\int_0^\infty \left(\frac{\partial u}{\partial x} \int_y^\infty u^2 dy \right) dy = - \int_0^\infty \left(\frac{\partial v}{\partial y} \int_y^\infty u^2 dy \right) dy$$

$$= - \left[v \int_y^\infty u^2 dy \right]_0^\infty - \int_0^\infty v u^2 dy$$

$$= - \int_0^\infty v u^2 dy. \tag{7}$$

Substituting Eq. (7) into Eq. (5) and the resultant equation into Eq. (4), we obtain

$$\frac{d}{dx} \int_0^\infty \left(u \int_y^\infty u^2 dy \right) dy = 0. \tag{8}$$

Substituting $u = \partial\psi/\partial y$ and $v = -\partial\psi/\partial x$ into the boundary-layer equation, Eq. (17.11) ($p = $ const), we have

$$\frac{\partial\psi}{\partial y} \cdot \frac{\partial^2\psi}{\partial y \partial x} - \frac{\partial\psi}{\partial x} \cdot \frac{\partial^2\psi}{\partial y^2} = \nu \frac{\partial^3\psi}{\partial y^3}, \tag{9}$$

in which $\psi = $ the stream function. The boundary conditions of Eq. (9) are

$$\psi = \frac{\partial\psi}{\partial y} = 0, \quad \text{at } y = 0, \tag{10}$$

$$\frac{\partial\psi}{\partial y} \to 0, \quad \text{as } y \to \infty. \tag{11}$$

Applying the similarity solution with the maximum velocity $\bar{u}(x)$ and assuming the following forms:

$$\psi(x,y) = [\nu x \bar{u}(x)]^{1/2} f(\eta), \tag{12}$$

$$\eta = \frac{y}{x} \sqrt{Re} = \frac{y}{x} [x\bar{u}(x)/\nu]^{1/2}, \tag{13}$$

we have

$$u(x,y) = \frac{\partial\psi}{\partial y} = \bar{u}(x) f'(\eta), \tag{14}$$

in which "′" indicates $d/d\eta$. Substituting these equations into Eq. (8), we obtain

$$\frac{d}{dx} \left[\bar{u}^3(x) \frac{\nu x}{\bar{u}(x)} \int_0^\infty f' \left(\int_y^\infty f'^2 d\eta \right) d\eta \right] = 0. \tag{15}$$

Since Eq. (15) is independent of x, we get

$$\bar{u}^2(x)x = \text{const} = C^2. \tag{16}$$

Therefore, we have

$$\bar{u}(x) = \frac{C}{\sqrt{x}}. \tag{17}$$

Therefore, we get the stream function ψ as

$$\psi(x, y) = \sqrt{\nu C} \cdot x^{1/4} f(\eta), \tag{18}$$

in which

$$\eta = \left(\frac{C}{\nu}\right)^{1/2} \frac{y}{x^{3/4}}. \tag{19}$$

Substituting Eq. (18) into Eq. (9) and multiplying $4x^2/(C^2\nu)$, we derive the following differential equation:

$$4f''' + ff'' + 2f'^2 = 0. \tag{20}$$

The boundary conditions of Eq. (20) are

$$f(0) = 0, \quad f'(0) = 0, \quad f'(\infty) = 0. \tag{21}$$

Problem 17.3. Solve Eq. (17.68) with the boundary conditions of Eqs. (17.69) and (17.70).

17.4. Boundary-Layer Induced by Wave Motion

Consider the boundary-layer at the bottom of water depth h generated by the wave motion. The unsteady boundary-layer equation is given by

$$\frac{\partial u}{\partial t} + u\frac{\partial u}{\partial x} + v\frac{\partial u}{\partial y} = -\frac{1}{\rho}\frac{\partial p}{\partial x} + \nu\frac{\partial^2 u}{\partial y^2}. \tag{17.75}$$

When the wave motion is infinitesimally small, the second and third terms on the left side are negligible. Thus, Eq. (17.75) becomes

$$\frac{\partial u}{\partial t} = -\frac{1}{\rho}\frac{\partial p}{\partial x} + \nu\frac{\partial^2 u}{\partial y^2}. \tag{17.76}$$

Since the boundary-layer is very thin, the pressure inside the boundary-layer is the same as the pressure outside. Thus, to obtain the pressure inside the boundary-layer, the pressure outside the boundary-layer is calculated by

$$-\frac{1}{\rho}\frac{\partial p}{\partial x} = \frac{\partial u_0}{\partial t}, \tag{17.77}$$

in which u_0 = the velocity in the x direction induced by the wave motion. From Eqs. (17.76) and (17.77), we get

$$\frac{\partial u}{\partial t} - \frac{\partial u_0}{\partial t} = \nu\frac{\partial^2 u}{\partial y^2}. \tag{17.78}$$

The velocity u_0 outside the boundary-layer is equal to the velocity at the bottom induced by the infinitesimally small amplitude wave motion and given by

$$u_0 = U_\infty \cos k(x - Ct), \tag{17.79}$$

in which

$$U_\infty = \frac{kCH}{2\sinh kh}, \tag{17.80}$$

in which H = wave height, k = an angular frequency, C = celerity, and h = water depth. The usage of Eq. (17.79) is the first approximation. Assume the velocity in Eq. (17.78) as follows:

$$u = \Re[f(y)e^{-ik(x - Ct)}], \tag{17.81}$$

in which $\Re$ indicates the real part. Substituting Eqs. (17.81) and (17.79) into Eq. (17.78) and rearranging it, we get

$$\frac{d^2 f}{dy^2} - \frac{ikC}{\nu}f = -\frac{ikC}{\nu}U_\infty. \tag{17.82}$$

Solving Eq. (17.82) with the boundary conditions $u = 0$ at $y = 0$, and $U = u_0$ at $y = +\infty$ we obtain

$$u = U_\infty[\cos k(x - Ct) - e^{-\beta y}\cos(kx - kCt + \beta y)], \tag{17.83}$$

in which $\beta = \sqrt{kC/(2\nu)}$. Equation (17.83) was first derived by Longuet-Higgins (1953).

Example 17.4. Derive Eq. (17.83) from Eq. (17.82).

Solution. Defining $f(y) = U_\infty + F(y)$ and substituting it into Eq. (17.82), we get

$$\frac{d^2 F}{dy^2} - \frac{ikC}{\nu} F = 0. \tag{1}$$

The solution of Eq. (1) is

$$F = C_1 e^{\sqrt{\frac{ikC}{\nu}} y} + C_2 e^{-\sqrt{\frac{ikC}{\nu}} y}. \tag{2}$$

Using the following relationship:

$$\sqrt{i} = e^{\frac{\pi i}{4}} = \frac{1+i}{\sqrt{2}}, \tag{3}$$

we get

$$F = C_1 e^{(1+i)\beta y} + C_2 e^{-(1+i)\beta y}. \tag{4}$$

Since $F \to 0$ for $y \to \infty$, we have $C_1 = 0$. Thus, we obtain

$$F = C_2 e^{-\beta y - i\beta y}, \tag{5}$$

$$f = F + U_\infty = C_2 e^{-\beta y - i\beta y} + U_\infty. \tag{6}$$

Substituting Eq. (6) into Eq. (17.81), we have

$$u = \Re[U_\infty e^{ik(Ct-x)}] + \Re[C_2 e^{-\beta y + i(kCt - kx - \beta y)}],$$

$$= U_\infty \cos k(Ct - x) + C_2 e^{-\beta y} \cos (kCt - kx - \beta y), \tag{7}$$

in which $u = 0$ at $y = 0$ shows $C_2 = -U_\infty$. Thus, Eq. (17.83) is derived.

Problem 17.4. We studied the problem of the oscillating plate in the fluid in Sec. 14.3. If $u = V_0 \cos \omega t$ at $y = 0$ in Eq. (14.17), compute Eq. (14.25) and compare it with Eq. (17.83).

17.5. Plane Boundary-Layer and Ekman Layer

When the two river flows in Fig. 17.5 are on the different water depth or different roughness, the plane boundary-layer is developed

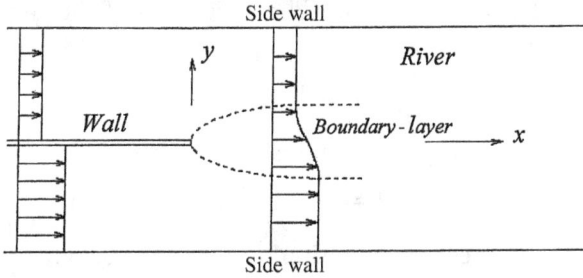

Figure 17.5. Plane boundary-layer.

in the lateral direction in the downstream after the confluence of the two rivers. The plane boundary-layer starts to develop from the confluence. Since the water depth is very small in comparison with the river width, there are large shear force acts in the direction to the normal to the water depth. Thus, the plane boundary-layer does not develop to the side walls and attains equilibrium state to some extent. Since the water depth is much smaller than the river width, the governing equations are the shallow water equations. Assuming the water depth is constant, we get

$$u\frac{\partial u}{\partial x} + v\frac{\partial u}{\partial y} = -g\frac{\partial H}{\partial x} + \epsilon\left(\frac{\partial^2 u}{\partial x^2} + \frac{\partial^2 u}{\partial y^2}\right) - \frac{f}{h}u\sqrt{u^2 + v^2}, \qquad (17.84)$$

$$u\frac{\partial v}{\partial x} + v\frac{\partial v}{\partial y} = -g\frac{\partial H}{\partial y} + \epsilon\left(\frac{\partial^2 v}{\partial x^2} + \frac{\partial^2 v}{\partial y^2}\right) - \frac{f}{h}v\sqrt{u^2 + v^2}, \qquad (17.85)$$

in which x = the coordinate system in the flow direction, y = the coordinate system in the lateral direction, u = the horizontal velocity in the x direction, v = the horizontal velocity in the y direction, H = the water level, h = the water depth, g = the gravitational acceleration, f = the friction coefficient, and ϵ = the horizontal eddy viscosity. Assuming that the velocity in the uniform flow is u_0 and $v = 0$ in Eq. (17.84), we get

$$gS_0 - \frac{f}{h}u_0^2 = 0, \qquad (17.86)$$

in which $S_0 =$ the bottom slope. The deviations of the variables from the uniform flow are as follows:

$$u = u_0 - u', \qquad (17.87)$$

$$v = v', \qquad (17.88)$$

$$H = -S_0 x + H', \qquad (17.89)$$

in which "$'$" indicates the deviation from the uniform flow. Assuming that H' is almost constant, by substituting Eqs. (17.87), (17.88), and (17.89) into Eqs. (17.84) and (17.85), rearranging them, and using the following boundary-layer assumption:

$$\left| \frac{\partial^2 u}{\partial x^2} \right| \ll \left| \frac{\partial^2 u}{\partial y^2} \right| \qquad (17.90)$$

and the following approximation (linearization) in Chap. 16:

$$\sqrt{u^2 + v^2} = \sqrt{(u_0 - u')^2 + v'^2} \cong u_0 \left(1 - \frac{u'}{u_0} \right), \qquad (17.91)$$

we get from Eq. (17.84)

$$-u_0 \frac{\partial u'}{\partial x} + u' \frac{\partial u'}{\partial x} + v' \frac{\partial u'}{\partial y} = -\epsilon \frac{\partial^2 u'}{\partial y^2} + \frac{f}{h}(2u_0 u' - u'^2). \qquad (17.92)$$

In the enough downstream, the thickness of the plane boundary-layer is constant, the dependent variables do not change, and the flow becomes uniform, $\frac{\partial u'}{\partial x} = \frac{\partial u'}{\partial y} = 0$. Then, Eq. (17.92) is written by

$$-\epsilon \frac{\partial^2 u'}{\partial y^2} + \frac{f}{h}(2u_0 u' - u'^2) = 0. \qquad (17.93)$$

In Eq. (17.93), defining U and η as follows:

$$\frac{u'}{u_0} = U, \quad \frac{y}{\delta} = \eta, \qquad (17.94)$$

we obtain

$$-k\ddot{U} + 2U - U^2 = 0, \qquad (17.95)$$

in which "$\cdot$" $= d/d\eta$ and $k = \epsilon h/(f u_0 \delta^2)$. Equation (17.95) is the nonlinear differential equation. Multiplying $\dot{U}$ to Eq. (17.95), we have

$$-k\dot{U}\ddot{U} + 2U\dot{U} - U^2\dot{U} = 0. \tag{17.96}$$

Integrating Eq. (17.96), we get

$$-\frac{k}{2}\dot{U}^2 + U^2 - \frac{U^3}{3} = C_1. \tag{17.97}$$

Assuming that $U = \dot{U} = 0$ at $\eta = \infty$, we have $C_1 = 0$. Then, Eq. (17.97) becomes

$$-\frac{k}{2}\dot{U}^2 + U^2 - \frac{U^3}{3} = 0. \tag{17.98}$$

Rearranging Eq. (17.98), we obtain

$$\frac{dU}{U\sqrt{1 - U/3}} = \sqrt{\frac{2}{k}}d\eta. \tag{17.99}$$

We use the integral formula as

$$\int \frac{dx}{x\sqrt{ax + b}} = \frac{1}{\sqrt{b}}\log\left|\frac{\sqrt{ax + b} - \sqrt{b}}{\sqrt{ax + b} + \sqrt{b}}\right|, \quad \text{for } b > 0. \tag{17.100}$$

Substituting $a = -1/3$ and $b = 1$, we get

$$\frac{\sqrt{1 - U/3} - 1}{\sqrt{1 - U/3} + 1} = Ce^{\sqrt{2/k}\eta}, \tag{17.101}$$

in which $C = $ an integral constant. Rearranging Eq. (17.101), we obtain

$$U = 3 - 3\left(\frac{1 + Ce^{\sqrt{2/k}\eta}}{1 - Ce^{\sqrt{2/k}\eta}}\right)^2. \tag{17.102}$$

The integral constant C is determined by $U = U_s$ at $\eta = 0$ as

$$C = \frac{\sqrt{1 - U_s/3} - 1}{\sqrt{1 - U_s/3} + 1}. \tag{17.103}$$

The substitution of $v = 0$ into Eq. (17.85) indicates $dH/dy = 0$.

Next, consider an Ekman layer. In 1905, Ekman analyzed flow in the sea and atmosphere induced by the spin of the earth, assuming that the Coriolis force is equal to the viscous force. Substituting the Coriolis force into the Navier–Stokes equations and conducting the Stokes approximation in Chap. 16, we have

$$-2\omega v = -\frac{\partial p}{\partial x} + \nu \nabla^2 u, \qquad (17.104)$$

$$2\omega u = -\frac{\partial p}{\partial y} + \nu \nabla^2 v, \qquad (17.105)$$

$$0 = -\frac{\partial p}{\partial z} + \nu \nabla^2 w, \qquad (17.106)$$

in which x and y = the horizontal coordinate system, z = the coordinate system in the vertical upward direction, ν = the kinematic viscosity, ω = an angular velocity due to the spin of the earth, p = the pressure, and u, v, and w = the velocities in the directions of x, y, and z, respectively. If $w = 0$, the pressure p is independent of z. If p is independent of x and y, Eqs. (17.104) and (17.105) are written by

$$-2\omega v = \nu \nabla^2 u, \qquad (17.107)$$

$$2\omega u = \nu \nabla^2 v. \qquad (17.108)$$

Ekman assumed that the velocities u and v are the function of z only. Applying the plane boundary-layer approximation $\partial^2/\partial z^2 \gg \partial^2/\partial x^2$ and $\partial^2/\partial z^2 \gg \partial^2/\partial y^2$ to Eqs. (17.107) and (17.108), we get

$$-2\omega v = \nu \frac{d^2 u}{dz^2}, \qquad (17.109)$$

$$2\omega u = \nu \frac{d^2 v}{dz^2}. \qquad (17.110)$$

Multiplying $i = \sqrt{-1}$ to Eq. (17.110) and adding it to Eq. (17.109), we have the following equation:

$$2\omega i (u + iv) = \nu \frac{d^2}{dz^2}(u + iv). \qquad (17.111)$$

The solution of Eq. (17.111) is easily obtained as

$$u + iv = (U + iV)\exp\left[\left(\frac{\omega}{\nu}\right)^{1/2}(1 + i)z\right], \qquad (17.112)$$

in which the boundary conditions in the sea are

$$u + iv = U + iV, \quad \text{at } z = 0, \tag{17.113}$$

$$u + iv = 0, \quad \text{at } z = -\infty, \tag{17.114}$$

in which U and V are the constant velocities in the water surface. Equation (17.112) indicates that the velocity spirally decreases as z becomes $-\infty$.

Example 17.5. Integrate Eq. (17.111).

Solution. Defining $w = u + iv$, we rewrite Eq. (17.111) as

$$2wiw = \nu \frac{d^2 w}{dz^2}. \tag{1}$$

Substituting $w = Ae^{az}$ into Eq. (1), we have

$$a^2 = \frac{2wi}{\nu}. \tag{2}$$

Since $i = e^{\pi i/2} + 2n\pi$ for $n = 0, 1, 2, \ldots$,

$$\sqrt{i} = e^{\pi i/4 + n\pi} = \frac{1 + i}{\sqrt{2}}. \tag{3}$$

Thus, based on Eq. (17.114) we get

$$a = \sqrt{\frac{\omega}{\nu}} (1 + i). \tag{4}$$

The solution e^{-az} does not satisfy the physical condition with Eq. (17.114). To satisfy Eq. (17.113), the solution is

$$w = u + iv = (U + iV) \exp \left[\sqrt{\frac{\omega}{\nu}} (1 + i) z \right]. \tag{5}$$

Problem 17.5. Solve the following ordinary differential equation:

$$\ddot{U} = U - U^3, \quad \text{for } -\sqrt{2} \le U \le \sqrt{2}. \tag{6}$$

Chapter 18

Variational Method

The variational method has indirect and direct ones. The indirect method is to solve an Euler's variational equation which is derived from the original differential equation in order that the original differential equation becomes extreme. The direct method is to obtain coefficients in a trial (composite) function which satisfies boundary conditions in order that the original differential equation with the coefficients has an extreme value. For the direct method, there exist the Ritz method and the Galerkin method. These are also applicable to nonlinear problems. We also introduce the least squares method, collocation method, and the method of moments as the direct method.

18.1. Ritz Method

Consider the following ordinary differential equation:

$$f(x, y, y') = 0, \quad \text{for } 0 \leq x \leq \ell, \tag{18.1}$$

in which $y' = dy/dx$. The boundary conditions are

$$y(0) = y_0 \quad \text{and} \quad y(\ell) = y_\ell. \tag{18.2}$$

This is called a two-point boundary-value problem (abbreviated by TPBVP). Instead of solving Eq. (18.1) directly, we consider an arbitrary function $F(x, y, y')$. Integrating $F(x, y, y')$ about x, we get

$$I(y) = \int_0^\ell F(x, y, y') dx. \tag{18.3}$$

Equation (18.3) is called a functional with an argument y. The function which makes Eq. (18.3) minimum is defined $\bar{y}(x)$. The function $\bar{y}(x)$ satisfies Eq. (18.2). A trial function $y_\epsilon(x)$ consists of $\bar{y}(x)$ and

$\eta(x)$ which satisfies

$$\eta(0) = 0 \quad \text{and} \quad \eta(\ell) = 0. \tag{18.4}$$

Introducing $\eta(x)$ and an arbitrary constant ϵ, we have

$$y_\epsilon(x) = \bar{y}(x) + \epsilon\eta(x). \tag{18.5}$$

The trial function $y_\epsilon(x)$ satisfies Eq. (18.2). The trial function $y_\epsilon(x)$ is called a function which is obtained by the variation of $I(y)$. After the substitution of $y = y_\epsilon(x)$ into Eq. (18.3), $I(y = y_\epsilon(x))$ is the function of ϵ and takes minimum at $\epsilon = 0$. Therefore, we get

$$\frac{\partial}{\partial\epsilon}I(y_\epsilon)\bigg|_{\epsilon=0} = 0. \tag{18.6}$$

We call $\bar{y}(x)$ a stationary function. Then, Eq. (18.6) is rewritten as

$$\frac{\partial}{\partial\epsilon}I(y_\epsilon)\bigg|_{\epsilon=0} = \frac{\partial}{\partial\epsilon}\int_0^\ell F\left[x, \bar{y}(x) + \epsilon\eta(x), \bar{y}'(x) + \epsilon\eta'(x)\right]dx\bigg|_{\epsilon=0}, \tag{18.7}$$

in which "$'$" $= d/dx$ Rearranging Eq. (18.7), we have

$$\int_0^\ell [F_y|_{\bar{y}(x)}\eta(x) + F_{y'}|_{\bar{y}(x)}\eta'(x)]dx = 0. \tag{18.8}$$

Considering Eq. (18.2) and integrating Eq. (18.8) partly, we obtain

$$\int_0^\ell \left[F_y|_{\bar{y}(x)} - \frac{d}{dx}F_{y'}\bigg|_{\bar{y}(x)}\right]\eta(x)dx = 0. \tag{18.9}$$

This indicates that $y = \bar{y}(x)$ satifies the following differential equation:

$$F_y - \frac{d}{dx}F_{y'} = 0. \tag{18.10}$$

Equation (18.10) is called the Euler's variational equation for Eq. (18.3). Since the left side of Eq. (18.1) is equal to Eq. (18.10), we get

$$f(x, y, y') = F_y - \frac{d}{dx}F_{y'} = 0, \tag{18.11}$$

in which

$$y' = \frac{dy}{dx}. \tag{18.12}$$

The method to solve Eq. (18.10) is called the indirect method. The solution of Eq. (18.10) is obtained by the indirect method. Equation (18.10) is called the Euler's variational equation. Equation (18.10) is a necessary condition in order that Eq. (18.3) has an extreme value. For example, let us consider the following equation:

$$F(x, y, y') = y'^2 + y^2 + 2xy. \tag{18.13}$$

The Euler's variational equation of Eq. (18.13) is given by

$$F_y - \frac{d}{dx} F_{y'} = 2(y + x) - 2y'' = 0. \tag{18.14}$$

Rearranging Eq. (18.14), we obtain

$$y'' = y + x. \tag{18.15}$$

To get the solution y as the variational problem, we must solve Eq. (18.15) again. Instead of the indirect method, Ritz suggested to obtain y under the condition, where Eq. (18.3) takes an extreme. This is called the direct method. When the boundary conditions are $y(0) = y_0$ and $y(\ell) = y_\ell$, the following trial function including a complete system of functions satisfies the boundary conditions $y = y_0$ at $x = 0$ and $y = y_\ell$ at $x = \ell$:

$$y(x) = \frac{x}{\ell} y_\ell + \frac{\ell - x}{\ell} y_0 + \sum_{k=1}^{\infty} a_k \sin \frac{k\pi x}{\ell}. \tag{18.16}$$

The sine functions belong to the complete system of functions. The coefficients a_k's are selected in order that the functional $I(y)$ in Eq. (18.3) becomes minimum. Namely, we assume that $y_n(x)$ of the $n+2$ terms can approximate the true solution $y(x)$. Thus, the approximate solution is written as

$$y_n(x) = \frac{x}{\ell} y_\ell + \frac{\ell - x}{\ell} y_0 + \sum_{k=1}^{n} a_k \sin \frac{k\pi x}{\ell}. \tag{18.17}$$

Substituting Eq. (18.17) into $I(y_n)$ in Eq. (18.3) and minimizing $I(y_n)$ about a_k, we get the following n equations:

$$\frac{\partial I(y_n)}{\partial a_1} = \frac{\partial I(y_n)}{\partial a_2} = \cdots = \frac{\partial I(y_n)}{\partial a_n} = 0. \tag{18.18}$$

Since Eq. (18.18) becomes n sequential linear equations, these linear equations show the coefficients $a_1, a_2, \ldots, a_k$. The principle is the same as that of the least squares method. Since the number of terms in Eq. (18.17) is finite, this is applicable to nonlinear differential equations. In Eq. (18.17), we use the sine functions as the complete system of functions. But we can use another complete system of functions which satisfies the boundary conditions $y = 0$ at $x = 0$ and $x = \ell$. For example, we can select the complete system of functions $(\ell - x)x^k$ for this problem. Then, the trial function is assumed to be

$$y(x) = \frac{x}{\ell} y_\ell + \frac{(\ell - x)}{\ell} y_0 + \sum_{k=1}^{\infty} a_k (\ell - x) x^k. \tag{18.19}$$

The coefficients may be determined by Eq. (18.18). The trick to the selection of the complete system of functions is the easy integral of the product of the complete system of functions.

Example 18.1. Solve the following partial differential equation by the Ritz method:

$$\frac{\partial^2 \phi}{\partial x^2} + \frac{\partial^2 \phi}{\partial y^2} = 1, \tag{1}$$

in which ϕ is zero at $x = 0$, $x = a$, $y = 0$, and $y = b$. The function ϕ is defined in the rectangular domain $0 \le x \le a$ and $0 \le y \le b$. The function ϕ minimizes the following integral:

$$I(\phi) = \int_0^a \int_0^b \left[\left(\frac{\partial \phi}{\partial x} \right)^2 + \left(\frac{\partial \phi}{\partial y} \right)^2 + 2\phi \right] dx dy. \tag{2}$$

Solution. Assume the following trial function:

$$\phi = \sum_{m=1}^{\infty} \sum_{n=1}^{\infty} a_{mn} \sin \frac{m\pi x}{a} \sin \frac{n\pi y}{b}, \tag{3}$$

in which ϕ satisfies the bounadry conditions. Computing $\partial \phi / \partial x$ and $\partial \phi / \partial y$ and substituting Eq. (3) into Eq. (2), we get

$$I = \sum_{m=1}^{\infty} \sum_{n=1}^{\infty} \left\{ \frac{\pi^2 ab}{4} a_{mn}^2 \left(\frac{m^2}{a^2} + \frac{n^2}{b^2} \right) \right.$$

$$\left. + \frac{2ab}{mn\pi^2} a_{mn} [(-1)^m - 1][(-1)^n - 1] \right\}. \tag{4}$$

Since $I(\phi)$ is the function of a_{mn}'s, a_{mn}'s are determined in order that $I(\phi)$ takes an extreme. That is

$$\frac{\partial I}{\partial a_{mn}} = \sum_{m=1}^{\infty}\sum_{n=1}^{\infty}\left\{\frac{\pi^2 ab}{2}a_{mn}\left(\frac{m^2}{a^2}+\frac{n^2}{b^2}\right)\right.$$

$$\left.+\frac{2ab}{mn\pi^2}[(-1)^m - 1][(-1)^n - 1]\right\} = 0,$$

$$\text{for } m = 1, 2, \ldots, \quad n = 1, 2, \ldots. \tag{5}$$

This indicates

$$a_{mn} = -\frac{4[(-1)^m - 1][(-1)^n - 1]}{mn\pi^4\left(\frac{m^2}{a^2}+\frac{n^2}{b^2}\right)},$$

$$\text{for } m = 1, 2, \ldots, \quad n = 1, 2, \ldots. \tag{6}$$

Problem 18.1. Show that when $y = \sin x$, the functional

$$I(y) = \int_0^{\pi/2} (y'^2 - y^2)dx, \quad y(0) = 0, \quad y(\pi/2) = 1 \tag{7}$$

takes an extreme.

18.2. Galerkin Method

The Galerkin method (Finlayson, 1972; Mura and Kaya, 1992) is essentially the same as the Ritz method. But the Galerkin method is applicable to the practical computation by the skillful technique. The governing equation is

$$F(x, y, y') = 0. \tag{18.20}$$

The boundary conditions are $y(0) = y_0$ and $y(\ell) = y_\ell$. Consider the problem which minimizes the following functional:

$$I(y) = \int_0^{\infty} F(x, y, y')dx. \tag{18.21}$$

The trial function is given by the linear combination of the complete system of functions $\omega_1, \omega_2, \ldots$. The trial function is assumed to be by ω_k as follows:

$$y = \frac{x}{\ell}y_\ell + \frac{\ell - x}{\ell}y_0 + a_1\omega_1 + a_2\omega_2 + \cdots + a_n\omega_n + \cdots. \tag{18.22}$$

Substituting Eq. (18.22) into Eq. (18.21), the coefficients a_n's are determined by the conditions

$$\frac{\partial I(y)}{\partial a_n} = 0, \quad \text{for } n = 1, 2, \dots . \tag{18.23}$$

Calculating Eq. (18.23), we obtain

$$\frac{\partial I(y)}{\partial a_n} = \int_0^\ell \frac{\partial F}{\partial a_n} dx = \int_0^\ell \left(\frac{\partial y}{\partial a_n} \frac{\partial F}{\partial y} + \frac{\partial y'}{\partial a_n} \frac{\partial F}{\partial y'} \right) dx$$

$$= \int_0^\ell \left(\omega_n \frac{\partial F}{\partial y} + \omega_n' \frac{\partial F}{\partial y'} \right) dx, \tag{18.24}$$

in which "′" indicates d/dx. Integrating Eq. (18.24) partly, we have

$$\frac{\partial I(x)}{\partial a_n} = \left[\omega_n \frac{\partial F}{\partial y'} \right]_0^\ell + \int_0^\ell \left[\frac{\partial F}{\partial y} - \frac{\partial}{\partial x} \left(\frac{\partial F}{\partial y'} \right) \right] \omega_n dx = 0. \tag{18.25}$$

Since ω_n is 0 at $x = 0$ and ℓ, Eq. (18.25) is rewritten as

$$\int_0^\ell \left[\frac{\partial F}{\partial y} - \frac{\partial}{\partial x} \left(\frac{\partial F}{\partial y'} \right) \right] \omega_n dx = 0. \tag{18.26}$$

Defining the inside of Eq. (18.26) as

$$L(u) = \frac{\partial F}{\partial y} - \frac{\partial}{\partial x} \left(\frac{\partial F}{\partial y'} \right). \tag{18.27}$$

The Euler's variational equation is defined by

$$L(u) = 0. \tag{18.28}$$

Solving Eq. (18.26) is the same as solving the Euler's variational equation of Eq. (18.27). Therefore, we have

$$\int_0^\ell L(u)\omega_n dx = 0, \tag{18.29}$$

in which the trial function is

$$u = u_0 + a_1 \omega_1 + a_2 \omega_2 + \cdots . \tag{18.30}$$

Substituting Eq. (18.30) into Eq. (18.29), we get the coefficients $a_1, a_2, \dots$. In order that Eq. (18.30) satisfies the boundary conditions, u_0 satisfies the non-homogeneous boundary conditions and a complete system of functions w_n satisfies the homogeneous boundary conditions. Otherwise, Eq. (18.26) is not simply formed, because Eq. (18.26) is not derived from Eq. (18.25). In solving processes on nonlinear and linear partial differential equations, boundary conditions play an important role.

Example 18.2. Consider the equation of the steady laminar flow in the circular pipe which is called a Hagen–Poiseuille flow. The Navier–Stokes equation in the cylindrical coordinate is

$$L(u) = \frac{d}{dr}\left(r\frac{du}{dr}\right) - \frac{r}{\mu}\frac{dp}{dx} = 0, \tag{1}$$

in which $r =$ the radial coordinate system, $x =$ the coordinate system in the flow direction, $u =$ velocity in the x direction, $\mu =$ the viscosity coefficient, and $p =$ pressure. We assume the velocity distribution is $u = a + br + cr^2$ and minimize the following function $I(u)$:

$$I(u) = \int_0^{D/2} [L(u)]^2 \, dr. \tag{2}$$

Then, obtain the constants a, b, and c, in which $D =$ the internal diameter of the circular pipe and $dp/dx =$ the pressure gradient.

Solution. Since the velocity in the pipe is maximum at $r = 0$, $b = 0$ is obtained. That is, we get $u = a + cr^2$. Substituting this equation into Eqs. (1) and (2), we have

$$I = \int_0^{D/2} \left[4cr - \frac{r}{\mu}\frac{dp}{dx}r\right]^2 \, dr \tag{3}$$

and differentiating Eq. (3) about c, we get

$$c = \frac{1}{4\mu}\frac{dp}{dx}. \tag{4}$$

Since $u = 0$ at $r = D/2$, from $u = a + cr^2$ we have

$$a = -\frac{1}{4\mu}\frac{dp}{dx}\left(\frac{D}{2}\right)^2. \tag{5}$$

Equations (4) and (5) show the velocity distribution as

$$u = -\frac{1}{4\mu}\frac{dp}{dx}\left(\frac{D^2}{4} - r^2\right). \tag{6}$$

Equation (6) is the same as the analytical solution of the Hagen–Poiseuille flow (Daily and Harleman, 1966).

Problem 18.2. Obtain the approximate solution of the ordinary differential equation

$$y'' + xy' + y = 2x \tag{7}$$

with the boundary conditions $y(0) = 1$ and $y(1) = 0$. Assuming

$$y = 1 - x + a_1 x(1 - x) + a_2 x^2 (1 - x) + \cdots . \tag{8}$$

and calculate a_1 and a_2 in Eq. (8).

18.3. Transform of Partial Differential Equation to Ordinary Differential Equation

The governing equation in hydraulic engineering is generally expressed by partial differential equations. One of the solving methods of the partial differential equations is the transform to ordinary differential equations. The similarity method, the method of separation of variables, and the functional transform which we studied in the previous chapters are representative. In this section, we apply the variational method to transform a partial differential equation to an ordinary differential equation. The governing equation is

$$L(u) = 0. \tag{18.31}$$

Assuming independent variables in a function u are x and y, we apply the Galerkin method. The boundary conditions are given by

$$u = 0, \quad \text{at } x = 0, a, \tag{18.32}$$

$$u = 0, \quad \text{at } y = 0, b. \tag{18.33}$$

Considering the following trial function which satisfies the boundary conditions, we have

$$u_n = w_0 + \sum_{k=1}^{n} w_k f_k(x). \tag{18.34}$$

The function $f_k(x)$ satisfies

$$\int_0^b L(u_n)\omega_k dy = 0, \quad \text{for } k = 1, 2, \ldots, n. \tag{18.35}$$

Usually, functions of polynomials of x or trigonometric functions of x are selected as the complete system of functions ω_k. As a result, $f_k(x)$ in Eq. (18.35) forms an ordinary differential equation. The obtained ordinary differential equations are numerically or analytically solved. If Eq. (18.31) is nonlinear, Eq. (18.35) are often nonlinear. If Eq. (18.35) is nonlinear, it is usually not solvable analytically. Mizumura and Yamasaka (2002) applied the Galerkin method to the flow pattern in the embayment along a river and compared it with the experimental result. Mizumura (2002a, 2003c, 2006a, 2009a) applied the Galerkin method to unconfined groundwater flow on a hillslope.

Example 18.3. Solve the Poisson equation, $\Delta u = -1$ with the following boundary conditions:

$$u = 0, \quad \text{at } 0 \le x \le a, \ y = 0, \tag{1}$$

$$u = 0, \quad \text{at } 0 \le x \le a, \ y = b, \tag{2}$$

$$u = 0, \quad \text{at } 0 \le y \le b, \ x = 0, \tag{3}$$

$$u = 0, \quad \text{at } 0 \le y \le b, \ x = a, \tag{4}$$

in which $\Delta = \frac{\partial^2}{\partial x^2} + \frac{\partial^2}{\partial y^2}$.

Solution. Assume the following trial functions:

$$u_2 = a_1(y)\sin\frac{\pi x}{a} + a_2(y)\sin\frac{2\pi x}{a}. \tag{5}$$

From Eq. (18.35), we have

$$\int_0^a (\Delta u + 1)\sin\frac{\pi x}{a}dx = 0 \tag{6}$$

and

$$\int_0^a (\Delta u + 1)\sin\frac{2\pi x}{a}dx = 0. \tag{7}$$

Equation (6) indicates

$$a_1 \left(\frac{\pi}{a}\right)^2 \frac{a}{2} + a_1'' \frac{a}{2} + 2 = 0,\tag{8}$$

in which "′" shows d/dy. The solution of Eq. (8) is

$$a_1 = C_1 e^{\frac{\pi y}{a}} + C_2 e^{-\frac{\pi y}{a}} + \frac{4}{a}\left(\frac{a}{\pi}\right)^2.\tag{9}$$

The boundary conditions (1) and (2) show

$$C_1 e^{\frac{\pi b}{a}} + C_2 e^{-\frac{\pi b}{a}} + \frac{4}{a}\left(\frac{a}{\pi}\right)^2 = 0, \quad \text{at } y = b.\tag{10}$$

and

$$C_1 + C_2 + \frac{4}{a}\left(\frac{a}{\pi}\right)^2 = 0, \quad \text{at } y = 0.\tag{11}$$

Equations (10) and (11) give

$$C_1 = -\frac{\frac{4}{a}\left(\frac{a}{\pi}\right)^2 \left(1 - e^{-\pi b/a}\right)}{2\sinh\frac{\pi b}{a}}\tag{12}$$

and

$$C_2 = \frac{\frac{4}{a}\left(\frac{a}{\pi}\right)^2 \left(1 - e^{\pi b/a}\right)}{2\sinh\frac{\pi b}{a}}.\tag{13}$$

Equation (7) indicates

$$a_2 = C_3 e^{2\pi y/a} + C_4 e^{-2\pi y/a}.\tag{14}$$

The integral constants C_3 and C_4 which satisfy the boundary conditions (1) and (2) are 0. Therefore, the function $a_2 = 0$ from Equation (14). This shows

$$u_2 = \left[-\frac{\frac{4}{a}\left(\frac{a}{\pi}\right)^2(1 - e^{-\pi b/a})e^{\pi y/a}}{2\sinh\frac{\pi b}{a}} \right.$$

$$\left. + \frac{\frac{4}{a}\left(\frac{a}{\pi}\right)^2(1 - e^{\pi b/a})e^{-\pi y/a}}{2\sinh\frac{\pi b}{a}} + \frac{4}{a}\left(\frac{a}{\pi}\right)^2 \right] \sin\frac{\pi x}{a}.\tag{15}$$

Problem 18.3. The unconfined groundwater flow on a hillslope is expressed by the Boussinesq equation in Chap. 14. When the bottom slope is zero, it is simplified as

$$\lambda \frac{\partial h}{\partial t} = k \frac{\partial}{\partial x} \left(h \frac{\partial h}{\partial x} \right), \tag{16}$$

in which λ = porosity, h = the water depth, t = time, x = the coordinate system in the flow direction, and k = the hydraulic conductivity. The boundary conditions are $\partial h / \partial x = 0$ at $x = 0$ and $h = 0$ at $x = \ell$. The trial function is assumed to be

$$h = \sum_{n=0}^{\infty} H_n(t) \cos \left[\frac{x}{\ell} \left(n\pi + \frac{\pi}{2} \right) \right]. \tag{17}$$

Then, derive the ordinary differential equation of $H_n(t)$ (Mizumura, 2002, 2003b, 2006a, 2009a).

18.4. Eigen Value Problem

The eigen value problem is transformed to the variational problem. Let us consider solving the variational problem by the direct method. As an example, we apply it to the differential equation of the Sturm–Liouville type. It is defined in the domain $0 \le x \le \ell$,

$$L(y) = \frac{d}{dx}(p(x)y') - q(x)y + \lambda y = 0, \quad p(x) > 0, \quad q(x) \ge 0, \tag{18.36}$$

in which "$'$" $= d/dx$ and $\lambda =$ an eigen value. The boundary conditions are

$$y(0) = y(\ell) = 0, \tag{18.37}$$

The functions $p(x)$ and $q(x)$ are differentiable in $0 \le x \le \ell$. To get the first eigen value, λ_1, consider the following function $I(y) - \lambda H(y)$:

$$J(y) = I(y) - \lambda H(y) = \int_0^\ell (p(x)y'^2 + q(x)y^2)dx - \lambda \int_0^\ell y^2 dx \tag{18.38}$$

must be minimized. Using the complete system of functions $\omega_k(x)$, we assume the trial function as

$$y = \sum_{k=1}^{n} a_k \omega_k(x). \tag{18.39}$$

The coefficients a_k's are determined in order that $J(y)$ in Eq. (18.38) is extreme. This corresponds to the Ritz method. To simplify the computation, let us apply the Galerkin method. Assuming that the first variation of $J(y)$ is zero,

$$\delta J = -2 \int_0^{\ell} [(p(x)y')' - q(x)y + \lambda y]\delta y dx = 0, \tag{18.40}$$

in which δ indicates the variation. Substituting Eq. (18.39) into Eq. (18.40) and defining $\delta y = \omega_r$ $(r = 1, 2, \ldots, n)$, we obtain

$$\int_0^{\ell} \left[\left(p(x) \sum_{k=1}^{n} a_k \omega_k' \right)' - q(x) \sum_{k=1}^{n} a_k \omega_k + \lambda \sum_{k=1}^{n} a_k \omega_k \right] \omega_r dx = 0,$$

$$\text{for } r = 1, 2, \ldots, n. \tag{18.41}$$

Equations (18.41) are the sequential equations of a_k's. These are the homogeneous algebraic equations. To get eigen values, the determinant of Eq. (18.41) is computed and it is the n-th degree equation. It has n solutions of λ's. Writing them from the small one to large one, we obtain

$$\lambda_1^{(n)}, \lambda_2^{(n)}, \ldots, \lambda_n^{(n)}. \tag{18.42}$$

These solutions correspond to the first, second, $\cdots$, and n-th eigen values.

Example 18.4. Consider the free vibration of a string which is fixed at $x = \pm 1$ and obtain eigen values. The governing equation is given by

$$\frac{\partial^2 u}{\partial x^2} = \frac{\partial^2 u}{\partial t^2}. \tag{1}$$

Substituting $u = e^{i\lambda t}v(x)$ into Eq. (1), we get

$$v''(x) + \lambda^2 v(x) = 0, \tag{2}$$

in which $``'" = d/dx$.

As the trial function, we assume

$$v(x) = (1 - x^2)(a_1 + a_2 x^2). \tag{3}$$

The complete system of functions is

$$\omega_1 = (1 - x^2), \quad \omega_2 = (1 - x^2)x^2. \tag{4}$$

Solution. Substituting v, ω_1, and ω_2 into Eq. (18.41), we have the following two equations:

$$\int_{-1}^{1} (v'' + \lambda^2 v)\omega_1 dx = (35 - 14\lambda^2)a_1 + (7 - 2\lambda^2)a_2 = 0, \tag{5}$$

$$\int_{-1}^{1} (v'' + \lambda^2 v)\omega_2 dx = (21 - 6\lambda^2)a_2 + (33 - 2\lambda^2)a_2 = 0. \tag{6}$$

That the determinant of Eqs. (5) and (6) is zero leads to the following quartic equation of λ:

$$\lambda^4 - 28\lambda^2 + 36 = 0. \tag{7}$$

The roots are easily

$$\lambda_1^2 = 3.14, \quad \lambda_2^2 = 24.86. \tag{8}$$

The analytical eigen values are

$$\lambda_1^2 = \left(\frac{\pi}{2}\right)^2 = 2.4674, \quad \lambda_2^2 = \pi^2 = 9.8676, \quad \lambda_3^2 = \left(\frac{3\pi}{2}\right)^2 = 22.207. \tag{9}$$

The eigen value for λ_2^2 from the variational method is 24.86. This corresponds to $\lambda_3{}^2$ from the analytical eigen value. Because we assumed the even function for $v(x)$, the variational method could not obtain the eigen value for the odd function. Therefore, assuming the following equation:

$$v(x) = (1 - x^2)(a_1 + a_2 x + a_3 x^2), \tag{10}$$

we get $\lambda_1{}^2$, $\lambda_2{}^2$, and $\lambda_3{}^2$ in the same way.

Problem 18.4. Show that we get Eq. (18.40) if the first variation of Eq. (18.38) is zero.

18.5. Application to Groundwater Flow

As the application of the variational method, consider the solution of Eq. (14.3), the Boussinesq equation in the finite domain (Chap. 14). This equation has the first-order in time and the second-order in space. This equation requires an initial condition and two boundary conditions. The initial condition corresponds to the present water table in the ground. One of the two boundary conditions is given at the upstream end of the computational domain. If it is impermeable, the velocity is zero or the hydraulic gradient is zero. The other boundary condition is given at the downstream end of the computational domain. It is dependent on the slope of the seepage face (Mizumura and Kaneda, 2010). The downstream boundary condition at the drawdown end is given by the unit hydraulic gradient (Mizumura, 2009b). Herein, we assume that the slope of the seepage face is vertical and the bottom slope of the aquifer is zero. The Boussinesq equation is

$$\frac{\partial h}{\partial t} = k' \frac{\partial}{\partial x}\left(h \frac{\partial h}{\partial x}\right), \tag{18.43}$$

in which $h =$ the water depth, $x =$ the coordinate system in the flow direction, $k' = k/\lambda$, $k =$ the hydraulic conductivity, and $\lambda =$ porosity. Define the following equation from Eq. (18.43):

$$L(\tilde{h}) = \frac{\partial \tilde{h}}{\partial t} - k' \frac{\partial}{\partial x}\left(\tilde{h} \frac{\partial \tilde{h}}{\partial x}\right), \tag{18.44}$$

in which $\tilde{h} =$ the approximate water depth of h. Then, we assume the following trial function which satisfies the boundary conditions:

$$\tilde{h} = \frac{\epsilon}{2} - \frac{1}{2\epsilon}(x + \epsilon - \ell)^2 U(x + \epsilon - \ell) + a_0 + \sum_{n=1}^{\infty} a_n \cos \frac{n\pi x}{\ell}, \tag{18.45}$$

in which $\epsilon =$ a constant, $\ell =$ the length of the aquifer, $U(\cdot) =$ the unit step function, and a_0 and a_n for $n = 1, 2, \ldots$ are coefficients of the Fourier cosine series and time-varying. The first and second terms in Eq. (18.45) are added to satisfy the boundary conditions. When we use the cosine functions as the complete system of functions, the cosine functions do not satisfy the downstream boundary condition. The downstream boundary condition is given by

$$\frac{\partial \tilde{h}}{\partial x} = -1. \tag{18.46}$$

Thus, the second term is added in Eq. (18.45) to satisfy the downstream boundary condition. The numerical value of ϵ is the effect of the boundary conditions. To restrict the influence range of the numerical value of ϵ, a smaller value is better. Substituting $\tilde{h} = D$ at $t = 0$ into Eq. (18.45), we get

$$a_0 = D - \frac{\epsilon}{2} + \frac{\epsilon^2}{6\ell}, \tag{18.47}$$

$$a_j = -\frac{(\epsilon - \ell)^2}{j\pi\epsilon} \sin\frac{j\pi(\ell - \epsilon)}{\ell} + \frac{(\epsilon - \ell)}{\epsilon} \int_{\ell-\epsilon}^{\ell} x \cos\frac{j\pi x}{\ell} dx$$

$$+ \frac{1}{2\epsilon} \int_{\ell-\epsilon}^{\ell} x^2 \cos\frac{j\pi x}{\ell} dx, \quad \text{for } j = 1, 2, \ldots, \tag{18.48}$$

in which D = the initial water depth. From Eq. (18.45) with $\tilde{h} = D$ the initial condition derives the following equation of a_0 and a_j for $j = 1, 2, \ldots$:

$$D - \frac{\epsilon}{2} + \frac{1}{2\epsilon}(x + \epsilon - \ell)^2 U(x + \epsilon - \ell) = a_0 + \sum_{n=1}^{\infty} a_n \cos\frac{n\pi x}{\ell}. \tag{18.49}$$

The initial values of a_0 and a_j for $j = 1, 2, \ldots$ are obtained from the Fourier cosine series in Chap. 10. Substituting Eq. (18.45) into Eq. (18.44), we apply the Galerkin method. Since the resultant equations are usually nonlinear, the numerical method such as the Runge–Kutta method in Chap. 2 is used to the resultant equations. Mizumura used the Runge–Kutta method and the result is satisfactory as shown in Fig. 18.1 (Mizumura, 2002, 2003b, 2009a). Hino (1975) used the variational method to compute the generation of rip currents on the shore using the Hermite functions.

Example 18.5. Derive Eqs. (18.47) and (18.48) from Eq. (18.49).

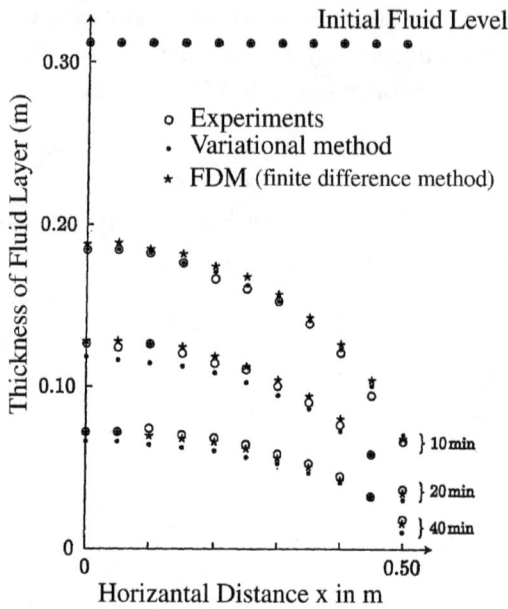

Figure 18.1.   Solution of Boussinesq equation.

Solution. Integrating Eq. (18.49) from 0 to ℓ, we get Eq. (18.47). Multiplying $\cos \frac{j\pi x}{\ell}$ to Eq. (18.49) and integrating the resultant equation from 0 to ℓ, we obtain Eq. (18.48).

Problem 18.5. Show that the following equation is 1 and 0 for $n = m$ and $n \neq m$, respectively.

$$\int_0^\ell \cos \frac{n\pi x}{\ell} \cos \frac{m\pi x}{\ell} dx = \begin{cases} 0 & \text{for } n \neq m \\ 1 & \text{for } n = m \end{cases}. \qquad (1)$$

18.6.   Least Squares Method and Collocation Method

Consider the following ordinary differential equation and boundary conditions:

$$\frac{d^2 u}{dx^2} + u = x^2, \quad u(0) = u(1) = 0. \qquad (18.50)$$

The analytical solution of Eq. (18.50) is

$$u = 2\cos x + \frac{1 - 2\cos 1}{\sin 1}\sin x + x^2 - 2$$

Assume the trial function that satisfies the boundary conditions

$$\tilde{u}(x) \cong x(1 - x)(a_1 + a_2 x). \tag{18.51}$$

Because $\tilde{u}(x)$ is an approximation to the exact solution $u(x)$, $\tilde{u}(x)$ satisfies Eq. (18.50) as follows:

$$\frac{d^2\tilde{u}}{dx^2} + \tilde{u} - x^2 = \epsilon(x). \tag{18.52}$$

The residual $\epsilon(x)$ in Eq. (18.52) is not zero. When both sides are squared and integrated over the domain, a new functional is given by

$$I(\tilde{u}) = \int_0^1 [\epsilon(x)]^2 dx$$

$$= \int_0^1 [-2a_1 + 2a_2 - 6a_2 x + a_1(x - x^2) + a_2(x^2 - x^3) - x^2]^2 dx. \tag{18.53}$$

Differentiating Eq. (18.53) about a_1 and a_2, respectively, we have

$$\frac{\partial I(\tilde{u})}{\partial a_1} = \frac{\partial I(\tilde{u})}{\partial a_2} = 0. \tag{18.54}$$

Equation (18.54) shows

$$3.367a_1 - 9.316a_2 = -0.6167, \tag{18.55}$$

$$1.683a_1 + 3.767a_2 = -0.8. \tag{18.56}$$

Then, we get

$$a_1 \cong -0.345, \quad a_2 \cong -0.0583. \tag{18.57}$$

This is the least squares method. The collocation method is the simplest direct method which can be considered as a variation of the Ritz method. The least squares method minimizes errors in the functional over the entire domain, but the collocation method aims at the vanishing error in the original ordinary differential equation at selected points in the domain. Although these points which are called collocation points can be arbitrarily selected, the accuracy of the solution depends heavily on the choice of these points. Substituting

Eq. (18.51) into Eq. (18.50), we have

$$(x^2 - x + 2)a_1 + (x^3 - x^2 + 6x - 2)a_2 = -x^2. \qquad (18.58)$$

Suppose the collocation points are selected at $x = 1/3$ and $x = 2/3$ in the domain $[0, 1]$. Substituting the two points into Eq. (18.58), we have

$$48a_1 - 2a_2 = -3, \qquad (18.59)$$

$$24a_1 + 25a_2 = -6. \qquad (18.60)$$

Then, the constants are

$$a_1 = -0.0697, \quad a_2 = -0.173. \qquad (18.61)$$

This is the collocation method.

Expanding x^2 on the right side of Eq. (18.50) by the Fourier series on the interval $[0, 1]$, we can express it as

$$x^2 = \sum_{n=0}^{\infty} \frac{2}{n^3 \pi^3} [2 - (n^2 \pi^2 - 2)(-1)^n] \sin n\pi x \qquad (18.62)$$

Substituting the Fourier series of u

$$u = \sum_{n=1}^{\infty} A_n \sin n\pi x \qquad (18.63)$$

into Eq. (18.50) and equating that the coefficient of $\sin n\pi x$ is 0, we obtain

$$A_n = \frac{2}{n^3 \pi^3 (1 - n^2 \pi^2)} \left[2 - (n^2 \pi^2 - 2)(-1)^n\right] \qquad (18.64)$$

This is the coefficients of the Fourier series of the solution in Eq. (18.50). The solution is

$$u = \sum_{n=1}^{\infty} \frac{2}{n^3 \pi^3 (1 - n^2 \pi^2)} \left[2 - (n^2 \pi^2 - 2)(-1)^n\right] \sin n\pi x \qquad (18.65)$$

Example 18.6. Solve the following differential equation by the collocation method:

$$\frac{d^2 y}{dx^2} + y = x, \quad y(0) = y(1) = 0, \qquad (1)$$

Solution. Assume
$$\tilde{y}(x) \cong x(1 - x)(a_1 + a_2 x). \tag{2}$$
Substituting Eq. (2) into Eq. (1), we get
$$-2a_1 + 2a_2 - 6a_2 x + x(1 - x)(a_1 + a_2 x) - x = 0. \tag{3}$$
Suppose the collocation points are selected at $x = 1/3$ and $x = 2/3$ in the domain $[0, 1]$. Substituting the two points into Eq. (3), we obtain
$$-48a_1 + 2a_2 = 9, \tag{4}$$
$$-48a_1 - 50a_2 = 18. \tag{5}$$
Then, the coefficients are
$$a_1 = -0.195, \quad a_2 = -0.173. \tag{6}$$

Problem 18.6. Approximate $x^{5/3}$ by $ax^2 + bx + c$ using the least squares method in the domain $[0, 1]$. The coefficients a, b, and c are constants to be determined. The irrational equation can be expressed by the polynomials.

18.7. The Method of Moments

When the governing differential equation is expressed by
$$L[y] = f(x) \quad \text{in } [a, b], \tag{18.66}$$
the residual $R(x)$ is written by
$$R(x) = L\left[u_0(x) + \sum_{i=1}^{\infty} a_i u_i(x)\right] - f(x), \tag{18.67}$$
in which the trial function is approximated by
$$\tilde{y} = u_0(x) + \sum_{i=1}^{n-1} a_i u_i(x). \tag{18.68}$$
The method of moments is designed in order that the residuals are orthogonal to x^i for $i = 0, 1, 2, \ldots$. They are written as
$$\int_a^b R(x) x^i dx = 0, \quad \text{for } i = 0, 1, 2, \ldots, n - 1. \tag{18.69}$$
When $R(x)$ satisfies Eqs. (18.69), $R(x)$ intersects the x axis at least n times between a and b. Thus, as $n \to \infty$, $R(x)$ intersects the x axis

∞ times between a and b. This indicates that $R(x) = 0$ is formed between a and b and Eq. (18.68) is the solution of Eq. (18.66) as $n \to \infty$. The merit of this method is the easiness of integrals of trial functions.

Example 18.7. Solve the following differential equation:

$$y'' + xy' + y = 2x, \quad y(0) = 1, \quad y(1) = 0, \tag{1}$$

in which "$'$" $= d/dx$.

Solution. Assuming that

$$y = 1 - x + a_1 x(1 - x) + a_2 x^2(1 - x), \tag{2}$$

we substitute Eq. (2) into Eq. (18.69) as

$$2a_1 + a_2 = -1, \quad \text{for } i = 0, \tag{3}$$

$$65a_1 + 63a_2 = -59, \quad \text{for } i = 1. \tag{4}$$

The solutions of Eqs. (3) and (4) are

$$a_1 = -\frac{13}{61} = -0.213, \quad a_2 = -\frac{35}{61} = -0.574. \tag{5}$$

The approximate solution is

$$y = 1 - x - 0.213x(1 - x) - 0.574x^2(1 - x). \tag{6}$$

Problem 18.7. Solve the following nonlinear differential equation and boundary conditions:

$$y'' + yy' + y = 2x, \quad y(0) = 1, \quad y(1) = 0, \tag{7}$$

by using the approximation

$$y = 1 - x + a_1 x(1 - x). \tag{8}$$

Chapter 19

Perturbation Methods

Consider the case that the differential equations are not easily and analytically solvable. When the differential equations or the non-dimensional forms of the differential equations contain a small parameter, the differential equations can be expanded in the Taylor series about the small parameter. Then, the solution can be obtained sequentially from the lower orders to higher orders of the small parameter. Since the differential equation for each order of the small parameter is usually linear, we can solve it analytically and form an approximate solution of the original solution. Then, we use a special technique to exclude permanent terms in the approximate solution. The permanent terms are defined to become infinite as independent variables are infinite.

19.1. Parameter Perturbation

Let us consider the famous Duffing equation to explain a nonlinear oscillation as follows:

$$\ddot{u} + u + \epsilon u^3 = 0, \quad u(0) = a, \quad \dot{u}(0) = 0, \tag{19.1}$$

in which "$\cdot$" indicates d/dt and ϵ is a small parameter. Integrating Eq. (19.1) once with the given boundary conditions, we get

$$\dot{u}^2 + u^2 + \epsilon \frac{u^4}{2} = \left(1 + \frac{\epsilon a^2}{2}\right) a^2, \tag{19.2}$$

in which we expand u in the power series of ϵ as

$$u = \sum_{m=0}^{\infty} \epsilon^m u_m(t). \tag{19.3}$$

Substituting Eq. (19.3) into Eq. (19.1), expand the resultant equation by the powers of ϵ, and rearranging it, we obtain

$$\ddot{u}_0 + u_0 = 0, \quad u_0(0) = a, \quad \dot{u}_0(0) = 0, \qquad (19.4)$$

$$\ddot{u}_1 + u_1 = -u_0^3, \quad u_1(0) = 0, \quad \dot{u}_1(0) = 0. \qquad (19.5)$$

The zero-th order solution of Eq. (19.4) is

$$u_0(t) = a \cos t. \qquad (19.6)$$

Substituting Eq. (19.6) into Eq. (19.5), we get

$$\ddot{u}_1 + u_1 = -a^3 \frac{\cos 3t + 3\cos t}{4}. \qquad (19.7)$$

Solving Eq. (19.7), we get the following first-order solution:

$$u_1(t) = -\frac{3a^3}{8} t \sin t + \frac{a^3}{32}(\cos 3t - \cos t). \qquad (19.8)$$

Using Eqs. (19.3), (19.6), and (19.8), we obtain the approximate solution as

$$u = a \cos t + \epsilon a^3 \left[-\frac{3}{8} t \sin t + \frac{1}{32}(\cos 3t - \cos t) \right] + O(\epsilon^2), \qquad (19.9)$$

in which $O(\epsilon^2)$ indicates terms of higher orders than ϵ^2. Mathematically, it shows

$$\lim_{\epsilon \to 0} \frac{O(\epsilon^2)}{\epsilon^2} < \infty, \quad \text{finite.} \qquad (19.10)$$

Equation (19.9) is not a reasonable solution, because the term $t \sin t$ becomes infinity except $\sin t = 0$ as t is ∞. Then, Eq. (19.9) does not converge as $t \to \infty$. This term is called the permanent term in the perturbation theory. To exclude a permanent term in the solution, an independent variable is defined as follows:

$$\zeta = t(1 + p_1\epsilon + p_2\epsilon^2 + \cdots), \qquad (19.11)$$

in which $p_1, p_2, \cdots$ = arbitrary small constants. The term $\ddot{u}$ in Eq. (19.1) is modified as

$$\ddot{u} = \frac{d^2u}{dt^2} = \frac{d^2u}{d\zeta^2}(1 + p_1\epsilon + p_2\epsilon^2 + \cdots)^2. \qquad (19.12)$$

Substituting Eq. (19.11) into Eq. (19.1), we find that the zero-th order equation is the same as Eq. (19.4). But Eq. (19.5) is transformed to

$$\ddot{u}_1 + u_1 = -u_0^3 - 2p_1\ddot{u}_0, \tag{19.13}$$

in which " $\ddot{}$ " indicates $d^2/d\zeta^2$. Therefore, Eq. (19.7) becomes the following one:

$$\ddot{u}_1 + u_1 = -a^3\frac{\cos 3\zeta + 3\cos \zeta}{4} + 2p_1 a \cos \zeta. \tag{19.14}$$

The solution of Eq. (19.14) is

$$u_1 = -\frac{3a^3}{8}\zeta \sin \zeta + \frac{a^3}{32}(\cos 3\zeta - \cos \zeta) + p_1 a\zeta \sin \zeta. \tag{19.15}$$

The permanent term $\zeta \sin \zeta$ on the right side of Eq. (19.15) becomes zero when $p_1 a - 3a^3/8 = 0$. Therefore, we get

$$p_1 = \frac{3}{8}a^2. \tag{19.16}$$

The approximate solution is

$$u = a \cos \zeta + \epsilon\frac{a^3}{32}(\cos 3\zeta - \cos \zeta) + O(\epsilon^2), \tag{19.17}$$

in which

$$\zeta = t\left(1 + \frac{3}{8}a^2\epsilon\right). \tag{19.18}$$

Expanding the independent variable by the powers of ϵ, we can exclude the permanent term.

Next, consider a slow flow past a sphere of radius a as shown in Fig. 19.1. The three-dimensional Navier–Stokes equation (Dyke, 1975) is given by

$$\nabla^4\psi = \frac{Re}{r^2\sin\theta}\left(\frac{\partial\psi}{\partial\theta}\frac{\partial}{\partial r} - \frac{\partial\psi}{\partial r}\frac{\partial}{\partial\theta} + 2\cot\theta\cdot\frac{\partial\psi}{\partial r} - \frac{2}{r}\frac{\partial\psi}{\partial\theta}\right)\nabla^2\psi, \tag{19.19}$$

in which $Re = Ua/\nu$, the Reynolds number, and Laplacian is given by

$$\nabla^2 = \frac{\partial^2}{\partial r^2} + \frac{\sin\theta}{r^2}\frac{\partial}{\partial\theta}\left(\frac{1}{\sin\theta}\cdot\frac{\partial}{\partial\theta}\right), \tag{19.20}$$

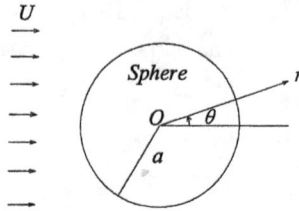

Figure 19.1.   Flow past a sphere.

in which $\psi(r, \theta)$ is the Stokes stream function. The flow pattern is axially symmetric. The velocity components in the r and θ directions u_r and u_θ are, respectively, given by

$$u_r = \frac{\partial \psi}{\partial \theta} \frac{1}{r^2 \sin \theta}, \qquad (19.21)$$

$$u_\theta = -\frac{\partial \psi}{\partial r} \frac{1}{r \sin \theta}. \qquad (19.22)$$

The boundary condition is described in order that the surrounding of the sphere forms the streamline. This indicates

$$\psi(1, \theta) = \frac{\partial \psi(1, \theta)}{\partial r} = 0. \qquad (19.23)$$

The condition that the flow is uniform at the enough upstream shows

$$\psi(r, \theta) \to \frac{1}{2} r^2 \sin^2 \theta, \quad \text{as } r \to \infty. \qquad (19.24)$$

For the small Reynolds number $Re \ll 1$, the stream function is assumed to be

$$\psi = \sum_{m=0}^{\infty} Re^m \, \psi_m(r, \theta). \qquad (19.25)$$

Substituting Eq. (19.25) into Eq. (19.19), we obtain the zero-th order equation as

$$\nabla^4 \psi_0 = 0. \qquad (19.26)$$

Equation (19.23) becomes

$$\psi_0(1, \theta) = \left. \frac{\partial \psi_0}{\partial r} \right|_{(1, \theta)} = 0. \qquad (19.27)$$

Equation (19.24) also becomes

$$\psi_0(r,\theta) \to \frac{1}{2}r^2\sin^2\theta, \quad \text{as } r \to \infty. \tag{19.28}$$

The first-order equation is

$$\nabla^4\psi_1 = \frac{Re}{r^2\sin\theta}\left(\frac{\partial\psi_0}{\partial\theta}\frac{\partial}{\partial r} - \frac{\partial\psi_0}{\partial r}\frac{\partial}{\partial\theta} + 2\cot\theta\cdot\frac{\partial\psi_0}{\partial r} - \frac{2}{r}\frac{\partial\psi_0}{\partial\theta}\right)\nabla^2\psi_0. \tag{19.29}$$

Equation (19.23) is

$$\psi_1(1,\theta) = \frac{\partial\psi_1(1,\theta)}{\partial r} = 0. \tag{19.30}$$

Equation (19.24) is

$$\psi_1(r,\theta) = o(r^2), \quad \text{as } r \to \infty, \tag{19.31}$$

in which o is defined as

$$\lim_{r\to\infty}\left|\frac{o(r^2)}{r^2}\right| = 0. \tag{19.32}$$

From Eq. (19.28), the solution of Eq. (19.26) forms the following stream function:

$$\psi_0 = f(r)\sin^2\theta. \tag{19.33}$$

Substituting Eq. (19.33) into Eq. (19.26), we have

$$f''' - \frac{4f''}{r^2} + \frac{8f'}{r^3} - \frac{8f}{r^4} = 0, \tag{19.34}$$

in which "$'$" $= d/dr$. The general solution of Eq. (19.34) is

$$f(r) = C_4 r^4 + C_2 r^2 + C_1 r + C_{-1}\frac{1}{r}, \tag{19.35}$$

in which C_{-1}, C_1, C_2, and C_4 are integral constants. Equation (19.28) indicates $C_4 = 0$ and $C_2 = 1/2$. Equation (19.27) shows $C_1 = -3/4$ and $C_{-1} = 1/4$. Thus, the stream function ψ_0 is given by

$$\psi_0 = \frac{1}{4}\left(2r^2 - 3r + \frac{1}{r}\right)\sin^2\theta. \tag{19.36}$$

Substituting Eq. (19.36) into Eq. (19.29), we have

$$\nabla^4\psi_1 = -\frac{9}{4}\left(\frac{2}{r^2} - \frac{3}{r^3} + \frac{1}{r^5}\right)\sin^2\theta\cos\theta. \tag{19.37}$$

Equations (19.30), (19.31), and (19.37) indicate ψ_1 to have the following form:

$$\psi_1 = g(r)\sin^2\theta\cos\theta. \tag{19.38}$$

Substituting Eq. (19.38) into Eq. (19.37), we get

$$g'''' - \frac{12g''}{r^2} + \frac{24g'}{r^3} = -\frac{9}{4}\left(\frac{2}{r^2} - \frac{3}{r^3} + \frac{1}{r^5}\right). \tag{19.39}$$

The boundary conditions are

$$g(1) = g'(1) = 0 \tag{19.40}$$

and

$$g(r) = o(r^2), \quad \text{as } r \to \infty. \tag{19.41}$$

The general solution of Eq. (19.39) is

$$g = b_{-2}r^{-2} + b_0 + b_3 r^3 + b_5 r^5 - \frac{3}{16}r^2 + \frac{9}{32}r + \frac{3}{32}r^{-1}, \tag{19.42}$$

in which b_{-2}, b_0, b_3, and b_5 are integral constants. Equation (19.41) shows $b_3 = b_5 = 0$. Equation (19.40) indicates $b_0 = b_{-2} = -3/32$. Therefore, we obtain

$$\psi_1 = -\frac{3}{32}\left(2r^2 - 3r + 1 - \frac{1}{r} + \frac{1}{r^2}\right)\sin^2\theta\cos\theta. \tag{19.43}$$

The stream function is

$$\psi = \frac{1}{4}\left(2r^2 - 3r + \frac{1}{r}\right)\sin^2\theta$$

$$- \frac{3}{32}Re\left(2r^2 - 3r + 1 - \frac{1}{r} + \frac{1}{r^2}\right)\sin^2\theta\cos\theta$$

$$+ O(Re^2), \quad \text{as } Re \to 0, \tag{19.44}$$

in which

$$\lim_{Re \to 0}\frac{O(Re^2)}{Re^2} < \infty, \quad \text{finite.} \tag{19.45}$$

But for a large r, Eq. (19.44) does not fit.

Example 19.1. Obtain the first-order solution of the following non-linear ordinary differential equation with a small parameter ϵ:

$$y' + y = \epsilon y^2, \quad y(0) = 1. \tag{1}$$

Solution. Assume that y is expressed by

$$y = y_0 + \epsilon y_1 + \epsilon^2 y_2 + \cdots . \tag{2}$$

Substituting Eq. (2) into Eq. (1), we obtain the equation of the zero-th order as

$$y_0' + y_0 = 0, \quad y_0(0) = 1. \tag{3}$$

The zero-th order solution is described by

$$y_0 = e^{-x}. \tag{4}$$

The equation of the first-order and the boundary condition are

$$y_1' + y_1 = y_0^2 = e^{-2x}, \quad y_1(0) = 0. \tag{5}$$

The first-order solution is

$$y_1 = e^{-x} - e^{-2x}. \tag{6}$$

The analytical solution of Eq. (1) from Eq. (5.15) is given by

$$y = \frac{1}{e^x + \epsilon(1 - e^x)} \tag{7}$$

When ϵ is very small, Eq. (7) is expanded as

$$y = e^{-x}[1 + \epsilon(1 - e^{-x}) + \epsilon^2(1 - e^{-x})^2 + \cdots] \tag{8}$$

The zero-th and first order solutions of Eq. (8) are the same as Eqs. (4) and (6).

Problem 19.1. Obtain the first-order solution of the following differential equation and the boundary condition:

$$y' + y = \epsilon y^{1/2}, \quad y(0) = 1, \tag{9}$$

in which ϵ is a small parameter.

19.2. Matched Asymptotic Expansion

According to Prandtl's boundary-layer theory, the flow consists of the boundary-layer near the body and the potential flow outside the boundary-layer. The matching of the two equations must be more precise than Prandtl's boundary-layer theory. The matching theory is the matched asymptotic expansion (Nayfeh, 1973). The matched asymptotic expansion gives higher order approximation than the boundary-layer theory. Let us consider a simple example. When the coefficient of the highest derivative is a small parameter ϵ, the following differential equation is given by

$$\epsilon y'' + y' + y = 0, \tag{19.46}$$

in which "$'$" $= d/dx$. The boundary conditions are

$$y(0) = \alpha, \quad y(1) = \beta. \tag{19.47}$$

The analytical solution of Eq. (19.46) is easily obtained, because Eq. (19.46) is linear. If we assume $\epsilon \to 0$ in Eq. (19.46) to apply the perturbation theory, Eq. (19.46) becomes

$$y' + y = 0. \tag{19.48}$$

Solving Eq. (19.48) with $y(1) = \beta$, we obtain

$$y = \beta e^{1-x}. \tag{19.49}$$

We call Eq. (19.49) an outer solution and define y^o. The outer solution does not satisfy $y(0) = \alpha$ as shown in Fig. 19.2. Next, define a new variable ζ as

$$\zeta = x/\epsilon, \tag{19.50}$$

enlarge the x coordinate and then Eq. (19.46) is

$$\frac{d^2 y}{d\zeta^2} + \frac{dy}{d\zeta} + \epsilon y = 0. \tag{19.51}$$

For the small ϵ Eq. (19.51) is written as

$$\frac{d^2 y}{d\zeta^2} + \frac{dy}{d\zeta} = 0 \tag{19.52}$$

The solution of Eq. (19.52) is given by

$$y = A + B e^{-\zeta}, \tag{19.53}$$

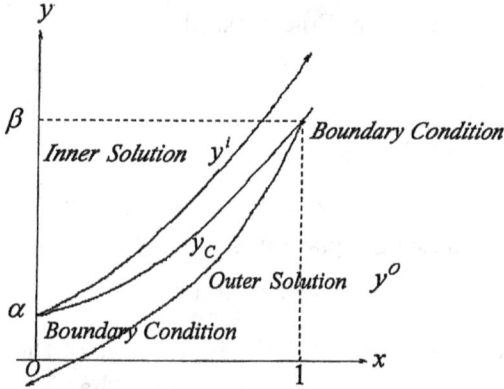

Figure 19.2.   Inner and outer solutions.

in which A and B are integral constants. From the condition of $y(0) = \alpha$ we get

$$y = A + (\alpha - A)e^{-\zeta}. \tag{19.54}$$

From the condition of $y(1/\epsilon) = \beta$ we get

$$A = \frac{\beta - \alpha e^{-1/\epsilon}}{1 - e^{-\epsilon}}. \tag{19.55}$$

The solution of Eq. (19.52) is

$$y = \alpha e^{-\zeta} + \frac{\beta - \alpha e^{-1/\epsilon}}{1 - e^{-\epsilon}}(1 - e^{-\zeta}), \tag{19.56}$$

This is the approximate solution for $\epsilon \cong 0$. Thus, it is rewritten by

$$y = \alpha e^{-\zeta} + \beta(1 - e^{-\zeta}). \tag{19.57}$$

This cannot match Eq. (19.49) smoothly. We call Eq. (19.54) an inner solution and show y^i. This does not satisfy $y(1) = \beta$. The outer solution Eq. (19.49) has a limit as x goes to zero,

$$\lim_{x \to 0} y^o = \beta e. \tag{19.58}$$

The inner solution Eq. (19.54) has a limit as ζ goes to infinity.

$$\lim_{\zeta \to \infty} y^i = A. \tag{19.59}$$

Since for a small x both Eqs. (19.58) and (19.59) are the same, we have

$$\beta e = A. \tag{19.60}$$

Therefore, Eq. (19.54) is modified as

$$y^i = \beta e + (\alpha - \beta e)e^{-\zeta}. \tag{19.61}$$

In general, the following equation is formed:

$$\lim_{x \to 0} y^o(x; \epsilon) = \lim_{\zeta \to \infty} y^i(x; \epsilon). \tag{19.62}$$

Equation (19.62) indicates that the inner limit of the outer solution equals the outer limit of the inner solution. That is,

$$(y^o)^i = (y^i)^o. \tag{19.63}$$

Erdelyi in 1961 showed the following solution which is applied to the inner and outer solutions y^c:

$$y^c = y^o + y^i - (y^o)^i = y^o + y^i - (y^i)^o. \tag{19.64}$$

Therefore, we have

$$(y^c)^o = y^o + (y^i)^o - (y^i)^o = y^o, \tag{19.65}$$

$$(y^c)^i = (y^o)^i + y^i - (y^o)^i = y^i. \tag{19.66}$$

Since $(y^i)^o = (y^o)^i = \beta e$ is formed, we have

$$y^c = \beta e^{1-x} + \beta e + (\alpha - \beta e)e^{-x/\epsilon} - \beta e + O(\epsilon)$$

$$= \beta e^{1-x} + (\alpha - \beta e)e^{-x/\epsilon} + O(\epsilon), \tag{19.67}$$

in which ϵ is very small. Thus,

$$e^{-1/\epsilon} \cong 0. \tag{19.68}$$

The approximate solution which satisfies $y = \alpha$ at $x = 0$ and $y = \beta$ at $x = 1$ is Eq. (19.67).

Referring to Example 17.1 as the application of the matched asymptotic expansion, the laminar boundary-layer equation is given by

$$2\frac{d^3 f}{d\eta^3} + f\frac{d^2 f}{d\eta^2} = 0, \tag{19.69}$$

in which $\eta = \sqrt{\frac{U}{\nu x}} \cdot y$, $\psi = \sqrt{\nu U x} \cdot f(\eta)$, and $\psi =$ stream function. To solve Eq. (19.69), we use the conditions which the velocity is 0 on the plate

$$f = f' = 0, \quad \text{at } \eta = 0 \tag{19.70}$$

and the flow is uniform outside the boundary layer as

$$f' = 1, \quad \text{at } \eta = \infty, \tag{19.71}$$

in which "$'$" $= d/d\eta$. We apply the matched asymptotic expansion to this problem (Rosenhead, 1963). Using Eq. (19.70), we expand the solution in the power series in Sec. 16.6.

$$f(\eta) = \sum_{n=0}^{\infty} \left(-\frac{1}{2}\right)^n \frac{\alpha^{n+1} C_n}{(3n+2)!} \eta^{3n+2}. \tag{19.72}$$

Substituting Eq. (19.72) into Eq. (19.69) and equating the coefficients of the various power of η, we get $C_0 = 1$, $C_1 = 1$, $C_2 = 11$, $C_3 = 375$, $C_4 = 27{,}897$, $C_5 = 3{,}817{,}137$, $\cdots$. An unknown α in Eq. (19.72) is determined from the condition of the uniform flow outside the boundary-layer. Since Eq. (19.69) is formed in the uniform flow, we apply the perturbation method to Eq. (19.69) at $\eta = \infty$. Consequently, we assume

$$f = f_0 + f_1 + f_2 + f_3 + \cdots, \tag{19.73}$$

in which $|f_0| \gg |f_1| \gg |f_2| \gg \cdots$ are supposed. Since Eq. (19.71) indicates $f' = 1$ at $\eta = \infty$, we can obtain

$$f_0 = \eta - \beta, \tag{19.74}$$

in which β is an integral constant. Equation (19.74) implies that $f_1'(\infty) = f_2'(\infty) = \cdots = 0$. The function f_0 in Eq. (19.74) is the zero-th approximation. The first-order approximation f_1 is obtained from Eq. (19.69) as

$$2f_1''' + f_0 f_1'' + f_1 f_0'' = 0, \tag{19.75}$$

in which "$'$" indicates $d/d\eta$. Substituting Eq. (19.74) into Eq. (19.75) and rearranging it, we have

$$\frac{f_1'''}{f_1''} = \frac{1}{2}(\beta - \eta). \tag{19.76}$$

Integrating Eq. (19.76) again, we get

$$\ln f_1'' = \frac{1}{2}\beta\eta - \frac{1}{4}\eta^2 + c, \tag{19.77}$$

in which c is an integral constant. Defining $c = -\beta^2/4 + \ln\gamma$, we obtain

$$f_1'' = \gamma \exp\left[-\frac{1}{4}(\eta - \beta)^2\right], \tag{19.78}$$

in which γ is an integral constant. Considering $f_1'(\infty) = 0$ and integrating Eq. (19.78), we have

$$f_1' = -\gamma \int_\eta^\infty \exp\left[-\frac{1}{4}(\eta - \beta)^2\right] d\eta. \qquad (19.79)$$

Integrating Eq. (19.79) again, we obtain

$$f_1 = \gamma \int_\eta^\infty d\eta \int_\eta^\infty \exp\left[-\frac{1}{4}(\eta - \beta)^2\right] d\eta. \qquad (19.80)$$

The approximate solution is given by the sum of f_0 and f_1 as follows:

$$f = f_0 + f_1 = \eta - \beta + \gamma \int_\eta^\infty d\eta \int_\eta^\infty \exp\left[-\frac{1}{4}(\eta - \beta)^2\right] d\eta. \qquad (19.81)$$

It is possible for us to improve the accuracy for the computation. Equations (19.72) and (19.81) are the solutions for small and large η's, respectively. Equation (19.72) has an unknown parameter α and Eq. (19.81) has two unknown parameters β and γ. To compute the three unknown parameters, f, f', and f'' must be smoothly continuous at a point outside the boundary-layer. Thus, the three unknown parameters α, β, and γ are, determined as

$$\alpha = 0.332, \quad \beta = 1.72, \quad \gamma = 0.231. \qquad (19.82)$$

The numerals in Eq. (19.82) are estimated by the comparison of the numerical computation of Eq. (19.69) with the Runge–Kutta method and iteration processes. The approximate solution with the three parameters fits the experimental data very well. The trials to improve the accuracy of the approximate solution have been conducted by many researchers.

Example 19.2. Solving the following differential equation:

$$\epsilon y'' + y' = 2x, \qquad (1)$$

with the boundary conditions $y(0) = \alpha$ and $y(1) = \beta$ and obtain the outer y^o, inner y^i solutions, and y^c.

Solution. The outer solution derived from $y^{o\prime} = 2x$ and $y^o(1) = \beta$ is

$$y^o = \beta - 1 + x^2. \tag{2}$$

Defining $\zeta = x/\epsilon$ in Eq. (1) and assuming a small ϵ, we obtain

$$\frac{d^2 y^i}{d\zeta^2} + \frac{dy^i}{d\zeta} = 0. \tag{3}$$

The inner solution is

$$y^i = C_2 e^{-\zeta} + C_1, \tag{4}$$

in which $y(0) = \alpha$. This shows

$$C_1 = \alpha - C_2. \tag{5}$$

Therefore, Eq. (19.64) is given by

$$y^c = y^o + y^i - (y^o)^i = \beta - 1 + x^2 + C_2 e^{-x/\epsilon} + C_1 - C_1. \tag{6}$$

Because the following limits are

$$\lim_{x \to 0} y^o = \beta - 1, \quad \lim_{\zeta \to \infty} y^i = C_1. \tag{7}$$

Sequently the approximate solution from Eq. (19.59) is obtained as

$$y^c = \beta - 1 + x^2 + (\alpha - \beta + 1)e^{-x/\epsilon} + O(\epsilon). \tag{8}$$

Problem 19.2. Obtain y^c of the differential equation

$$\epsilon y'' + y' = \sin x$$

with the boundary conditions $y(0) = 0$ and $y(1) = 1$.

19.3. Free Surface Flow over Wavy Bed

Consider the two-dimensional free surface flow over a wavy bed (Mizumura, 1995a). The definition sketch is represented in Fig. 19.3. The governing equation is

$$\frac{\partial^2 \phi}{\partial x^2} + \frac{\partial^2 \phi}{\partial y^2} = 0, \tag{19.83}$$

in which ϕ = the velocity potential, x = the horizontal coordinate system on the free water surface and indicates the flow direction, and y = the vertically upward coordinate system. When the bed form is flat, the water depth is d and the velocity is U. The velocity potential ϕ is caused by the existence of the wavy bed. The effect of the wavy bed η is given by ζ on the free water surface. The kinematic boundary condition at the free water surface is

$$\left(U + \frac{\partial \phi}{\partial x}\right) \frac{\partial \zeta}{\partial x} = \frac{\partial \phi}{\partial y}, \quad \text{at } y = \zeta. \tag{19.84}$$

The dynamic boundary condition at the free water surface is also

$$g\zeta + \frac{1}{2}\left[\left(U + \frac{\partial \phi}{\partial x}\right)^2 + \left(\frac{\partial \phi}{\partial y}\right)^2\right] = \frac{U^2}{2}, \quad \text{at } y = \zeta, \tag{19.85}$$

in which g = the gravitational acceleratin. The boundary condition at the wavy bed $y = \eta - d$ is given by

$$\frac{\partial \phi}{\partial y} = \left(U + \frac{\partial \phi}{\partial x}\right) \frac{\partial \eta}{\partial x}, \quad \text{at } y = \eta - d, \tag{19.86}$$

in which the form of the wavy bed is described by

$$\eta = a \sin kx, \tag{19.87}$$

in which a = the amplitude of the wavy bed and k = an angular wave number. To solve the problem, the variables are non-dimensionalized as follows:

$$x = x^*L, \quad y = y^*L, \quad \zeta = \zeta^*a, \quad \phi = \phi^*Ua, \quad \eta = \eta^*a, \tag{19.88}$$

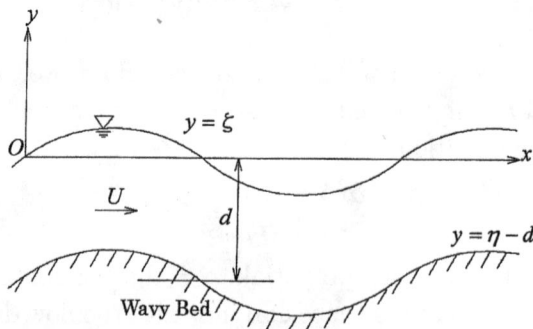

Figure 19.3.   Shallow flow over wavy bed.

in which $L = 2\pi/k$, L = the wave length of the wavy bed, and $*$ indicates a non-dimensional variable. Equation (19.83) becomes

$$\frac{\partial^2 \phi^*}{\partial x^{*2}} + \frac{\partial^2 \phi^*}{\partial y^{*2}} = 0.$$ (19.89)

Equation (19.84) becomes

$$\left(1 + \frac{a}{L}\frac{\partial \phi^*}{\partial x^*}\right)\frac{\partial \zeta^*}{\partial x^*} = \frac{\partial \phi^*}{\partial y^*}, \quad \text{at } y^* = \frac{a}{L}\zeta^*.$$ (19.90)

Equation (19.85) becomes

$$\frac{ga}{U^2}\zeta^* + \frac{1}{2}\left[\left(1 + \frac{a}{L}\frac{\partial \phi^*}{\partial x^*}\right)^2 + \left(\frac{a}{L}\frac{\partial \phi^*}{\partial y^*}\right)^2\right] = \frac{1}{2}, \quad \text{at } y^* = \frac{a}{L}\zeta^*.$$ (19.91)

Equation (19.86) becomes

$$\frac{\partial \phi^*}{\partial y^*} = \left(1 + \frac{a}{L}\frac{\partial \phi^*}{\partial x^*}\right)\frac{\partial \eta^*}{\partial x^*}, \quad \text{at } y^* = \frac{a}{L}\eta^* - \frac{d}{L}.$$ (19.92)

Although Eq. (19.89) is linear, the boundary conditions Eqs. (19.90), (19.91), and (19.92) are nonlinear. Therefore, we assume dependent variables ϕ^*, ζ^*, and η^* are expanded in the power series of a small parameter α.

$$\phi^* = \phi_0^* + \alpha\phi_1^* + \alpha^2\phi_2^* + \cdots,$$ (19.93)

$$\zeta^* = \zeta_0^* + \alpha\zeta_1^* + \alpha^2\zeta_2^* + \cdots,$$ (19.94)

$$\eta^* = \eta_0^* + \alpha\eta_1^* + \alpha^2\eta_2^* + \cdots,$$ (19.95)

in which $\alpha = a/L$. Applying the Taylor series to ϕ^* at the free surface and wavy bed, we obtain

$$\phi_{y^*=\alpha\zeta^*}^* = \phi^*|_{y^*=0} + \alpha\zeta^*\left.\frac{\partial \phi^*}{\partial y^*}\right|_{y^*=0} + \frac{(\alpha\zeta^*)^2}{2!}\left.\frac{\partial^2 \phi^*}{\partial y^{*2}}\right|_{y^*=0} + \cdots,$$ (19.96)

$$\phi_{y^*=\alpha\eta^*-d/L}^* = \phi^*|_{y^*=-d/L} + \alpha\eta^*\left.\frac{\partial \phi^*}{\partial y^*}\right|_{y^*=-d/L}$$

$$+ \frac{(\alpha\eta^*)^2}{2!}\left.\frac{\partial^2 \phi^*}{\partial y^{*2}}\right|_{y^*=-d/L} + \cdots.$$ (19.97)

The wavy bed is given by

$$\eta_1^* = \sin kx = \sin k^* x^*, \quad \eta_2^* = \eta_3^* = \cdots = 0, \qquad (19.98)$$

in which $k^* = kL = 2\pi$. The governing equation of the zero-th order solution is from Eq. (19.89) as

$$\frac{\partial^2 \phi_0^*}{\partial x^{*2}} + \frac{\partial^2 \phi_0^*}{\partial y^{*2}} = 0. \qquad (19.99)$$

From Eq. (19.90) we get

$$\frac{\partial \zeta_0^*}{\partial x^*} = \frac{\partial \phi_0^*}{\partial y^*}, \quad \text{at } y^* = 0. \qquad (19.100)$$

From Eq. (19.91) we have

$$\frac{ga}{U^2} \zeta_0^* + \frac{\partial \phi_0^*}{\partial x^*} = 0, \quad \text{at } y^* = 0. \qquad (19.101)$$

From Eq. (19.92) we obtain

$$\frac{\partial \phi_0^*}{\partial y^*} = \frac{\partial \eta_0^*}{\partial x^*}, \quad \text{at } d^* = -\frac{d}{L}. \qquad (19.102)$$

By using the linearized boundary conditions Eqs. (19.100), (19.101), and (19.102), we can solve Eq. (19.99). The solutions are

$$\phi_0^* = \frac{Pe^{k^* y^*} + e^{-k^* y^*}}{Pe^{-k^* d^*} - e^{k^* d^*}} \cos k^* x^*, \qquad (19.103)$$

$$\zeta_0^* = \frac{U^2 k^*}{ga} \frac{P+1}{Pe^{-k^* d^*} - e^{k^* d^*}} \sin k^* x^*, \qquad (19.104)$$

in which

$$P = \frac{ga + k^* U^2}{ga - k^* U^2}. \qquad (19.105)$$

In the same way, the first-order solution satisfies

$$\frac{\partial^2 \phi_1^*}{\partial x^{*2}} + \frac{\partial^2 \phi_1^*}{\partial y^{*2}} = 0. \qquad (19.106)$$

The boundary conditions are, respectively,

$$\frac{\partial \zeta_1^*}{\partial x^*} + \frac{\partial \phi_0^*}{\partial x^*} \frac{\partial \zeta_0^*}{\partial x^*} = \frac{\partial \phi_1^*}{\partial y^*} + \zeta_0^* \frac{\partial^2 \phi_0^*}{\partial y^{*2}}, \quad \text{at } y^* = 0, \qquad (19.107)$$

$$\frac{ga}{U^2}\zeta_1^* + \frac{\partial \phi_1^*}{\partial x^*} + \zeta_0^* \frac{\partial^2 \phi_0^*}{\partial x^* \partial y^*}$$

$$+ \frac{1}{2}\left[\left(\frac{\partial \phi_0^*}{\partial x^*}\right)^2 + \left(\frac{\partial \phi_0^*}{\partial y^*}\right)^2\right] = 0, \quad \text{at } y^* = 0, \quad (19.108)$$

$$\frac{\partial \phi_1^*}{\partial y^*} + \eta_0^* \frac{\partial^2 \phi_0^*}{\partial y^{*2}} = \frac{\partial \phi_0^*}{\partial x^*}\frac{\partial \eta_0^*}{\partial x^*}, \quad \text{at } y^* = -d^*. \quad (19.109)$$

After the computation of ϕ_1 (Mizumura, 1995a), the first-order solution is simply expressed by

$$\phi_1^* \sim \text{(Function of } y^*) \times \sin 2k^* x^*, \quad (19.110)$$

$$\zeta_1^* \sim \text{(Constant)} \times \cos 2k^* x^*. \quad (19.111)$$

Therefore, the water surface profile for the zero-th order is a sine function by Eq. (19.104). But the water surface profile for the first-order solution has the wave length of half the original wave length of the wavy bed. Due to the existence of the cosine function in Eq. (19.111), the peak of the water surface profile becomes flatter and the trough is steeper than the sine function for the subcritical flow. Due to the existence of the cosine function, the peak of the water surface profile becomes steeper and the trough is flatter than the sine function for the supercritical flow (Mizumura, 1995a). Stillmore, Mizumura (1998) obtained the water surface profile and the velocity distribution for the flow over the wavy porous media. Using the perturbation method, Mizumura (1993) and Mizumura and Yamasaka (1998) succeeded in the linear and nonlinear analysis of water rivulets on a smooth plane. Figure 19.4 represents meandering water rivulet on the smooth plane. Hino in 1975 analyzed the generation of rip currents on the shore using the perturbation method. The perturbation method is very powerful for an analysis of nonlinear problems.

Example 19.3. Derive Eqs. (19.99) and (19.100).

Solution. Substituting Eq. (19.93) into Eq. (19.89), we have

$$\frac{\partial^2 \phi_0^*}{\partial x^{*2}} + \frac{\partial^2 \phi_0^*}{\partial y^{*2}} + \alpha\left(\frac{\partial^2 \phi_1^*}{\partial x^{*2}} + \frac{\partial^2 \phi_1^*}{\partial y^{*2}}\right) + \cdots = 0. \quad (1)$$

Equation (1) indicates Eq. (19.99) by equating the coefficients of the various power of α. Substituting Eqs. (19.93), (19.94), and (19.95)

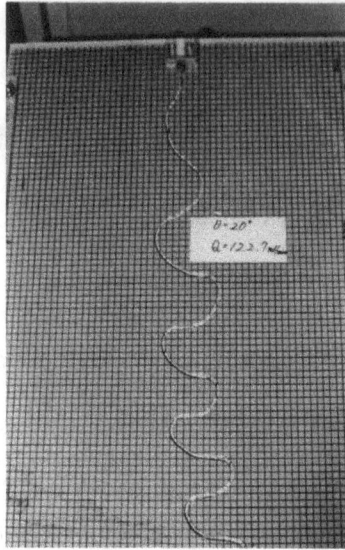

Figure 19.4.   Water rivulet.

into Eq. (19.90), we get

$$\left[1 + \alpha\left(\frac{\partial \phi_0^*}{\partial x^*} + \alpha\frac{\partial \phi_1^*}{\partial x^*} + \cdots\right)\right]\left[\frac{\partial \zeta_0^*}{\partial x^*} + \alpha\frac{\partial \zeta_1^*}{\partial x^*} + \cdots\right]$$
$$= \frac{\partial \phi_0^*}{\partial y^*} + \alpha\frac{\partial \phi_1^*}{\partial y^*} + \cdots. \tag{2}$$

The first-order approximation is

$$\frac{\partial \zeta_0^*}{\partial x^*} = \frac{\partial \phi_0^*}{\partial y^*}. \tag{3}$$

This is Eq. (19.100).

Problem 19.3. Obtain Eqs. (19.91) and (19.92).

19.4. Application to Kinematic Wave Method

When we analyze overland flow on the steep slope in hydrology, we use the kinematic wave method. Mizumura (1992a, 2006b) approximated Manning's formula by the parabolic curve and obtained the analytical solution in the closed form. The governing equation is

$$\frac{\partial h}{\partial t} + \frac{\partial q}{\partial x} = R(t), \tag{19.112}$$

in which h = the water depth, q = discharge per unit width, t = time, x = the coordinate system in the flow direction, and $R(t)$ = time-varying rainfall intensity. The momentum equation is the Manning's formula as

$$q = \frac{1}{n} S_o^{1/2} h^{5/3}, \tag{19.113}$$

in which n = Manning's roughness coefficient and S_o = the bottom slope. For simplicity, Eq. (19.113) is approximated by

$$q = kh^2, \tag{19.114}$$

in which $k = S_o^{1/2}/n$. Substituting Eq. (19.114) into Eq. (19.112), we have

$$\frac{\partial h}{\partial t} + 2kh\frac{\partial h}{\partial x} = R(t). \tag{19.115}$$

Because Eq. (19.115) is nonlinear, we apply the perturbation method to Eq. (19.115). We apply the method for the stability analysis of the nonlinear system. The method assumes that a dependent variable consists of two parts. One is the constant or known part and the other is the small part. The small part is analyzed for time changes. When the constant rainfall continues, we assume that the time-varying small rainfall starts from $t = 0$. Thus, the water depth h changes on the slope according to the small time-varying rainfall. Thus, we assume the following rainfall:

$$R(t) = R_o + R'(t), \quad R_o \gg R'(t), \tag{19.116}$$

in which R_o = the constant rainfall and $R'(t)$ = the time-varying small rainfall. The water depth is also given by

$$h(x,t) = h_o(x) + h'(x,t), \quad h_o(x) \gg h'(x,t), \tag{19.117}$$

in which "o" indicates constant in time and "$'$" shows time-varying. Substituting Eqs. (19.116) and (19.117) into Eq. (19.115) and assuming that the constant rainfall corresponds to the spatial water depth, we get

$$2kh_o\frac{dh_o}{dx} = R_o. \tag{19.118}$$

Neglecting squared terms of small variables, we have

$$\frac{\partial h'}{\partial t} + 2kh_o\frac{\partial h'}{\partial x} + 2kh'\frac{dh_o}{dx} = R'(t). \tag{19.119}$$

The solution of Eq. (19.118) is easily

$$h_o(x) = \sqrt{\frac{R_o x}{k}}, \tag{19.120}$$

in which $h_o(0) = 0$ is assumed. The initial and boundary conditions of h' in Eq. (19.119) are as follows:

$$h'(x,\ t=0) = 0, \tag{19.121}$$

$$h'(x=0,\ t) = 0. \tag{19.122}$$

Assuming $R_o \gg A$, we consider the time-varying rainfall of the sine function.

$$R'(t) = A\sin\omega t. \tag{19.123}$$

Substituting Eqs. (19.120) and (19.123) into Eq. (19.119), we get the following linear partial differential equation:

$$\frac{\partial h'}{\partial t} + 2\sqrt{kR_o x}\frac{\partial h'}{\partial x} + \sqrt{\frac{R_o k}{x}}h' = A\sin\omega t. \tag{19.124}$$

Equation (19.124) is the wave equation and has characteristics on the x and t plane. The characteristic that intersects the origin is called the limiting characteristic in Fig. 19.5. Consider the solution $h_o + h'_1$ after the limiting characteristic reaches. To obtain h'_1, let us assume the following form:

$$h'_1 = B_0(x) + B_1(x)\cos\omega t + B_2(x)\sin\omega t. \tag{19.125}$$

Substituting Eq. (19.125) into Eq. (19.124), we obtain $B_0(x)$, $B_1(x)$, and $B_2(x)$. They are

$$B_0(x) = 0, \tag{19.126}$$

$$B_1(x) = \frac{A}{\omega^2}\sqrt{\frac{kR_o}{x}}\sin\left(\omega\sqrt{\frac{x}{kR_o}}\right) - \frac{A}{\omega}, \tag{19.127}$$

$$B_2(x) = \frac{A}{\omega^2}\sqrt{\frac{kR_o}{x}}\left[1 - \cos\left(\omega\sqrt{\frac{x}{kR_o}}\right)\right]. \tag{19.128}$$

Substituting Eqs. (19.126), (19.127), and (19.128) into Eq. (19.125), we get the solution h'_1. Thus, the solution $h_o + h'_1$ is

$$h_o + h'_1 = \sqrt{\frac{R_o x}{k}} + \left[\frac{A}{\omega^2}\sqrt{\frac{kR_o}{x}}\sin\left(\omega\sqrt{\frac{x}{kR_o}}\right) - \frac{A}{\omega}\right]\cos\omega t$$

$$+ \frac{A}{\omega^2}\sqrt{\frac{kR_o}{x}}\left[1 - \cos\left(\omega\sqrt{\frac{x}{kR_o}}\right)\right]\sin\omega t. \tag{19.129}$$

Equation (19.129) satisfies Eq. (19.122), but does not satisfy the initial condition Eq. (19.121). The solution before the limiting characteristic reaches is given by $h_o + h'_1 + h'_2$ because Eq. (19.119) is linear. The existence of h'_2 helps $h_o + h'_1 + h'_2$ satisfy the initial condition Eq. (19.121). The function h'_2 satisfies

$$\frac{\partial h'_2}{\partial t} + 2\sqrt{kR_o x}\frac{\partial h'_2}{\partial x} + \sqrt{\frac{R_o k}{x}}h'_2 = 0. \tag{19.130}$$

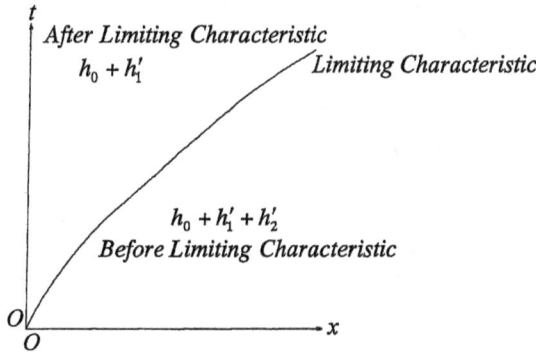

Figure 19.5.   Rainfall and runoff analysis.

In order that $h'_1 + h'_2$ satisfies the initial condition Eq. (19.121), h'_2 is equal to $-h'_1$ at $t = 0$. In the equation, we write

$$h'_2(x, \ t = 0) = \frac{A}{\omega} - \frac{A}{\omega^2}\sqrt{\frac{kR_o}{x}}\sin\left(\omega\sqrt{\frac{x}{kR_o}}\right) \qquad (19.131)$$

is the initial condition for h'_2. Referring to Chap. 8, the characteristic equation of Eq (19.130) is

$$\frac{dt}{1} = \frac{dx}{2\sqrt{kR_ox}} = -\frac{dh'_2}{h'_2\sqrt{kR_o/x}}. \qquad (19.132)$$

The solution is

$$h'_2\sqrt{x} = F\left(t - \sqrt{\frac{x}{kR_o}}\right), \qquad (19.133)$$

in which $F(\cdot)$ is an arbitrary function. Introducing a dummy variable ξ and comparing Eq. (19.131), we get

$$F(\xi) = -\frac{A\sqrt{kR_o}}{\omega}\xi + \frac{A\sqrt{kR_o}}{\omega^2}\sin\omega\xi. \qquad (19.134)$$

The solution before the limiting characteristic reaches is given by

$$h_o + h'_1 + h'_2 = \sqrt{\frac{R_ox}{k}} + \frac{A}{\omega}(1 - \cos\omega t) + \frac{A}{\omega^2}\sqrt{\frac{kR_o}{x}}\sin\omega t$$

$$- \frac{A}{\omega}\sqrt{\frac{kR_o}{x}}t. \qquad (19.135)$$

Equation (19.135) satisfies Eq. (19.121), but it does not satisfy Eq. (19.122). The third term in Eq. (19.135) becomes infinity as $t \to \infty$ and the permanent term. But this is the solution before the limiting characteristic reaches and t is finite in this solution. Thus, this is not a problem (Mizumura 1992a, 2006b). The limiting characteristic is given by the condition that Eq. (19.129) equals Eq. (19.135) as follows:

$$\sin w\left(\sqrt{\frac{x}{kR_0}} - t\right) = w\left(\sqrt{\frac{x}{kR_0}} - t\right) \qquad (19.136)$$

Because $x = t = 0$ satisfies Eq. (19.136), Eq. (19.136) passes through the origin.

Example 19.4. Derive Eq. (19.133) from Eq. (19.132).

Solution. We get from Eq. (19.132),

$$t - \sqrt{\frac{x}{kR_o}} = C_1, \tag{1}$$

$$h'_2\sqrt{x} = C_2. \tag{2}$$

The general solution from Eqs. (1) and (2) is

$$h'_2\sqrt{x} = F\left(t - \sqrt{\frac{x}{kR_o}}\right), \tag{3}$$

in which $F(\cdot)$ is an arbitrary function.

Problem 19.4. Show that h'_1 in Eq. (19.125) satisfies Eq. (19.122) at $x = 0$.

19.5. Averaging Method

When the analytical solution of a nonlinear differential equation of order zero is assumed to be obtained, consider the nonlinear differential equation that includes a small parameter ϵ as follows:

$$\ddot{X} + \omega^2 X = \epsilon f(X, \dot{X}), \tag{19.137}$$

in which " $\cdot$ " indicates d/dt and $f(X, \dot{X})$ is an arbitrary function of X and $\dot{X}$. If $\epsilon = 0$ in Eq. (19.137), Eq. (19.137) shows the simple oscillation equation. The solution is

$$X(t) = a\cos(\omega t + \phi), \tag{19.138}$$

in which $a =$ an amplitude and $\phi =$ a phase lag. They are constant and determined by the initial condition. In the averaging method, Eq. (19.138) is assumed not to significantly change for the small parameter ϵ and $a = a(t)$ and $\phi = \phi(t)$ slowly vary for changing t. Then, the solution is supposed to be

$$X(t) = a(t)\cos[\omega t + \phi(t)]. \tag{19.139}$$

The derivative of Eq. (19.139) is

$$\dot{X}(T) = -a(t)\omega\sin[\omega t + \phi(t)]. \tag{19.140}$$

The assumption of Eq. (19.140) is the same as that of the following equation to be formed as

$$\dot{a}(t)\cos[\omega t + \phi(t)] - a(t)\dot{\phi}(t)\sin[\omega t + \phi(t)] = 0. \qquad (19.141)$$

From Eq. (19.140) $\ddot{X}$ is

$$\ddot{X} = -\dot{a}(t)\omega\sin[\omega t + \phi(t)] - a(t)\omega^2\cos[\omega t + \phi(t)]$$
$$- a(t)\dot{\phi}(t)\omega\cos[\omega t + \phi(t)]. \qquad (19.142)$$

Substituting Eqs. (19.139) and (19.142) into Eq. (19.137), we obtain

$$-\omega\dot{a}(t)\sin[\omega t + \phi(t)] - \omega a(t)\dot{\phi}(t)\cos[\omega t + \phi(t)]$$
$$= \epsilon f\{a(t)\cos[\omega t + \phi(t)], -a(t)\omega\sin[\omega t + \phi(t)]\}. \qquad (19.143)$$

Solving $\dot{a}$ and $\dot{\phi}$ by Eqs. (19.141) and (19.143), we get

$$\dot{a} = -\frac{\epsilon}{\omega}f(a\cos\psi, -a\omega\sin\psi)\sin\psi, \qquad (19.144)$$

$$\dot{\phi} = -\frac{\epsilon}{\omega a}f(a\cos\psi, -a\omega\sin\psi)\cos\psi, \qquad (19.145)$$

in which $\psi = \omega t + \phi(t)$. If the terms on the right side in Eqs. (19.144) and (19.145) are expanded in the Fourier series, we have

$$f(a\cos\psi, -a\omega\sin\psi)\sin\psi = \frac{1}{2}\alpha_0 + \sum_{n=1}^{\infty}(\alpha_n\cos n\psi + \beta_n\sin n\psi),$$
$$(19.146)$$

$$f(a\cos\psi, -a\omega\sin\psi)\cos\psi = \frac{1}{2}\alpha_0' + \sum_{n=1}^{\infty}(\alpha_n'\cos n\psi + \beta_n'\sin n\psi),$$
$$(19.147)$$

in which

$$\alpha_n = \frac{1}{\pi}\int_0^{2\pi} f(a\cos\psi, -a\omega\sin\psi)\sin\psi\cos n\psi\, d\psi, \qquad (19.148)$$

$$\beta_n = \frac{1}{\pi}\int_0^{2\pi} f(a\cos\psi, -a\omega\sin\psi)\sin\psi\sin n\psi\, d\psi, \qquad (19.149)$$

$$\alpha_n' = \frac{1}{\pi}\int_0^{2\pi} f(a\cos\psi, -a\omega\sin\psi)\cos\psi\cos n\psi\, d\psi, \qquad (19.150)$$

$$\beta_n' = \frac{1}{\pi}\int_0^{2\pi} f(a\cos\psi, -a\omega\sin\psi)\cos\psi\sin n\psi\, d\psi, \qquad (19.151)$$

for $n = 0, 1, 2, \ldots, \infty$. As the zero-th order solution, considering the constant terms in the Fourier series $\frac{1}{2}\alpha_0$ and $\frac{1}{2}\alpha'_0$, we get

$$\dot{a}(t) \cong -\frac{\epsilon}{2\pi\omega} \int_0^{2\pi} f(a\cos\psi, -a\omega\sin\psi)\sin\psi\, d\psi, \qquad (19.152)$$

$$\dot{\phi}(t) \cong -\frac{\epsilon}{2\pi\omega a} \int_0^{2\pi} f(a\cos\psi, -a\omega\sin\psi)\cos\psi\, d\psi. \qquad (19.153)$$

This is called a method of Kryloff and Bogoliuboff (Aggarwal, 1992). This method is applicable for the nonlinear oscillation phenomena.

Example 19.5. Obtain the zero-th order solution of the following Rayleigh equation by the averaging method:

$$\ddot{X} + X = \epsilon\left(\dot{X} - \frac{\dot{X}^3}{3}\right), \qquad (1)$$

in which ϵ is a small parameter.

Solution. From Eq. (19.140) we get

$$\dot{X} = -a\omega\sin\psi, \qquad (2)$$

$$f = -a\omega\sin\psi + \frac{a^3\omega^3}{3}\sin^3\psi. \qquad (3)$$

The integral of Eq. (19.152) is

$$\int_0^{2\pi}\left(-a\omega\sin\psi + \frac{a^3\omega^3}{3}\sin^3\psi\right)\sin\psi\, d\psi = -a\omega\pi + \frac{\pi a^3\omega^3}{4}. \qquad (4)$$

The integral of Eq. (19.153) is

$$\int_0^{2\pi}\left(-a\omega\sin\psi + \frac{a^3\omega^3}{3}\sin^3\psi\right)\cos\psi\, d\psi = 0. \qquad (5)$$

Thus, Eq. (19.152) is

$$\dot{a} \cong -\frac{\epsilon}{2}a\left(1 - \frac{a^2\omega^2}{4}\right). \qquad (6)$$

From Eq. (19.153) we obtain

$$\dot{\phi} \cong 0. \qquad (7)$$

Using the initial conditions a_0 and ϕ_0, we obtain

$$a(t) = \frac{a_0 e^{\epsilon t/2}}{\left[1 + \frac{1}{4}a_0^2\omega^2(e^{\epsilon t} - 1)\right]^{1/2}}, \qquad (8)$$

$$\phi(t) = \phi_0. \qquad (9)$$

Equations (8) and (9) indicate that $a(t)$ and $\phi(t)$ are slowly varying.

Problem 19.5. Obtain the zero-th order solution of the following van del Pol equation by the averaging method:

$$\ddot{X} + \epsilon \dot{X}(X^2 - 1) + \omega^2 X = 0, \tag{10}$$

in which ϵ is very small.

19.6. Stability Analysis in Fluid Flow

Let us apply the perturbation method to the stability analysis of the two-dimensional laminar boundary-layer on the plate, respectively (Yih, 1969). When the x and y directions are flow direction and normal direction to the plate, the thickness of the laminar boundary-layer increases in the x direction and the velocity in the y direction in the laminar boundary-layer is negligible in comparison with the velocity in the x direction. Therefore, substituting the following equations into the Navier–Stokes equations:

$$u = U(y) + u(x, y, t), \tag{19.154}$$

$$v = v(x, y, t), \tag{19.155}$$

$$p = P + p(x, y, t). \tag{19.156}$$

Since $U(y) \gg u(x, y, t)$ and $P \gg p(x, y, t)$, the small components of subscript 1 in the Navier–Stokes equations satisfy the following equations:

$$\frac{\partial u}{\partial t} + U\frac{\partial u}{\partial x} + v\frac{\partial U}{\partial y} = -\frac{1}{\rho}\frac{\partial p}{\partial x} + \nu\nabla^2 u, \tag{19.157}$$

$$\frac{\partial v}{\partial t} + U\frac{\partial v}{\partial x} = -\frac{1}{\rho}\frac{\partial p}{\partial y} + \nu\nabla^2 v, \tag{19.158}$$

$$\frac{\partial u}{\partial x} + \frac{\partial v}{\partial y} = 0. \tag{19.159}$$

If the variation of infinitesimally small amplitude occurs, the small components of the velocities and pressure are assumed to be

$$u(x, y, t) = \bar{u}(y)e^{(\alpha x - \beta t)}, \tag{19.160}$$

$$v(x, y, t) = \bar{v}(y)e^{(\alpha x - \beta t)}, \tag{19.161}$$

$$p(x, y, t) = \bar{p}(y)e^{(\alpha x - \beta t)}, \tag{19.162}$$

in which α = the real angular wave number and real and β = an complex angular frequency. Therefore, we can write

$$\beta = \beta_r + i\beta_i. \qquad (19.163)$$

Thus, we rewrite

$$e^{i(\alpha x - \beta t)} = e^{i(\alpha x - \beta_r t)} \cdot e^{\beta_i t}. \qquad (19.164)$$

When $\beta_i > 0$, the amplitude increases as time goes. The flow becomes unstable and turbulent. When $\beta_i < 0$, the amplitude decreases as time goes. The flow becomes stable and laminar. $\beta_i = 0$ indicates a neutral condition. Substituting Eq. (19.159) into Eq. (19.158) and eliminating $\bar{u}$ and $\bar{p}$, we obtain the following equation about $\phi(y) = \bar{v}/U_\infty$:

$$(U - c)(\phi'' - \alpha^2 \phi) - U''\phi = -\frac{i}{\alpha Re}(\phi'''' - 2\alpha^2 \phi'' + \alpha^4 \phi), \qquad (19.165)$$

in which $c = \beta/\alpha$, $Re = U_\infty \delta/\nu$, $U_\infty = U(\delta)$, $\eta = y/\delta$, "$'$" $= d/d\eta$, and δ = the thickness of the laminar boundary-layer. Equation (19.165) is the fundamental equation to determine whether the varying flow of disturbance increases or decreases. Equation (19.165) is called an Orr–Sommerfeld equation. Orr in 1907 and Sommerfeld in 1908 derived Eq. (19.165) independently. The boundary conditions for Eq. (19.165) are

$$u = v = 0 \quad \text{at } y = 0 \to \phi = \phi' = 0 \quad \text{at } \eta = 0, \qquad (19.166)$$

$$u = v = 0 \quad \text{at } y = \infty \to \phi = \phi' = 0 \quad \text{at } \eta = \infty. \qquad (19.167)$$

Equations (19.165), (19.166), and (19.167) are the boundary value problem of the fourth-order differential equation. When α and Re in Eq. (19.165) are known, the complex angular frequency β is the eigen value.

Example 19.6. Show that the flow of the perfect fluid is stable if a point of inflection is in the velocity distribution (Rayleigh unstable theorem (Yih, 1969)).

Solution. The Orr–Sommerfeld equation in the perfect fluid for $Re \to \infty$ is given by

$$\phi'' - \left(\alpha^2 + \frac{U''}{U-c}\right)\phi = 0. \tag{1}$$

The boundary conditions are $v = 0$ ($\phi = 0$) at $y = y_1$ and $y = y_2$. Multiplying the conjugate function $\bar{\phi}$ of ϕ to Eq. (1) and integrating from $y = y_1$ to y_2, we have

$$\int_{y_1}^{y_2} \phi''\bar{\phi}\,dy = [\phi'\bar{\phi}]_{y_1}^{y_2} - \int_{y_1}^{y_2} \phi'\bar{\phi}'\,dy = -\int_{y_1}^{y_2} |\phi'|^2 dy. \tag{2}$$

Since $\phi(y) = 0$ and $\bar{\phi}(y) = 0$ at $y = y_1$ and y_2. Therefore, Eq. (1) is rewritten by Eq. (2) as

$$\int_{y_1}^{y_2} \left\{|\phi'|^2 + \left[\alpha^2 + \frac{U''(\overline{U-c})}{|U-c|^2}\right]|\phi|^2\right\} dy = 0. \tag{3}$$

If the celerity of the disturbance c is a complex, we have

$$c = c_r + ic_i. \tag{4}$$

The imaginary part of Eq. (3) is

$$c_i \int_{y_1}^{y_2} \frac{U''}{|U-c|^2}|\phi|^2 dy = 0. \tag{5}$$

To satisfy Eq. (5), $c_i = 0$ or U'' changes the sign between $[y_1, y_2]$. The latter indicates that the velocity distribution has a point of inflection. In this case, there possibly exists the celerity of disturbance $c_i > 0$.

Problem 19.6. When the energy slope of the open channel flow on the uniform slope is approximated by the water surface slope, then the velocity and water depth are slightly varied from the uniform flow. The equation that perturbed velocity and water depth becomes the diffusion equation. Is it true?

Chapter 20

Nonlinear Systems Analysis

When the second-order differential equation is nonlinear, we study the behavior of the solution on the phase plane. A linearized equation near a singular point forms a node, saddle, focus, or center. The special form of the center is a limit cycle. As the application of the nonlinear system, there exist chaos and fractals. The computational technique is useful for the study of the estimation and control theory.

20.1. Vector Calculation

Consider the following vector variable:

$$\underline{X} = \begin{bmatrix} x_1 \\ x_2 \\ \vdots \\ x_m \end{bmatrix}. \tag{20.1}$$

Then, a vector function $\underline{f}(\underline{X})$ is defined as follows:

$$\underline{f}(\underline{X}) = \begin{bmatrix} f_1(\underline{X}) \\ f_2(\underline{X}) \\ \vdots \\ f_m(\underline{X}) \end{bmatrix} = \begin{bmatrix} f_1(x_1, x_2, \ldots, x_m) \\ f_2(x_1, x_2, \ldots, x_m) \\ \vdots \\ f_m(x_1, x_2, \ldots, x_m) \end{bmatrix}. \tag{20.2}$$

In this chapter a vector is customarily defined as $\underline{X}$. For example, consider

$$f_1(x_1, x_2) = ax_1 + bx_2, \tag{20.3}$$

$$f_2(x_1, x_2) = cx_1 + dx_2. \tag{20.4}$$

These two equations Eqs. (20.3) and (20.4) are written by

$$\underline{f}(\underline{X}) = A\underline{X},\qquad (20.5)$$

in which

$$\underline{f}(\underline{X}) = \begin{bmatrix} f_1(x_1, x_2) \\ f_2(x_1, x_2) \end{bmatrix}, \quad A = \begin{bmatrix} a & b \\ c & d \end{bmatrix}, \quad \underline{X} = \begin{bmatrix} x_1 \\ x_2 \end{bmatrix}. \qquad (20.6)$$

Differentiating $\underline{f}(\underline{X})$ about $\underline{X}$, we get

$$\frac{d\underline{f}(\underline{X})}{d\underline{X}} = \begin{bmatrix} \dfrac{\partial f_1}{\partial x_1} & \dfrac{\partial f_1}{\partial x_2} & \cdots & \dfrac{\partial f_1}{\partial x_m} \\[2mm] \dfrac{\partial f_2}{\partial x_1} & \dfrac{\partial f_2}{\partial x_2} & \cdots & \dfrac{\partial f_2}{\partial x_m} \\[2mm] \vdots & \vdots & \ddots & \vdots \\[2mm] \dfrac{\partial f_m}{\partial x_1} & \dfrac{\partial f_m}{\partial x_2} & \cdots & \dfrac{\partial f_m}{\partial x_m} \end{bmatrix}. \qquad (20.7)$$

Equation (20.7) is a matrix. Especially if $\underline{X}$ is a scalar, because $\underline{X} = x_1$, we get

$$\frac{d\underline{f}(\underline{X})}{d\underline{X}} = \frac{d\underline{f}(\underline{X})}{dx_1} = \begin{bmatrix} \dfrac{\partial f_1}{\partial x_1} \\[2mm] \dfrac{\partial f_2}{\partial x_1} \\[2mm] \vdots \\[2mm] \dfrac{\partial f_m}{\partial x_1} \end{bmatrix} = \begin{bmatrix} \dfrac{df_1}{dx_1} \\[2mm] \dfrac{df_2}{dx_1} \\[2mm] \vdots \\[2mm] \dfrac{df_m}{dx_1} \end{bmatrix}. \qquad (20.8)$$

This is a column vector. If $\underline{f}$ is a scalar, because $\underline{f} = f_1$, we obtain

$$\frac{df}{d\underline{X}} = \frac{df_1}{d\underline{X}} = \begin{bmatrix} \dfrac{\partial f_1}{\partial x_1}, & \dfrac{\partial f_1}{\partial x_2}, & \cdots, & \dfrac{\partial f_1}{\partial x_m} \end{bmatrix}. \qquad (20.9)$$

This is a row vector. For example, we have the following general formulas:

$$\frac{d}{d\underline{X}}(A\underline{f}(\underline{X})) = A\frac{d\underline{f}(x)}{d\underline{X}}, \tag{20.10}$$

$$\frac{d}{d\underline{X}}[\underline{f}^T(\underline{X})A\underline{g}(\underline{X})] = \underline{f}^T(\underline{X})A\frac{d\underline{g}(\underline{X})}{d\underline{X}} + \underline{g}^T(\underline{X})A^T\frac{d\underline{f}(\underline{X})}{d\underline{X}}, \tag{20.11}$$

in which superscript T indicates a transpose matrix or vector. Equation (20.7) shows

$$\frac{d\underline{X}}{d\underline{X}} = \begin{bmatrix} 1 & 0 & \cdots & 0 \\ 0 & 1 & \cdots & 0 \\ \vdots & \vdots & \ddots & \vdots \\ 0 & 0 & \cdots & 1 \end{bmatrix} = I, \tag{20.12}$$

in which I is a unit matrix.

Example 20.1. Show the following equation:

$$\frac{d}{d\underline{X}}(\underline{X}^T A\underline{X}) = \underline{X}^T(A + A^T). \tag{1}$$

Solution. Substituting $\underline{f}(\underline{X}) = \underline{X}$ and $\underline{g}(\underline{X}) = \underline{X}$ in Eq. (20.11), we have Eq. (20.12)

$$\frac{d\underline{X}}{d\underline{X}} = I \tag{2}$$

and obtain Eq. (1).

Problem 20.1. Consider the equation of a pendulum oscillation with a finite amplitude. It is

$$\ddot{x} + \sin x = 0. \tag{3}$$

If we define

$$\dot{x} = y \quad \text{and} \quad \dot{y} = -\sin x, \tag{4}$$

Eq. (3) is expressed by the vector form

$$\underline{X} = \begin{bmatrix} x \\ y \end{bmatrix}, \tag{5}$$

$$\underline{f}(\underline{X}) = \begin{bmatrix} y \\ -\sin x \end{bmatrix}. \tag{6}$$

Then, obtain $d\underline{f}/d\underline{X}$, in which " · " indicates d/dt.

20.2. Phase Plane Analysis

When an independent variable is t and the nonlinear differential equation of the second-order is an autonomous system (Aggarwal, 1972; Strintzis, 1974), the phase plane analysis can properly investigate the original nonlinear differential equation. For the linear system, the analytical solution is generally obtained. If " · " shows d/dt, the autonomous system is written as

$$\dot{x} = x^2, \tag{20.13}$$

The autonomous system does not contain t explicitly. Thus,

$$\dot{x} = 2tx^2 \tag{20.14}$$

is not autonomous. Equations (20.13) and (20.14) are nonlinear. The general form of the second-order differential equation is given by

$$\ddot{x} = f(\dot{x}, x). \tag{20.15}$$

Introducing a new variable y such as

$$\dot{x} = y, \tag{20.16}$$

we obtian from Eq. (20.15)

$$\dot{y} = f(y, x). \tag{20.17}$$

Combining Eqs. (20.16) and (20.17), we get the following vector form:

$$\begin{bmatrix} \dot{x} \\ \dot{y} \end{bmatrix} = \begin{bmatrix} y \\ f(y, x) \end{bmatrix}. \tag{20.18}$$

This is the vector form of the first-order differential equation. The second-order differential equation becomes a two-dimensional vector form of the first-order. The third-order differential equation forms a three-dimensional vector form of the first-order. The higher order differential equation is also expressed by the vector form of the first-order. Next, let us analyze the autonomous differential equation of the second-order. The equations are written as

$$\dot{x}_1 = f_1(x_1, x_2), \quad x_1(0) = \xi_1, \tag{20.19}$$

$$\dot{x}_2 = f_2(x_1, x_2), \quad x_2(0) = \xi_2, \tag{20.20}$$

in which ξ_1 and ξ_2 are arbitrary variables and " $\cdot$ " indicates d/dt. Combining Eqs. (20.19) and (20.20), we get

$$\frac{dx_2}{dx_1} = \frac{f_2(x_1, x_2)}{f_1(x_1, x_2)}. \tag{20.21}$$

Equation (20.21) shows the derivative on the $x_1 - x_2$ plane. Let us consider the following equations:

$$f_1(\hat{x}_1, \hat{x}_2) = f_2(\hat{x}_1, \hat{x}_2) = 0. \tag{20.22}$$

The point $(\hat{x}_1, \hat{x}_2)$ that satisfies Eq. (20.22) is called a singular point. Expanding Eqs. (20.19) and (20.20) about the singular point by the Taylor series, we have

$$\dot{x}_1 = f_1(x_1, x_2) = f_1(\hat{x}_1, \hat{x}_2) + \frac{\partial f_1(\hat{x}_1, \hat{x}_2)}{\partial x_1}(x_1 - \hat{x}_1)$$

$$+ \frac{\partial f_1(\hat{x}_1, \hat{x}_2)}{\partial x_2}(x_2 - \hat{x}_2) + \cdots, \tag{20.23}$$

$$\dot{x}_2 = f_2(x_1, x_2) = f_2(\hat{x}_1, \hat{x}_2) + \frac{\partial f_2(\hat{x}_1, \hat{x}_2)}{\partial x_1}(x_1 - \hat{x}_1)$$

$$+ \frac{\partial f_2(\hat{x}_1, \hat{x}_2)}{\partial x_2}(x_2 - \hat{x}_2) + \cdots. \tag{20.24}$$

In Eqs. (20.23) and (20.24), we define

$$y_1(t) = x_1(t) - \hat{x}_1, \quad y_2(t) = x_2(t) - \hat{x}_2, \tag{20.25}$$

$$a_{11} = \frac{\partial f_1(\hat{x}_1, \hat{x}_2)}{\partial x_1}, \quad a_{12} = \frac{\partial f_1(\hat{x}_1, \hat{x}_2)}{\partial x_2}, \qquad (20.26)$$

$$a_{21} = \frac{\partial f_2(\hat{x}_1, \hat{x}_2)}{\partial x_1}, \quad a_{22} = \frac{\partial f_2(\hat{x}_1, \hat{x}_2)}{\partial x_2}. \qquad (20.27)$$

Note that

$$\frac{d}{dt}(x_1 - \hat{x}_1) = \dot{x}_1 = \dot{y}_1, \qquad (20.28)$$

$$\frac{d}{dt}(x_2 - \hat{x}_2) = \dot{x}_2 = \dot{y}_2. \qquad (20.29)$$

The two equations Eqs. (20.23) and (20.24) are written as

$$\dot{y}_1 = a_{11}y_1 + a_{12}y_2 + H.O.T., \qquad (20.30)$$

$$\dot{y}_2 = a_{21}y_1 + a_{22}y_2 + H.O.T., \qquad (20.31)$$

in which *H.O.T.* indicates higher order terms of $x_1 - \hat{x}_1$ or $x_2 - \hat{x}_2$. Then, the singular point of Eqs. (20.30) and (20.31) is (0, 0). In Eqs. (20.30) and (20.31), neglecting the higher order terms and defining $z_1 = y_1$ and $z_2 = y_2$, we transform Eqs. (20.30) and (20.31) to

$$\dot{z}_1 = a_{11}z_1 + a_{12}z_2, \qquad (20.32)$$

$$\dot{z}_2 = a_{21}z_1 + a_{22}z_2. \qquad (20.33)$$

Combining Eqs. (20.32) and (20.33), we have

$$\underline{\dot{Z}} = A\underline{Z}, \qquad (20.34)$$

in which

$$\underline{Z} = \begin{bmatrix} z_1 \\ z_2 \end{bmatrix}, \quad A = \begin{bmatrix} a_{11} & a_{12} \\ a_{21} & a_{22} \end{bmatrix}. \qquad (20.35)$$

Using a transformation matrix T, we transform $\underline{Z}$ to $\underline{W}$ as follows:

$$\underline{W} = T\underline{Z}. \qquad (20.36)$$

Then, we get

$$\underline{\dot{W}} = T\underline{\dot{Z}} = TA\underline{Z}. \qquad (20.37)$$

Equation (20.36) indicates

$$\underline{Z} = T^{-1}\underline{W}. \tag{20.38}$$

Equation (20.37) is also transformed

$$\underline{\dot{W}} = TAT^{-1}\underline{W}. \tag{20.39}$$

Transforming an arbitrary matrix A to $\Lambda = TAT^{-1}$, we have Λ as

$$\Lambda = \begin{bmatrix} \lambda_1 & 0 \\ 0 & \lambda_2 \end{bmatrix}, \quad \text{or} \quad \begin{bmatrix} \mu & 1 \\ 0 & \mu \end{bmatrix}, \quad \text{or} \quad \begin{bmatrix} \alpha & -\beta \\ \beta & \alpha \end{bmatrix}. \tag{20.40}$$

Thus, Eq. (20.39) is written as

$$\underline{\dot{W}} = \begin{bmatrix} \dot{w}_1 \\ \dot{w}_2 \end{bmatrix} = \Lambda\underline{W}. \tag{20.41}$$

According to Eq. (20.40), the property of the singular points is determined. The property of the singular points shows the trajectories of the solution. The trajectories represent the solution on the phase plane.

Example 20.2. Obtain the singular point and the matrix Λ in Eq. (20.40) of the following Duffing equation:

$$\ddot{u} + u + \epsilon u^3 = 0. \tag{1}$$

Solution. Defining $\dot{u} = v$, we transform Eq. (1) to

$$\dot{v} = -u - \epsilon u^3, \tag{2}$$

$$\dot{u} = v. \tag{3}$$

The singular point of Eqs. (2) and (3) is $(v, u) = (0, 0)$. Defining $f_1 = -u - \epsilon u^3$ and $f_2 = v$, we get

$$\frac{\partial f_1}{\partial v} = 0, \quad \frac{\partial f_1}{\partial u} = -1 - 3\epsilon u^2, \tag{4}$$

$$\frac{\partial f_2}{\partial v} = 1, \quad \frac{\partial f_2}{\partial u} = 0. \tag{5}$$

The derivatives at the singular point in Eqs. (4) and (5) are obtained by substituting $u = v = 0$. Assuming

$$\underline{Z} = \begin{bmatrix} v \\ u \end{bmatrix}. \tag{6}$$

Equation (20.34) becomes

$$\begin{bmatrix} \dot{v} \\ \dot{u} \end{bmatrix} = \begin{bmatrix} 0 & -1 \\ 1 & 0 \end{bmatrix} \begin{bmatrix} v \\ u \end{bmatrix}. \tag{7}$$

Therefore, we have the matrix A as follows:

$$\begin{bmatrix} 0 & -1 \\ 1 & 0 \end{bmatrix} = A. \tag{8}$$

The eigen values of the matrix A are given by

$$|A - \lambda I| = \begin{vmatrix} -\lambda & -1 \\ 1 & -\lambda \end{vmatrix} = \lambda^2 + 1 = 0. \tag{9}$$

Since from Eq. (9) we get $\lambda = \pm i$, the matrix Λ is

$$\Lambda = \begin{pmatrix} 0 & -i \\ i & 0 \end{pmatrix}. \tag{10}$$

Problem 20.2. Calculate singular points in Problem 20.1 and obtain the matrix Λ in Eq. (20.40).

20.3. Property of Singular Point

We showed that the matrix Λ is classified into three types in Eq. (20.40) in the previous section. We compute the eigen values of the matrix A. The eigen values are obtained from

$$|A - \lambda I| = \lambda^2 - (a_{11} + a_{22})\lambda + a_{11}a_{22} - a_{21}a_{12} = 0. \tag{20.42}$$

When Eq. (20.42) has two real roots λ_1 and λ_2, we have a diagonal form

$$\Lambda = \begin{bmatrix} \lambda_1 & 0 \\ 0 & \lambda_2 \end{bmatrix}. \tag{20.43}$$

When Eq. (20.42) has a repeated eigen value μ, we have a Jordan form

$$\Lambda = \begin{bmatrix} \mu & 1 \\ 0 & \mu \end{bmatrix}. \tag{20.44}$$

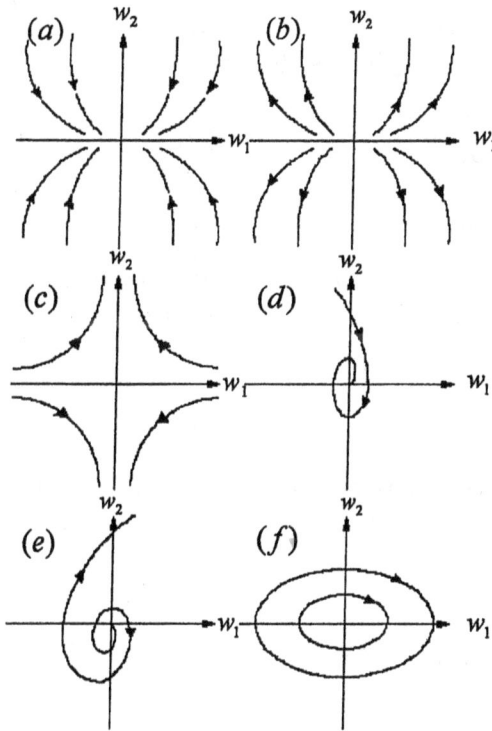

Figure 20.1. Singular points

When Eq. (20.42) has complex roots $\alpha \pm i\beta$, we have a complex conjugate form

$$\Lambda = \begin{bmatrix} \alpha & -\beta \\ \beta & \alpha \end{bmatrix}. \tag{20.45}$$

When λ_1 and λ_2 are negative and positive in Eq. (20.43), the singular point forms a stable node in Fig. 20.1(a) and an unstable node in Fig. 20.1(b), respectively. When the signs of λ_1 and λ_2 in Eq. (20.43) are different, the singular point forms a saddle in Fig. 20.1(c). When μ is negative and positive in Eq. (20.44), the singular point forms a stable and unstable node, respectively. Equation (20.45) represents a stable focus in Fig. 20.1(d) and unstable focus in Fig. 20.1(e), respectively, according to the negative and positive α. If α is zero, the singular point forms a center as shown in Fig. 20.1(f). For example, let us consider the nonlinear differential equation (16.125) that expresses the form of the saline wedge. Equation (16.125) can be written in the phase plane as follows:

$$\frac{d\xi}{dt} = 2(1 + S_0\xi - \eta)(\eta^3 - \eta_c^3) = f_1(\xi, \eta), \tag{20.46}$$

$$\frac{d\eta}{dt} = f_i(1 + S_0\xi) = f_2(\xi, \eta), \tag{20.47}$$

in which t is a dummy variable. The number of singular points of Eqs. (20.46) and (20.47) is two. They are

$$\xi = -1/S_0, \quad \eta = \eta_c, \tag{20.48}$$

$$\xi = -1/S_0, \quad \eta = 0. \tag{20.49}$$

Taking the partial derivatives of f_1 and f_2, we obtain

$$\frac{\partial f_1}{\partial \xi} = 2S_0(\eta^3 - \eta_c^3), \tag{20.50}$$

$$\frac{\partial f_1}{\partial \eta} = -2(\eta^3 - \eta_c^3) + 6(1 + S_0\xi - \eta)\eta^2, \tag{20.51}$$

$$\frac{\partial f_2}{\partial \xi} = f_i S_0, \tag{20.52}$$

$$\frac{\partial f_2}{\partial \eta} = 0. \tag{20.53}$$

The dot "." indicates d/dt. At the singular point of Eq. (20.48), Eq. (20.41) is

$$\begin{bmatrix} \dot{\xi} \\ \dot{\eta} \end{bmatrix} = \begin{bmatrix} 0 & -6\eta_c^3 \\ f_i S_0 & 0 \end{bmatrix} \begin{bmatrix} \xi \\ \eta \end{bmatrix}. \tag{20.54}$$

At the singular point of Eq. (20.49), Eq. (20.41) is

$$\begin{bmatrix} \dot{\xi} \\ \dot{\eta} \end{bmatrix} = \begin{bmatrix} -2S_0\eta_c^3 & 2\eta_c^3 \\ f_i S_0 & 0 \end{bmatrix} \begin{bmatrix} \xi \\ \eta \end{bmatrix}. \tag{20.55}$$

The eigen values of Eq. (20.54) are given by

$$\begin{vmatrix} -\lambda & -6\eta_c^3 \\ f_i S_0 & -\lambda \end{vmatrix} = \lambda^2 + 6f_i S_0\eta_c^3 = 0. \tag{20.56}$$

Therefore, we get

$$\lambda = \pm i \sqrt{6 f_i S_0 \eta_c^3}. \tag{20.57}$$

Since α is zero in Eq. (20.45), Eq. (20.48) indicates a center. The eigen values of Eq. (20.55) are

$$\begin{vmatrix} -2S_0\eta_c^3 - \lambda & 2\eta_c^3 \\ f_i S_0 & -\lambda \end{vmatrix} = \lambda(2 f_i S_0 \eta_c^3 + \lambda) - f_i S_0 \eta_c^3 = 0. \tag{20.58}$$

This shows

$$\lambda = S_0\eta_c^3 \pm \sqrt{(S_0\eta_c^3)^2 + 2 f_i S_0 \eta_c^3}. \tag{20.59}$$

This indicates that one of the two eigen values is positive and the other is negative. Thus, Eq. (20.49) forms a saddle point. The resultant phase plane is explained in Fig. 20.2. The solution near the saddle of Eq. (20.49) shows the saline wedge.

The following example is the nonlinear differential equation (17.95) to describe the plane boundary-layer. It is

$$k\ddot{U} = 2U - U^2, \quad \text{for } k > 0, \tag{20.60}$$

in which "$\cdot$" indicates $d/d\eta$. Defining $y_1 = U$ and $y_2 = \dot{U}$, we apply the phase plane analysis as follows:

$$\frac{dy_1}{d\eta} = y_2, \tag{20.61}$$

$$\frac{dy_2}{d\eta} = \frac{2y_1 - y_1^2}{k}. \tag{20.62}$$

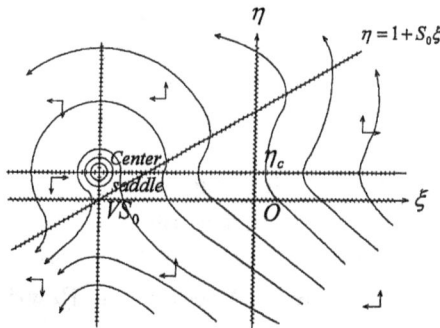

Figure 20.2.   Singular points in Eqs. (20.46) and (20.47).

The number of the singular points of Eqs. (20.61) and (20.62) is two. They are

$$y_1 = 0, \quad y_2 = 0, \tag{20.63}$$

$$y_1 = 2, \quad y_2 = 0. \tag{20.64}$$

According to the phase plane analysis, the singular point of Eq. (20.63) is

$$\begin{bmatrix} \dot{y}_1 \\ \dot{y}_2 \end{bmatrix} = \begin{bmatrix} 0 & 1 \\ 2/k & 0 \end{bmatrix} \begin{bmatrix} y_1 \\ y_2 \end{bmatrix}. \tag{20.65}$$

The singular point of Eq. (20.64) is

$$\begin{bmatrix} \dot{y}_1 \\ \dot{y}_2 \end{bmatrix} = \begin{bmatrix} 0 & 1 \\ -2/k & 0 \end{bmatrix} \begin{bmatrix} y_1 \\ y_2 \end{bmatrix}. \tag{20.66}$$

The eigen values of the singular point of Eq. (20.63) are obtained from

$$\begin{vmatrix} \lambda & -1 \\ -2/k & \lambda \end{vmatrix} = \lambda^2 - \frac{2}{k} = 0. \tag{20.67}$$

It is

$$\lambda = \pm\sqrt{\frac{2}{k}}. \tag{20.68}$$

Equation (20.68) indicates that the singular points of Eq. (20.63) forms the saddle. The eigen values of the singular point of Eq. (20.64) are obtained from

$$\begin{vmatrix} \lambda & -1 \\ 2/k & \lambda \end{vmatrix} = \lambda^2 + \frac{2}{k} = 0. \tag{20.69}$$

The eigen values are

$$\lambda = \pm i\sqrt{\frac{2}{k}}. \tag{20.70}$$

Equation (20.70) indicates that the singular point of Eq. (20.64) forms the center. The result of the phase plane analysis is represented in Fig. 20.3. The solution near the saddle shows the plane boundary-layer. Next is the KdV equation. Rewriting Eq. (16.58) with $A = 0$, we have

$$-Cu + 3u^2 + \ddot{u} = 0, \quad \text{for } C > 0, \tag{20.71}$$

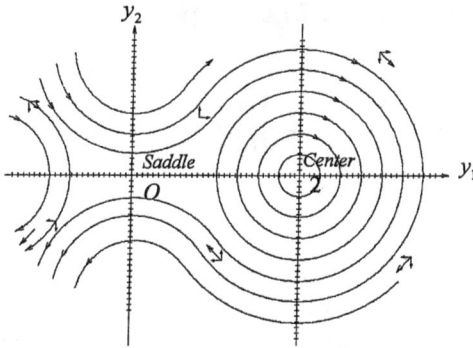

Figure 20.3.   Singular points of Eq. (20.60).

in which " $\cdot$ " $= d/d\xi$. Defining a dependent variable u and its derivative $\dot{u}$ by x_1 and x_2, respectively, we have

$$x_1 = u, \tag{20.72}$$

$$x_2 = \dot{u}. \tag{20.73}$$

Rewriting Eq. (20.71) by Eqs. (20.72) and (20.73), we obtain

$$\dot{x}_1 = f_1(x_1, x_2) = x_2, \tag{20.74}$$

$$\dot{x}_2 = f_2(x_1, x_2) = Cx_1 - 3x_1^2. \tag{20.75}$$

The singular points of Eqs. (20.74) and (20.75) are

$$x_1 = 0, \quad x_2 = 0 \tag{20.76}$$

and

$$x_1 = \frac{C}{3}, \quad x_2 = 0. \tag{20.77}$$

To describe the solution near the singular points, take the partial derivative of Eqs. (20.74) and (20.75) as

$$\frac{\partial f_1}{\partial x_1} = 0, \quad \frac{\partial f_1}{\partial x_2} = 1, \tag{20.78}$$

$$\frac{\partial f_2}{\partial x_1} = C - 6x_1, \quad \frac{\partial f_2}{\partial x_2} = 0. \tag{20.79}$$

Thus, Eqs. (20.74) and (20.75) are expressed near the singular point Eq. (20.76) as follows:

$$\begin{bmatrix} \dot{x}_1 \\ \dot{x}_2 \end{bmatrix} = \begin{bmatrix} 0 & 1 \\ C & 0 \end{bmatrix} \begin{bmatrix} x_1 \\ x_2 \end{bmatrix}. \tag{20.80}$$

Equations (20.74) and (20.75) are expressed near the singular point Eq. (20.76) as follows:

$$\begin{bmatrix} \dot{x}_1 \\ \dot{x}_2 \end{bmatrix} = \begin{bmatrix} 0 & 1 \\ -C & 0 \end{bmatrix} \begin{bmatrix} x_1 \\ x_2 \end{bmatrix}. \tag{20.81}$$

The property of the singular point Eq. (20.76) is

$$|A - \lambda I| = \left| \begin{pmatrix} 0 & 1 \\ C & 0 \end{pmatrix} - \begin{pmatrix} \lambda & 0 \\ 0 & \lambda \end{pmatrix} \right| = \begin{vmatrix} -\lambda & 1 \\ C & -\lambda \end{vmatrix} = 0. \tag{20.82}$$

The eigen values are

$$\lambda = \pm\sqrt{C}. \tag{20.83}$$

The eigen values are positive and negative reals. Thus, it indicates the saddle. The property of the singular point Eq. (20.77) is

$$|A - \lambda I| = \left| \begin{pmatrix} 0 & 1 \\ -C & 0 \end{pmatrix} - \begin{pmatrix} \lambda & 0 \\ 0 & \lambda \end{pmatrix} \right| = \begin{vmatrix} -\lambda & 1 \\ -C & -\lambda \end{vmatrix} = 0. \tag{20.84}$$

The eigen values are

$$\lambda = \pm i\sqrt{C}. \tag{20.85}$$

The eigen values are imaginary. Thus, it indicates the center. The solution near the center forms oscillatory motion such as water waves. Consider a simplified prey–predator equation by Lotka–Volterra (Thompson and Stewart, 1986). Foxes prey on rabbits. The model is written by

$$\dot{x} = K_1 x (N_y - y) = f_1(x, y), \tag{20.86}$$

$$\dot{y} = K_2 y (x - N_x) = f_2(x, y), \tag{20.87}$$

in which "." $= d/dt$, $x =$ the number of rabbits, $y =$ the number of foxes, and K_1, K_2, N_x, and $N_y =$ positive constants. The steady solution of Eqs. (20.86) and (20.87) is

$$x = N_x, \tag{20.88}$$

$$y = N_y. \tag{20.89}$$

This indicates that the numbers of rabbits and foxes are constant. To study the stability of this solution, adding small disturbances, we assume

$$x = x' + N_x, \tag{20.90}$$

$$y = y' + N_y, \tag{20.91}$$

in which $x' \ll N_x$ and $y' \ll N_y$. Substituting Eqs. (20.90) and (20.91) into Eqs. (20.86) and (20.87), we get

$$\dot{x}' = -K_1 N_x y', \tag{20.92}$$

$$\dot{y}' = K_2 N_y x'. \tag{20.93}$$

The substitution of Eqs. (20.93) into (20.92) is expressed by

$$\ddot{x}' + K_1 K_2 N_x N_y x' = 0. \tag{20.94}$$

The solution of Eq. (20.94) shows a harmonic oscillation. The singular points of Eqs. (20.86) and (20.87) are $(0, 0)$ and (N_x, N_y). The partial derivatives of the right side of Eqs. (20.86) and (20.87) are

$$\frac{\partial f_1}{\partial x} = K_1(N_y - y), \tag{20.95}$$

$$\frac{\partial f_1}{\partial y} = -K_1 x, \tag{20.96}$$

$$\frac{\partial f_2}{\partial x} = K_2 y, \tag{20.97}$$

$$\frac{\partial f_2}{\partial y} = K_2(x - N_x). \tag{20.98}$$

Therefore, the eigen values of the matrix A at the singular point $(0, 0)$ are obtained by

$$
\begin{aligned}
|A - \lambda I| &= \left| \begin{pmatrix} K_1 N_y & 0 \\ 0 & -K_2 N_x \end{pmatrix} - \begin{pmatrix} \lambda & 0 \\ 0 & \lambda \end{pmatrix} \right| \\
&= \begin{vmatrix} K_1 N_y - \lambda & 0 \\ 0 & -K_2 N_x - \lambda \end{vmatrix} \\
&= (K_1 N_y - \lambda)(K_2 N_x + \lambda) = 0.
\end{aligned} \tag{20.99}
$$

Thus, $\lambda = K_1 N_y$ and $-K_2 N_x$ indicate the saddle. The eigen values of the matrix A at (N_x, N_y) are obtained by

$$|A - \lambda I| = \left| \begin{pmatrix} 0 & -K_1 N_x \\ K_2 N_y & 0 \end{pmatrix} - \begin{pmatrix} \lambda & 0 \\ 0 & \lambda \end{pmatrix} \right|$$

$$= \begin{vmatrix} -\lambda & -K_1 N_x \\ K_2 N_y & -\lambda \end{vmatrix}$$

$$= \lambda^2 + K_1 K_2 N_x N_y = 0. \tag{20.100}$$

Thus, $\lambda = \pm i \sqrt{K_1 K_2 N_x N_y}$ indicate a center. Near the center x and y oscillate and Eq. (20.94) approximately shows the solution of the center. Finally, consider a limit cycle. Substituting $\epsilon = 1$ in the modified van der Pol equation, we get

$$\ddot{x} + (\dot{x}^2 + x^2 - 1)\dot{x} + x = 0. \tag{20.101}$$

Herein, defining

$$x_1 = x, \tag{20.102}$$

$$x_2 = \dot{x}, \tag{20.103}$$

we transform Eq. (20.86) to

$$\dot{x}_1 = x_2, \tag{20.104}$$

$$\dot{x}_2 = -(x_2^2 + x_1^2 - 1)x_2 - x_1. \tag{20.105}$$

The singular point is $x_1 = x_2 = 0$. The matrix A at the singular point is written as

$$A = \begin{bmatrix} 0 & 1 \\ -2x_1 x_2 - 1 & -3x_2^2 \quad x_1^2 + 1 \end{bmatrix}_{\text{singular point}} = \begin{bmatrix} 0 & 1 \\ -1 & 1 \end{bmatrix}. \tag{20.106}$$

The eigen values of the matrix A are $\frac{1}{2} \pm i\frac{\sqrt{3}}{2}$. These indicate the unstable focus. Next, consider

$$\frac{d}{dt}(x_1^2 + x_2^2) = 2x_1\dot{x}_1 + 2x_2\dot{x}_2 = 2x_1 x_2 + 2x_2[-(x_2^2 + x_1^2 - 1)x_2 - x_1]$$

$$= -2x_2^2(x_1^2 + x_2^2 - 1). \tag{20.107}$$

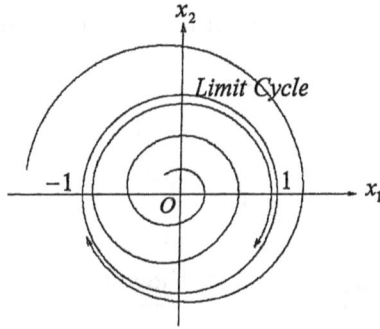

Figure 20.4. Limit cycle.

When $x_1^2 + x_2^2 < 1$, $\frac{d}{dt}(x_1^2 + x_2^2) > 0$ is formed. When $x_1^2 + x_2^2 > 1$, $\frac{d}{dt}(x_1^2 + x_2^2) < 0$ is also formed. This indicates that the trajectories always go to a circle $x_1^2 + x_2^2 = 1$. This indicates that the solution approaches from outside and inside a circle

$$x_1^2 + x_2^2 = 1. \tag{20.108}$$

This circle is called a stable limit cycle. The phase is described in Fig. 20.4. The limit cycle contains a center, stable limit cycle, unstable limit cycle, and semi-limit cycle. The limit cycle appears in the nonlinear system. Finally, consider the following equation (Thompson and Stewart, 1986):

$$\ddot{x} - \mu\dot{x} + x + \dot{x}^3 = 0, \tag{20.109}$$

in which "." $= d/dt$. Defining

$$x_1 = x, \tag{20.110}$$

$$\dot{x}_1 = x_2, \tag{20.111}$$

we get

$$\dot{x}_2 = \mu x_2 - x_1 - x_2^3. \tag{20.112}$$

The singular point is $x_1 = x_2 = 0$. The matrix A of the singular point is written as

$$A = \begin{bmatrix} 0 & 1 \\ -1 & \mu - 3x_2^2 \end{bmatrix}_{\text{singular point}} = \begin{bmatrix} 0 & 1 \\ -1 & \mu \end{bmatrix}. \tag{20.113}$$

The eigen values of the matrix A are derived from

$$|A - \lambda I| = \lambda^2 - \mu\lambda + 1 = 0. \tag{20.114}$$

Thus, we have

$$\lambda = \frac{\mu \pm \sqrt{\mu^2 - 4}}{2}. \tag{20.115}$$

The singular point is classified by the value of μ.

Parameter	Singular point
$\mu \leq -2$	Stable node
$-2 < \mu < 0$	Stable focus
$\mu = 0$	Center
$0 < \mu < 2$	Unstable focus
$\mu \geq 2$	Unstable node

Defining a circle $r^2 = x_1^2 + x_2^2$, we have

$$\frac{1}{2}\frac{d}{dt}r^2 = x_1\dot{x}_1 + x_2\dot{x}_2 = x_2^2(\mu - x_2^2). \tag{20.116}$$

If $\mu \leq 0$,

$$\frac{1}{2}\frac{d}{dt}r^2 < 0. \tag{20.117}$$

The trajectories go to the origin. There does not exist a limit cycle. If $\mu > 0$,

$$\frac{1}{2}\frac{d}{dt}r^2 > 0, \quad \text{for } x_2^2 < \mu, \tag{20.118}$$

$$\frac{1}{2}\frac{d}{dt}r^2 < 0, \quad \text{for } x_2^2 > \mu. \tag{20.119}$$

The trajectories go to a circle $x_1^2 + x_2^2 = r^2$. The equation $x_2 = \sqrt{\mu}$ indicates that $r^2 \geq \mu$. There is a stable limit cycle of which radius is more than $\sqrt{\mu}$. This indicates that the trajectories of the solution approach from outside and inside a limit cycle. In fluid mechanics,

Lorenz in 1963 investigated the following equation:

$$\frac{dx}{dt} = \sigma(y - x), \tag{20.120}$$

$$\frac{dy}{dt} = \rho x - y - xz, \tag{20.121}$$

$$\frac{dz}{dt} = -\beta z + xy, \tag{20.122}$$

in which σ = the Prandtl number, ρ = Rayleigh number, and $\beta = 8/3$. Then, he found a Lorenz attractor (Lorenz, 1963). This indicates that the solution is very sensitively dependent on the initial condition and the limitation of weather forecasting, although the original equations are deterministic. Ueda in 1971 also found chaos from the following equations:

$$\frac{dx}{dt} = y, \tag{20.123}$$

$$\frac{dy}{dt} = -ky - x^3 + B\cos t, \tag{20.124}$$

in which k and B = constants. As an example of chaos a logistic model is also described by

$$x_{n+1} = \mu x_n(1 - x_n), \quad \text{for } \mu > 0. \tag{20.125}$$

According to May's phase plane analysis for x_n and x_{n+1}, Eq. (20.125) shows a stable point, a limit cycle, and chaos for $\mu = 2.5$, 3.1, and 3.8, respectively (Thompson and Stewart, 1986). The roots of Eq. (20.125) for $x_n = x_{n+1}$ are $x_n = 0$ and $(\mu - 1)/\mu$. The derivative of the right side at $x_n = (\mu - 1)/\mu$ is given by

$$\frac{d}{dx_n}[\mu x_n(1 - x_n)] = \mu - 2\mu x_n = 2 - \mu. \tag{20.126}$$

The solution at $x_n = (\mu - 1)/\mu$ is unstable near $\mu = 3$. Since the derivative at $x_n = (\mu - 1)/\mu$ is -1, Eq. (20.125) does not converge. Li and Yorke (1975) found that the period 3 implies chaos from Eq. (20.126).

Example 20.3. Obtain the singular points and analyze the phase plane of the following equation:

$$\ddot{x} + 2\left(\frac{\dot{x}^3}{3} - \dot{x}\right) + x = 0, \tag{1}$$

in which ".$" = d/dt$.

Solution. Defining

$$x_1 = x, \tag{2}$$

$$x_2 = \dot{x}_1, \tag{3}$$

we get

$$\dot{x}_2 = -2\left(\frac{x_2^3}{3} - x_2\right) - x_1. \tag{4}$$

The singular point of Eqs. (3) and (4) is $x_1 = x_2 = 0$. The property near the singular point is expressed by

$$\begin{bmatrix} \dot{x}_1 \\ \dot{x}_2 \end{bmatrix} = \begin{bmatrix} 0 & 1 \\ -1 & 2 \end{bmatrix} \begin{bmatrix} x_1 \\ x_2 \end{bmatrix}. \tag{5}$$

Since the eigen value of Eq. (5) is 1, the singular point is the unstable node. The slope S of the solution is given by

$$S = \frac{-x_1 + 2x_2}{x_2}. \tag{6}$$

The result of the phase plane analysis is given in Fig. 20.5.

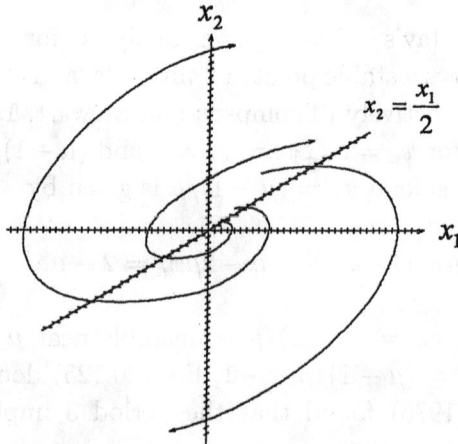

Figure 20.5.   Example 20.3.

Problem 20.3. Analyze the phase plane for the following equations:
(i) a linear oscillation equation

$$\ddot{x} + x = 0, \tag{7}$$

(ii) a nonlinear oscillation equation

$$\ddot{x} + x + x^3 = 0, \tag{8}$$

and (iii) a finite oscillation equation

$$\ddot{x} + \sin x = 0. \tag{9}$$

References

I used many books and papers to write this book. Most representative references are given. Readers who need more information for their studies or research, please refer to the following books and papers:

Aggarwal, J. K. (1972): *Note on Nonlinear Systems*, Van Nostrand Reinhold Com., New York.

Ahlfors, L. V. (1966): *Complex Analysis*, McGraw-Hill, New York.

Ames, W. F. (1965): *Nonlinear Partial Differential Equations in Engineering*, Academic Press, New York.

Ames, W. F. (1969): *Numerical Methods for Partial Differential Equations in Engineering*, Academic Press, New York.

Ames, W. F. (1972): *Nonlinear Partial Differential Equations in Engineering*, Vol. II, Academic Press, New York.

Aris, R. (1962): *Vectors, Tensors, and the Basic Equations of Fluid Mechanics*, Dover, New York.

Bluman, G. W. and Cole, J. D. (1974): *Similarity Methods for Differential Equations*, Springer-Verlag, New York.

Brutsaert, W. (2005): *Hydrology, An Introduction*, Cambridge Univ. Press., U.K.

Chiu, C. L., Lin, H. C., and Mizumura, K. (1976): Simulation of hydraulic process in open channels, *J. Hydraul. Div.*, 102(HY2).

Chow, V.-T. (1959): *Open Channel Hydraulics*, McGraw-Hill, New York.

Churchill, R. V. (1941): *Fourier Series and Boundary Value Problems*, McGraw-Hill, New York.

Churchill, R. V. (1958): *Operational Mathematics*, McGraw-Hill, New York.

Daily, J. W. and Harleman, D. R. F. (1966): *Fluid Dynamics*, Addison Wesley, New York.

Debnath, L. (1994): *Nonlinear Water Waves*, Academic Press Inc., San Diego.

Dyke, M. V. (1975): *Perturbation Methods in Fluid Mechanics*, The Parabolic Press, CA.

Erdelyi, A. (1961): An expansion procedure for singular perturbations, *Atti. Acad. Sci. Torino, CI. Sci. FIS. Mat. Nat.*, **95**.

Finlayson, B. A. (1972): *The Method of Weighted Residuals and Variational Principles*, Academic Press, New York.

Graf, W. H. (1984): *Hydraulics of Sediment Transport*, Water Resources Pub., Littleton, CO.

Hancock, H. (1958): *Lecture on the Theory of Elliptic Functions*, Dover, New York.

Harr, M. E. (1962): *Groundwater and Seepage*, McGraw-Hill, New York.

Hino, M. and Oonishi, S. (1967): Analysis on stratified flow into a point sink, *Technical Report*, Dept. of Civil Eng., Tokyo Inst. of Tech., No. 5 (in Japanese).

Hino, M. (1975): Theory on the formation of shore-current and shore-topography based on the response concept, *JSCE*, 237 (in Japanese).

Ince, E. L. (1956): *Ordinary Differential Equations*, Dover, New York.

Kennedy, J. F. (1963): The mechanism of dunes and antidunes in erodible bed channels, *J. Fluid Mech.*, 16(4).

Kikkawa, H., Yamada, T., and Mizutani, T. (1978): Study on unsteady selective withdrawal, *Proc. 22nd Japanese Conf. on Hydraulics* (in Japanese).

Lamb, H. (1945): *Hydrodynamics*, Cambridge Univ. Press, U.K.

Li, T. Y. and Yorke, J. A. (1975): Period three implies chaos, *Am. Math. Mon.*, 82.

Liggett, J. A. (1994): *Fluid Mechanics*, McGraw-Hill, New York.

Longuet-Higgins, M. S. (1953): Mass transport in water waves, *Philos. Trans. Roy. Soc. Ser. A*, 245(903).

Lorenz, E. N. (1963): Deterministic non-periodic flow, *J. Atmos. Sci.*, 20.

Madsen. O. S. and White, S. M. (1976): Energy dissipation on a rough slope, *J. Waterways, Port, Coastal, and Ocean Engineering*, 102(WW1).

Milne-Thomson (1968): *Theoretical Hydrodynamics*, Macmillan, New York.

Mizumura, K. (1984): Flow analyses in idealized rubble-mound breakwater, *J. Waterways, Port, Coastal, and Ocean Engineering*, 110(WW3).

Mizumura, K. (1989): On generation of dunes and antidunes, *IAHR Congress*, Sec. B.

Mizumura, K. and Yamamoto, M. (1991): Wave reflection and transmission through sloped rubble-mound breakwater, *Computational Hydraulics and Hydrology, Computational Method in Water Resources II* (eds. Brebbia *et al.*), Computational Mechanics Pub., Southampton.

Mizumura, K. (1992a): Nonlinear analysis of rainfall and runoff process, in *Catchment Runoff and Rational Formula* (ed. B.-C. Yen), Water Resources Publications, Littleton, CO.

Mizumura, K. (1992b): *Coastal and Ocean Engineering*, Kyouritsu Shuppan, Tokyo (in Japanese).

Mizumura, K. (1993): Meandering water rivulet, *J. Hydraul. Eng.*, 119(11).

Mizumura, K. (1995a): Free surface profile of open channel flow with wavy boundaries, *J. Hydraul. Eng.*, 121(7).

Mizumura, K. (1995b): Runoff prediction by a simple tank model using recession curves, *J. Hydraul. Eng.*, 121(11).

Mizumura, K. (1997): *Hydraulic Engineering*, Kyouritsu Shuppan, Tokyo (in Japanese).

Mizumura, K. (1998): Free surface flow over permeable wavy bed, *J. Hydraul. Eng.*, 124(9).

Mizumura, K. and Yamasaka, M. (1998): Analysis of meandering water rivulets of finite amplitude, *J. Hydraul. Eng.*, 123(11).

Mizumura, K. (2002): Drought flow from a hillslope, *J. Hydrolog. Eng.*, 7(2).

Mizumura, K. and Yamasaka, M. (2002): Flow in open-channel embayments, *J. Hydraul. Eng.*, 128(12).

Mizumura, K. (2003a): Chloride ion in groundwater near disposal of solid wastes in landfills, *J. Hydrolog. Eng.*, 8(4).

Mizumura, K., Yamasaka, M., and Adachi. J. (2003b): Side outflow to supercritical channel flow, *J. Hydraul. Eng.*, 129(10).

Mizumura, K. (2003c): Closure to "drought flow of hillslope", *J. Hydrolog. Eng.*, 8(6).

Mizumura, K. (2005a): Analyses of flow mechanism based on master recession curves, *J. Hydrolog. Eng.*, 10(6).

Mizumura, K. (2005b): Discharge ratio of side outflow to supercritical channel flow, *J. Hydraul. Eng.*, 131(9).

Mizumura, K. (2006a): Analysis of drought flow in recession period, *J. Hydrolog. Eng.*, 11(2).

Mizumura, K. (2006b): Analytical solution of kinematic wave model, *J. Hydrolog. Eng.*, 11(6).

Mizumura, K. (2008a): *Fundamentals in Hydrology*, Denkidaigaku Shuppankai, Tokyo (in Japanese).

Mizumura, K. (2008b): *Mathematics in Hydrology*, Denkidaigaku Shuppankai, Tokyo (in Japanese).

Mizumura, K. (2009a): Approximate solution of nonlinear Boussinesq equation, *J. Hydrolog. Eng.*, 14(10).

Mizumura, K. (2009b): Theoretical boundary condition groundwater flow at the drawdown end, *J Hydrolog. Eng.*, 14(10).

Mizumura, K. and Kaneda, T (2010): Boundary condition of ground water flow through sloping seepage face, *J. Hydrolog. Eng.*, 15(9).

Modi, P. N., Apiel, P. D., and Dandekar, M. M. (1981): Conformal mapping for channel junction flow, *J. Hydraul. Div.*, 107(12).

Mura T. and Koya, T. (1992): *Variational Methods in Mechanics*, Oxford Univ. Press, U.K.

Nayfeh, A. H. (1973): *Perturbation Methods*, Wiley-Interscience Pub., New York.

Pipes, L. A. and Harrill, L. R. (1971): *Applied Mathematics for Engineers and Physicists*, McGraw-Hill, New York.

Prager, W. (1961): *Introduction to Mechanics of Continua*, Ginn and Company, Boston.

Rosenhead, L. (1963): *Laminar Boundary Layer*, Oxford Univ. Press, U.K.

Schlichting, H. (1966): *Boundary Layer Theory*, McGraw-Hill, New York.

Shames, I. H. (1962): *Mechanics of Fluids*, McGraw-Hill, New York.

Singh, V. P. (1996): *Kinematic Wave Modeling in Water Resources*, John Wiley & Sons, Inc., NJ.

Sneddon, I. N. (1957): *Elements of Partial Differential Equations*, McGraw-Hill, New York.

Sokolnikoff, I. S. (1951): *Tensor Analysis*, John Wiley & Sons, New York.

Stoker, J. J. (1957): *Water Waves*, Interscience Pub., New York.

Streeter, V. L. (1958): *Fluid Mechanics*, McGraw-Hill, New York.

Strintzis, M. G. (1974): *Lecture Note on Nonlinear System Theory*, Univ. of Pittsburgh, Pittsburgh, PA.

Thompson, J. M. T. and Stewart, H. B. (1986): *Nonlinear Dynamics and Chaos — Geometrical Methods for Engineers and Scientists*, John Wiley & Sons, New York.

Toda, M. (1983): *Nonlinear Waves and Soliton*, Nippon Hyoronsha, Tokyo (in Japanese).

Turner, J. S. (1973): *Buoyancy Effects in Fluids*, Cambridge Univ. Press, U.K.

Wadachi, M. (1992): *Nonlinear Waves*, Iwanamishoten, Tokyo (in Japanese).

Yano, K. (1957): *Geometry*, Iwanami Series, Iwanamishoten, Tokyo (in Japanese).

Yih, C.-S. (1958): On the flow of stratified fluid, *Proc. 3rd U.S. Nat. Cong. Applied Mechanics*, ASME.

Yih, C.-S. (1969): *Fluid Mechanics*, McGraw-Hill, New York.

Yih, C.-S. (1982): On the nonexistence of solution of a differential system governing axisymmetric flow of a stratified fluid, *Q. Appl. Math.*, April.

Yoshida, K. (1950): *Integral Equations*, Iwanamishoten, Tokyo (in Japanese).

Index